PROGRESS IN COLLOID & POLYMER SCIENCE

Editors: F. Kremer (Leipzig) and G. Lagaly (Kiel)

Volume 111 (1998)

Structure, Dynamics and Properties of Disperse Colloidal Systems

Guest Editors:

H. Rehage and G. Peschel (Essen)

Springer

ISBN 978-3-662-15608-7
ISSN 0340-255 X

Die Deutsche Bibliothek –
CIP-Einheitsaufnahme

Progress in colloid & polymer science. –

Früher Schriftenreihe
Vol. 111. Structure, dynamics and
properties of disperse colloidal systems.
– 1998
**Structure, dynamics and properties of
disperse colloidal systems** / guest ed.:
H. Rehage and G. Peschel.
(Progress in colloid & polymer science ;
Vol. 111)
ISBN 978-3-662-15608-7
ISBN 978-3-7985-1652-6 (eBook)
DOI 10.1007/978-3-7985-1652-6

© 1998 by Springer-Verlag Berlin Heidelberg
Originally published by Dr. Dietrich Steinkopff
Verlag GmbH & Co. KG, Darmstadt in 1998
Softcover reprint of the hardcover 1st edition 1

Chemistry Editor:
Dr. Maria Magdalene Nabbe;
Production: Holger Frey, Ajit Vaidya.

Typesetting and Copy-Editing:
Macmillan Ltd., Bangalore, India

Prog Colloid Polym Sci (1998) V
© Steinkopff Verlag 1998

This special issue contains of a selection of papers and posters which were presented at the 38th General Meeting of the German Colloid Society. This conference was held at the University of Essen, Germany, from September 29th to October 2nd, 1997. The main topic of this congress was focused on "Structure, Dynamics, and Properties of Disperse Colloidal Systems". The lectures and posters covered a broad range of colloidal particles, such as emulsions, microemulsions, foams, dispersions, membranes, vesicles, microcapsules, nanoparticles, surfactants, polymers and liquid crystalline phases. On October 2nd, an additional symposium was organized on the application of colloidal systems in environmental science. Special topics of this session were focused on new processes of soil and ground water remediation.

The conference was attended by over 360 participants coming from 12 different countries. It is interesting to note that about 40 % of the participants work in industry. The scientific program was composed of 132 contributions, covering theoretical, experimental, and technical aspects. Lively discussions took place during the lectures and around the posters.

The first session was opened by the President of the German Colloid Society, Prof. Dr. M. J. Schwuger. Two well-known and famous scientists received awards from the German Colloid Society. In honor of his excellent and outstanding work on surfactants, emulsions, and microemulsions, Prof. Dr. Stig Friberg, Potsdam, New York, received the Wolfgang Ostwald Prize.

The Thomas Graham Medal was conferred to Prof. Dr. Armin. Weiß. Munich, for his detailed investigations of clay minerals and for his exceptional merits in colloid science. The Richard Zsigmondy Scholarship for successful young colloid scientists was awarded to Dr. Andreas Pohlmeier, Forschungszentrum Jülich.

In addition to these distinctions two students were honored for their excellent poster presentations. These prizes were awarded to: M. Dreja, B. Tieke (Köln): Polymerization of styrene in ternary microemulsions using cationic gemini surfactants, and H. Lippold, K. Quitsch (Leipzig): Micellar incorporation without solubilizing effect: Experimental studies of the surfactant system water/phenol/MEGA-10.

Review lectures on main topics of the conference were presented by S. E. Friberg (Clarkson University), P. G. de Gennes (Paris), D. Horn (BASF AG), G. Rossmy (Th. Goldschmidt AG), W. Umbach (Henkel KGaA), and A. Weiß (Univ. München). These plenary lectures were focused on emulsions, liquid films, advanced experimental techniques, amphiphilic solid particles, the importance of colloid chemistry in industrial practice, and the stability of suspensions. In addition to these contributions, 18 invited lectures, 48 contributed lectures, and 60 posters were presented. These activities provided a platform to review recent developments and to exchange new results and ideas. They also served to improve our current understanding of microscopic structures and macroscopic properties. The different contributions can be divided into the following sessions, which appear roughly in the order as they were presented in the meeting:

- Technical Applications
- Advanced Experimental Techniques
- Thin Films and Interfaces
- Suspensions and Microcapsules
- Emulsions, Microemulsions, and Foams
- Macromolecules
- Association Colloids
- Colloidal Systems in Environmental Science

The conference covered both fundamental aspects as well as technological applications. The scientific program was established under the guidance of an efficient program committee comprising P. Belouschek, G. Peschel, H. Rehage, W. v. Rybinski, and M. J. Schwuger. Our thanks go to all who contributed to the cordial scientific atmosphere of the meeting. Generous donations, which helped to finance the meeting, were made by BASF AG, Bayer AG, Clariant GmbH, Henkel KGaA, Industrie und Handelskammer Essen, Th. Goldschmidt AG, and RWE AG. We also gratefully acknowledge financial support from the University of Essen which kindly allowed us to use the facilities, where the conference took place. On behalf of the organizing comittee, we would like to thank all the participants and lecturers for their brilliant contributions and stimulating discussions. Finally, it is also a great pleasure to thank the authors of this volume for their excellent presentations. We sincerely do hope that the conference will stimulate new activities for future research in the whole area of colloid and polymer science.

Heinz Rehage
Gerhard Peschel

Prog Colloid Polym Sci (1998) VII
© Steinkopff Verlag 1998

CONTENTS

Progr Colloid Polym Sci (1998) 111:1–2
© Steinkopff Verlag 1998

M.J. Schwuger

Opening address at the 38th General Conference of the German Colloid Society in Essen

Prof. Dr. M.J. Schwuger (✉)
Forschungszentrum Jülich GmH
Institut für Angewandte
Physikalische Chemie
D-52425 Jülich
Germany

Ladies and Gentlemen,

Today I would first like to discuss the situation of chemistry and especially that of colloid chemistry in Germany. These two disciplines differ considerably. You are all aware that in the industrial sector chemistry is flourishing, companies are achieving high returns; nevertheless the employment situation for chemistry graduates is not very encouraging. In comparison, the situation for good colloid and interfacial chemists is distinctly better. This is due to the fact that interfacial and colloid chemistry is a discipline of great significance in industrial processes, a fact which is becoming increasingly apparent to companies. The same is also true for the environmental research field. Processes in the terrestrial environment, in soil, water and also in the gas phase, are largely of a colloid and interfacial chemical nature. These processes are today involved in the assessment of the environmental compatibility of new substances and their transport behavior. Some new environmental technologies are also based on such processes. To this extent, the significance of colloid chemistry in the environmental field has also increased in the past few decades. This naturally has a positive effect on the employment situation in our branch of science.

In the late 1960s and early 1970s, the image of colloid chemistry in Germany was very unfavorable. A large number of professorships ceased to exist and scientists went over to other so-called modern disciplines of physical chemistry. This meant that the industry had great difficulty in obtaining trained staff, and in some cases personnel had to be trained in the individual companies. This was recognized and we have a much more satisfactory situation in Germany today. A number of new research institutions have been established, both at universities and at national research centers. Special mention should be made of the new institutes of the Max Planck Society in Berlin and the Helmholtz Association in Jülich. In this respect, large research institutions are also now devoting themselves to colloid chemistry as an important research field. Such a significant discipline also needs a strong association and representation. We have, therefore, attempted in the course of the past four years to invigorate the Kolloid-Gesellschaft and provide it with new impetus.

Until four years ago, we essentially held only the general meetings of the Kolloid-Gesellschaft every other year. There was an optional possibility of holding other events under the umbrella of the Kolloid-Gesellschaft, but this was not sufficiently exploited. We have made a radical change here.

First of all, the structure of the general meetings was revamped. At the beginning of each conference, leading personalities from politics, industry, and science are to be invited to deal with topics of overlapping significance in general lectures. Furthermore, we want to enhance the attractiveness of the events by holding a special symposium within the framework of the general meetings. We organized such a symposium for the first time in Essen concerning the significance of colloid chemistry for innovative environmental technologies. In this respect, we are trying to make the events as interesting as possible in order to attract colleagues from industry, public authorities, and universities to the Kolloid-Gesellschaft. We shall see how this new conception for the general meetings is received. However, it can already be said that a record was achieved with about 350 delegates in Essen since previous events all had fewer attendees.

As a further innovation we instituted the Wolfgang Ostwald Colloquia to be held once or twice a year: once in the year of the general meeting and twice in the years between the general meetings. In one or two day workshops special, as a rule closely defined, topics are to be addressed and brought to the attention of a broad group of interested persons by invited speakers. Five such colloquia have been held to date. They were all a success. The number of participants ranged between 90 and 140, which represents an optimum size for events of this type.

We have recently also opened up another field intended to appeal to young researchers. Meetings for young scientists in colloid and interfacial research have been introduced. Being specially tailored to students, taking their final examinations, PhD students, and those working on their *Habilitation*, these meetings are aimed at giving young people the opportunity to present their work to a wider

public for the first time. We are still at the beginning here and hope that this idea will gain acceptance in the future.

Finally, we also intend to hold international conferences on an occasional basis in the gaps between the general meetings taking place every other year, either alone or together with other societies. This means that we have introduced different types of events, or are planning to do so, and hope that we can, thus, increase interest in the Kolloid-Gesellschaft and colloid chemistry in Germany.

Four conferences are planned for 1998. In order to avoid an overload this year, the young scientists' meetings will be organized separately but held at the same venue in Jülich together with the 6th Wolfgang Ostwald Colloquium so that both events can profit from each other. We shall also organize a 1st International Conference in Dresden on "Self-Assembly of Amphiphilic Systems" and hold the 7th Wolfgang Ostwald Colloquium in Berlin.

As far as cooperation with other societies is concerned, we have already had a joint Wolfgang Ostwald Colloquium in Dresden with the Society of German Chemists (GDCh) and a joint Wolfgang Ostwald Colloquium is planned for 1999 with Swiss colleagues. In the year 2000 an international conference is to be held with our Hungarian colleagues in Hungary.

Our Society's prizes are also an opportunity for promotion. As of now we have four prizes which are awarded as the occasion arises: the Wolfgang Ostwald Prize for lifework, the Thomas Graham Prize for lifework and commitment to colloid chemistry in an international and national framework, the Steinkopff Prize for applying colloid chemistry to environment and technology, and the Zsigmondy Scholarship for young scientists. We find that one instrument is still lacking in this context we shall institute a new prize for scientists in the medium age range between 40 and 55. In American societies there

are prizes especially for this group, and we would also like to introduce one at the Kolloid-Gesellschaft. As chairman, I do not wish to merely fulfill a representative function so I have decided to donate this prize from my own private funds. After lengthy discussion, we agreed to name this prize after Raphael Eduard Liesegang. Perhaps some of the younger generation are not immediately familiar with Raphael Eduard Liesegang. He was born in Wuppertal-Elberfeld in 1869 and became professor at the University of Frankfurt am Main in 1908. He died at a ripe old age in 1947. Liesegang's particular achievements were in "classical" colloid chemistry. Reactions in gelatinous systems, i.e., in gels and sols, the Liesegang rings, are also familiar to all young chemists. The reason why he appears particularly important is that he was one of the first to recognize the significance of colloids in technology. His books "Colloids in Technology", "Colloid Chemical Technology" as well as "Biological Technology" were ground-breaking for what we today regard – undoubtedly from a different standpoint – as particularly significant for our discipline. I believe that this will be a worthy namesake for our new prize.

The Society's activities and active recruitment have led to membership having almost doubled in the past few years. Nevertheless, the potential for colloid chemistry in German-speaking countries is much greater than expressed by this level of membership. We are therefore recruiting new members, but at the same time we regret to announce the death of respected colleagues who have accompanied us throughout the years. In the past two years, the death occurred of Professor Hans Sonntag, who was awarded the last Ostwald Prize in Dresden, Professor Erika Cremer from Innsbruck, and Dr. Moll from Walsrode.

I hope this conference will be a great success.

Progr Colloid Polym Sci (1998) 111:3–5
© Steinkopff Verlag 1998

E. Matijević

Wolfgang-Ostwald-Prize 1997: Awarded to Stig Friberg

E. Matijević
Distinguished University Professor
Clarkson University
Potsdam, New York 13699-5814
USA

Professor Stig E. Friberg, Potsdam, NY, USA, received the Wolfgang Ostwald Prize of the Kolloid-Gesellschaft for his investigations on the applications of amphiphilic association structres in dispersions ranging from colloidal to macrosize systems, with special emphasis on liquid dispersions.

His research has led to a large number of scientific firsts in the area of amphiphilic association structures, such as the formation of non-aqueous lyotropic liquid crystals, microemulsions, and stable foams from low-surface free-energy liquids. He also developed new stabilization mechanisms for emulsions, as well as catalytic activity by inverse micelles and by the surfaces in lyotropic liquid crystals.

Curriculum vitae

Stig E. Friberg was born in the small village of Rimforsa in the heavily wooded, hilly southern part of Öster-götland in Sweden. His unusual intellectual capacity was already realized in the little country school, where he received his first education. Later, a combination of part-time employment and scholarships allowed him to study physical chemistry at the Uni-

versity of Stockholm with the extremely gifted (and temperamental) Arne Ölander as his "hands off" mentor.

One year after obtaining his PhD he was appointed successor to Per Ekwall as director of the Laboratory for Surface Chemistry in Stockholm, a research institute, which under his leadership developed into an internationally known establishment with some sixty scientists supported by the Swedish government and industries from Sweden, Norway, and Denmark.

For his contributions to colloid science and the technical development in Sweden, Stig Friberg was elected as a member of the Swedish National Academy for Engineering Science in 1974, at the time the youngest person to be so honored.

Managing the institute limited his research activities, and although several scientific firsts were published from Stockholm, it was obvious that Stig Friberg wanted to devote more time for research. Thus, in 1976 he accepted the position as Chairman and Professor in the Chemistry Department, University of Missouri at Rolla, Missouri, USA. His research efforts became appreciated internationally, as shown by the Award for Excellence in Research by the Japanese Colloid Division in 1978 and nationally with several awards from the American Cosmetic Chemists Society, the American Chemical Society Award in Colloid and Surface Science, and the University of Missouri at Rolla Faculty Medal for the Curators Distinguished Professorship.

In 1987, Stig Friberg moved to Clarkson University, Potsdam, NY, USA as Chairman of the Chemistry Department, and a member of the University's Center for Advanced Materials Processing. During his tenure at Clarkson, he has received honorary doctorates from Lund University, Sweden, Yokohama National University, Japan, and Åbo Academy University, Finland. His recent interest in the fundamentals of skin care was rewarded with the highest honor from the American Society of Cosmetic Chemists, the Maison G. De-Navarre Medal Award.

He is an outstanding lecturer both in the classroom and in professional forums. His activity in continuing education led to the American Chemical Society's Award for Exceptional Achievement in this area.

Research activities

Microemulsions and emulsions

Stig Friberg realized very early the fundamental difference between emulsions and microemulsions and introduced phase diagrams as a tool for the formulation of the latter, an approach that is overwhelmingly used for these systems today.

He discovered multilayer emulsion stabilization, a wide spread phenomenon in pharmaceutical, food, and personal care systems. His analysis of the multilayer stabilization mechanism yielded a surprising fundamental result: the van der Waals forces showed a strong discontinuity in the transition from flocculation to coalescence. This discontinuity results in more than 90% reduction of the

force, a significant contribution to the enhanced stabilization by the multilayer adsorption.

The problems with the aerosol package formulations for foams led Friberg to investigate potential mechanisms for the stability of foams from hydrocarbons. The problem is fundamentally simple: hydrocarbon based surfactants do not reduce their surface free energy. Hence, there is no adsorption at the air/hydrocarbon interface and no foam stabilization may be obtained.

Stig Friberg could show that the phase change from an inverse micellar solution to a lamellar liquid crystal led to a reduction of surface free energy due to a restricted conformation of the surfactant in the latter phase. The surface of the lamellar liquid crystal contains predominantly methyl groups instead of a mixture of methyl and methylene groups in the disordered liquid. The difference in the surface free energy is at the level of $6–10 \, nM/m$, which is sufficient to give a preferred organization hydrocarbon/liquid crystal/air instead of liquid crystal/oil/air in a three-phase system. Hence, the foam bubbles are covered with a macroscopic layer of a liquid crystal, providing excellent stabilization.

Ever since the Winsor analysis in the 1950s, it was generally accepted that water was a necessary solvent in order to form micelles or lyotropic liquid crystals. Stig Friberg demonstrated this conclusion to be premature by discovering the first example of a waterless lamellar liquid crystal in 1979, which led to a series of publications, clarifying the structure of such liquid crystals. His contributions led several European laboratories to engage in this area of research.

Sol–gel process

The sol–gel process uses alcohol solutions to hydrolyze and subsequently

condense alkoxymetal compound combinations to form glasses of extreme purity and of molecularly controlled homogeneous composition. Stig Friberg recognized the potential for using water-in-oil microemulsions as solvents, thereby opening the field for formulating glasses with a high content of inorganic electrolytes. Transparent glasses containing inorganic salts, or even concentrated sulfuric acid in excess of 50%, were prepared and tested for application in laser technology.

Vapor pressure variation in colloidal systems

Recently Stig Friberg has become interested in the variation of vapor pressures in colloidal systems for applications in personal care products, coatings, and high-speed printing inks. The new results have demonstrated that the association structures have a decisive influence on the vapor pressure of solubilized volatile compounds in microemulsions, liquid crystals, and emulsions.

Some personal observations

The writer of this laudation met Professor Stig Friberg nearly 30 years ago and countless times both in Sweden and in the USA even before he joined Clarkson's faculty. These contacts made it possible to add some personal notes to the more formal description of his vitae and accomplishments. In summary, Stig is a true scholar as an academic person need be. His interests transgress the area of research, which he certainly enjoys doing, because he is equally able to discuss and argue topics in philosophy, civilization, politics, religion, art – and extra curricular activities.

Friberg can also readily adapt to a new environment or different situations. When I first met Stig in his

Progr Colloid Polym Sci (1998) 111:3–5
© Steinkopff Verlag 1998

native country, he always wore a dark suit and followed a rigid working schedule. A few years later I found him in Rolla, Missouri attired with cowboy boots, a shoelace tie, and a five gallon hat! But there was a reason for it. In Sweden his outdoor activities included sailing; indeed, he is an expert seaman who sailed the waters of the Baltic Sea and the Atlantic Ocean. Well, there is little chance to sail in Missouri, so Stig turned his interest to horses. Within a short time he knew more about these animals than most people who owned stables

for a long time. He still rides daily in his "corral" in Potsdam.

Stig's ability to adapt pays off professionally. He is one of the academic people who has the ability to translate a research problem (or project) into useful application. For this reason, he has been a valued consultant to many corporations in the USA and abroad.

Stig enjoys life. He likes fine foods and good wines, especially in the company of friends, who are conducive to interesting conversations.

Finally, Stig Friberg never denies his Swedish roots. However, he has

accepted the USA as his home to which he is most loyal. One of his proud moments was when he became a citizen of this country. With his move to the United States we have gained much, but the "Old World" has not suffered a loss, because Stig is scientifically a – World Citizen.

Egon Matijević
Distinguished University Professor
Clarkson University, Potsdam, NY,
USA

Progr Colloid Polym Sci (1998) 111:6–8
© Steinkopff Verlag 1998

K. Beneke
G. Lagaly
G. Schön

Thomas Graham Prize 1997 awarded to Armin Weiss

K. Beneke · Prof. Dr. Dr. h. c. G. Lagaly (✉)
Prof. Dr. G. Schön
Institut für Anorganische Chemie
Universität Kiel
D-24098 Kiel
Germany

The Kolloid–Gesellschaft awards the Thomas Graham Prize for exceptional merits in colloid science, for encouragement of international cooperation in colloid science and advancements in interdisciplinary research. Armin Weiss is deserving of this prize because of his activities during the last 30 years.

In 1966, Armin Weiss became the editor of Kolloid Z. Z. Polymere. During this time, the journal (founded 1906 by Wolfgang Ostwald) was already an international journal. Nevertheless, the transformation of it into Colloid Polymer Science in 1974 was indicative of the strong international orientation of Armin Weiss. In the 20 years in this position, he made strong efforts to publish outstanding contributions in almost every area of colloid and interface science.

During this period Armin Weiss was also President of Kolloid–Gesellschaft. He held this position for over 20 years (1967–1987), almost as long as Wolfgang Ostwald (1922–1943). At the beginning of this period the destiny of the Kolloid–Gesellschaft was oscillating between a quietly running German group and a vivid association with an international reputation. Armin Weiss was responsible for gaining the international reputation of this society. In the 1960s, this was not as easy as one imagines today.

The charming, fascinating personality of Armin Weiss and his high scientific reputation outside Germany promoted this metamorphosis of the society into one with an international standing. He also created many contacts between the Kolloid–Gesellschaft and industrial research and application groups; he always stressed the importance of diffusion through the interface between the society and the industrial sector.

Armin Weiss' activity for the Kolloid–Gesellschaft was recognised world wide: 1979–1983 he was appointed Vicepresident of the International Association of Colloid and Surface Scientists. In this period, 1978–1982, he also served as Vicepresident of the Association International pour l'Étude des Argiles.

The success of Armin Weiss as editor of Colloid Polymer Science and as President of the Kolloid–Gesellschaft was certainly not based on red-type behavior (he always stays in an atmosphere of "ordered chaos") but solely on his exceptional ability to persuade people of his ideas and on his outstanding scientific reputation.

The oeuvre of Armin Weiss is of greatest importance. In his doctoral thesis (1953, Technical University Darmstadt, Professor Ulrich Hofmann) he described reactions in the interlayer space of layered crystals.

Progr Colloid Polym Sci (1998) 111: 6–8
© Steinkopff Verlag 1998

After two years (!) he passed the Habilitation with a contribution about cyanides of transition elements. In the 1950s he published studies on the structure and properties of various chalcogenides, silicides, germanides, cyanides, mercury compounds, thiosalts, etc. (many papers together with his brother, Alarich Weiss, G. Nagorsen, and H. Schäfer). An highlight was the preparation and crystal structure determination of a new modification of silicium dioxide which consists of infinite chains of SiO_4 tetrahedra sharing opposite edges.

The first publication (1951) was "Batavit", a vermiculite-like 2:1 clay mineral. As a student in the group of Ulrich Hofmann (Diploma examination in Munich, 1951) he acquired a long lasting interest focused on the study of clay minerals. He described fibrous vermiculite, wolchonskoite and saponite, but his main interest lay in the study of ion exchange properties and the intracrystalline reactivity of the clay minerals, in particular towards organic compounds. Thus, numerous intercalation compounds were first described by Armin Weiss and coworkers. In this field of research he also contributed to our knowledge of mechanisms of petroleum formation, and, for some time, he was engaged in the discussions about the role of clay minerals in the origin of life. We have enjoyed and preserved a vivid memory to endless discussions on this topic in Munich. Unfortunately, these brilliant ideas were never published in the convincing original framework.

A significant impact on colloid science resulted from studies on the thixotropy of clay mineral dispersion. The current understanding of ceramic processes is based on the principles of colloid chemistry of clay minerals worked out by Ulrich Hofmann and Armin Weiss.

More recently, Armin Weiss and coworkers studied alteration of clay minerals by phosphate and formation of taranakite and other alumophosphates, an important aspect to be considered when clay minerals are used in barriers.

A milestone (1961) was the observation of the intercalation capability of kaolinite, when kaolinite was reacted with urea. The idea behind these experiments was that the hydrogen bonds between the kaolinite layers may be opened by molecules such as urea, which are known to break hydrogen bonds.

This discovery not only resulted in practical applications but provided an explanation of the secret of Chinese procelain. Everyone who was lucky to attend one of his lectures on this topic was deeply impressed by his fascinating bridging between old eastern cultural heritage, empirical knowledge and modern science.

A few years later, Armin Weiss observed the intercalation of organic molecules into titanium disulfide and initiated a new field of research. Since then, the number of new intercalation compounds had increased rapidly.

In parallel with the studies of 2:1 clay minerals he investigated intracrystalline reactions of various layer compounds such as titanates, phosphates, vanadates, uranium micas (often together with K. Hartl and E. Michel), graphite oxide and, more recently, iron oxychloride and earth alkali quadrates (together with C. Robl). He also described, for the first time, the two-dimensional swelling of many chain-like compounds: polyphosphates, polyvanadates, mercury amidosulfonates, alginic acid, pectic acid and deoxyribonucleic acid.

A fascinating aspect of his research was the study of silicosis. In 1958 he analyzed solid materials enriched in human lungs and isolated a swelling, toxical iron phosphate silicate, which grows within the air cells and finally destroys them.

Besides these studies Armin Weiss was also interested in complex chemistry and published several papers on silicon complexes with octahedral coordination. More recently, he (together with S. Dick) modified the clay mineral surface with binuclear iron complexes to prepare enzyme models.

Armin Weiss continuously covered new fields of interest. He initiated solid state chemistry at high temperature (with K. Hartl) and high pressure (with K. J. Range). Dielectric measurements on layer compounds (with G. Schön) laid the foundations for dielectric spectroscopy and modern impedance spectroscopy in material research. Detection of kink-type conformational changes of the alkyl chains in bimolecular interlayer films (with G. Lagaly) bordered on polymer chemistry and biomembrane science. He also studied the nerve myelin structure and excitation of nerve cells.

In 1974 he initiated (together with H.-P. Boehm and G. Lagaly) an advanced course "stability of dispersions and emulsions". This training of scientists from industry and practice became one of the most effective courses of Fortbildungskurse der Gesellschaft Deutscher Chemiker. From 1967 to 1969 he performed 27 telecourses on general and inorganic chemistry.

Armin Weiss was always open to discussions of environmental problems. An increasing activity in this field was not unexpected as Armin Weiss always engaged himself fully in all areas he touched. From 1986 to 1990 he acted as a member of the Bayerischer Landtag (parliament of Bavaria). Many of his colleagues could not understand his consequent morale since this was connected with many troubles and hostilities.

As annex we report biographical data:

1927 born in Stefling (a small village near Regensburg), 1943–1945 Luftwaffenhelfer in the second world war. 1945 Vorexamen in pharmacy, 1947–1951 study of chemistry in Regensburg, Würzburg and München. 1951 diploma examination in Munich,

1953 thesis (Technical University, Darmstadt, Prof. Dr. Ulrich Hofmann), 1955 Habilitation (Technical University Darmstadt). 1961–1965 Professor (Extraordinarius) of Inorganic Chemistry, University Heidelberg, 1965–1996 Professor (Ordinarius), University Munich.

Several of his coworkers hold or held positions as professor of inorganic chemistry: G. Lagaly (Univ. Kiel), K.-J. Range (Univ. Regensburg), C. Robl (Univ. Jena), H. Schäfer (Univ. Darmstadt, ✠ 1986), R. Schöllhorn (Techn. Univ. Berlin), G. Schön (Univ. Essen).

Armin Weiss was honoured with several awards:

1963, Medal d'Hommage, Université Libre Bruxelles;
1979, Member of New York Academy of Sciences;
1981, Liebig-Preis der Gesellschaft Deutscher Chemiker;
1984, Preis für Verfahrenstechnik der Textilveredelung;
1985, Dr. rer. nat. h. c., Lorand-Eötvös-University Budapest;
1987, Distinguished member of Sociedad Española de Arcillas.

He was visiting professor at Université Libre, Bruxelles (1963), Unileyer Visiting Professor, University Bristol (1974), and guest at Institute Laue-Langevin in Grenoble (1980).

A few basic papers concerning clay mineral and colloid chemistry are listed:

Weiss A, Mehler A, Koch G, Hofmann U (1956) Über das Anionenaustauschvermögen der Tonmineralien. Z Anorg Allg Chem 284:247

Weiss A (1958) Die innerkristalline Quellung als allgemeines Modell für Quellungsvorgänge. Chem Ber 91:487

Weiss A (1958) Über das Kationenaustauschvermögen der Tonminerale. I. Vergleich der Untersuchungsmethoden. Z Anorg Allg Chem 297:232

Weiss A (1958) Über das Kationenaustauschvermögen der Tonminerale. II. Der Kationenaustausch bei den Mineralen der Glimmer-, Vermikulit- und Montmorillonitgruppe. Z Anorg Allg Chem 297:257

Weiss A (1959) Über das Kationenaustauschvermögen der Tonminerale. III. Der Kationenaustausch bei Kaolinit. Z Anorg Allg Cehm 299:92

Weiss A, Michel E, Weiss Al (1959) Über den Einfluß von Wasserstoffbrückenbindungen auf ein- und zweidimensionale innerkristalline Quellungsvorgänge. In: Hydrogen Bonding, S. 495. Pergamon Press, London

Weiss A (1961) Eine Schichteinschlußverbindung von Kaolinit mit Harnstoff. Angew Chem 73:736

Weiss A (1962) Neuere Untersuchungen über die Struktur thixotroper Gele. Rheologica Acta 2:299

Weiss A (1963) Organische Derivate der glimmerartigen Schichtsilicate. Angew Chem 75:113

Weiss A (1963) Ein Geheimnis des chinesischen Porzellans. Angew Chem 75:755

Weiss A (1963) Isolierung und konstitutionsermittlung des quellungsfähigen Phosphatosilicates aus Lungen Silikosekranker. Beitr zur Silikoseforschg Sonderband Grundfragen der Silikoseforschg 5:93

Weiss A, Roloff G (1964) Die Rolle organischer Derivate von glimmerartigen Schichtsilikaten bei der Bildung von Erdöl. Internat Clay Conf, 1963, Vol 2, S. 373. Pergamon Press, London, 1964

Range K-J, Weiss A (1968) Über das Verhalten von Kaolinit bei hohen Drucken. Ber Dtsch Keram Ges 46:231

Weiss A, Range K-J (1968) Zur Existenz von Kaolinithydraten. Z Naturf 23b:1144

Weiss (1969) Organic derivatives of clay minerals, zeolites and related minerals. Organic Geochemistry, p 737. Springer Verlag, New York, Heidelberg

Schön G, Weiss A (1973) Nachweis von strukturellen L-Defekten im innerkristallinen Wasser von hydratisierten Vermikuliteinkristallen mittels dielektrischer Messungen. Z Naturf 28b:140

Lagaly G, Fitz S, Weiss A (1975) Kink block structures in clay organic complexes. Clays Clay Minerals, Vol 23. p 45, Pergamon Press

Lagaly G, Beneke K, Weiss A (1975) Magadiite and H-magadiite. Sodium magadiite and some of its derivatives. Am Min 60:642; H-magadiite and its intercalation compounds. Am Min 60:651

Lagaly G, Weiss A (1976) The layer charge of smectitic layer silicates. Proc Internat Clay Conf 1975, p 157

Lagaly G, Weiss A, Stuke E (1977) Effect of double-bonds on bimolecular films in membrane models. Biochim Biophys Acta 470:331

Weiss A (1977) Replication, evolution and differentiation in clay minerals – A model of protolife. Proc 3rd Europ Clay Conf, Oslo, 1977, p 228

Weiss A, Gossner U, Robl C (1995) Transformation of clay minerals into taranakite and the crystal structure of taranakite. Proc 10th Internat Clay Conf, Adelaide, 1993, CSR10 Publ, Melbourne, p 253

Weiss A, Dick S (1997) Interaction of montmorillonite with binuclear hydroxo-bridged iron complexes and their peroxo adducts. Clay Min 32:135

Progr Colloid Polym Sci (1998) 111:9–16
© Steinkopff Verlag 1998

TECHNICAL APPLICATIONS

W. Umbach

The importance of colloid chemistry in industrial practice

Dr. W. Umbach
Henkel KGaA
Henkelstraße 67
D-40191 Düsseldorf
Germany

Abstract Colloid and interfacial chemistry has been of outstanding importance for the development of products and processes for a long time. As there is a close interaction between basic research and applied industrial research in colloid and interface science, the demands which influence the innovation process in companies also affect the fundamental research. A summary of new ideas and strategies is given regarding the innovation process using Henkel KGaA as an example. New developments in the fields of detergents and cleansers, cosmetics, surface technologies, and adhesives show the importance of colloid and interfce science as well as the necessity of efficient project control within the context of the innovation process.

Key words Colloids – surfactants – innovation process – detergents – cosmetics – computer simulation

Introduction

Colloid and interfacial research is firmly established today as a basis for the development of methods and products. The areas of application which are shaped by colloid systems and interfacial effects are numerous and almost impossible to survey. They range from paints and lacquers which are, incidentally, one of the oldest examples for the practical implementation of colloid and interface know-how, via detergents and cleansers, adhesives, metal processing and textile treatment up to pharmaceuticals, cosmetics and food. Accordingly, colloid and interfacial chemistry also has a great economic importance. This is shown for Germany in Fig. 1 for selected groups of products which concern the bottom-line activities of Henkel KGaA. The consumer prices or ex-factory prices are shown for 1996.

In the raw materials sector the importance of colloid and interfacial chemistry is also obvious. The raw materials group – surfactants – can be regarded as being one of the most important classes of interface-modifying substances. Figure 2 shows this clearly on the basis of the market volumes of different types of surfactants. The information refers to the volume of active ingredient used in Germany. It must be pointed out that anionic and nonionic surfactants in particular are manufactured in very large quantities and used for very different applications.

In colloid and interfacial chemistry there has always been a close interaction between basic research and applied industrial research in the development of products and methods. At Henkel KGaA this has a long tradition. The work of Götte [1], Kling [2] and Lange [3] was concerned with the investigation of the interfacial chemistry of surfactants and the physical chemistry of the washing process, and laid the foundations for numerous new products and application areas. Systematization of the research investigations was given a high priority.

Today the role that chemical research plays has fundamentally changed. The companies face complex demands with respect to the year 2000 and beyond. This also affects research in the chemical industry and thus the colloid and interfacial chemistry sectors. The way in which the chemical industry has reacted is described below using Henkel KGaA as an example.

- Product groups* -

▪ Detergents and cleansers	7.3 billion DM
▪ Cosmetics	15.8 billion DM
▪ Adhesives	1.7 billion DM
▪ Metal processing additives	0.7 billion DM

(automotive industry, industry, water treatment)

Fig. 1 Economic importance of colloid chemistry. *Turnover in Germany, 1996

- Surfactants* -

Anionic surfactants	312000 t
Nonionic surfactants	365000 t
Cationic surfactants	52000 t
Amphoteric surfactants	18000 t

Fig. 2 Market volumes of surfactants. * Market volume in Germany according to active ingredient, 1996

The innovation process

An important criterion for the economic future ability of companies is their potential for innovation. Shorter product life cycles and aggressive competition in increasingly saturated markets have led to the ability to innovate becoming an essential competitive factor. This leads to direct demands being placed on research and develop-ment: work which is close to the market, flexible and rapid is required.

In recent years Henkel has strategically reacted to these alterations with a partial decentralization of research and development (R + D). A share of approx. 80% of the R + D budget is directly allocated to the operative business and is used for market-near developments. Of the remaining budget approx. 25% is used under the direction of central research for exploratory research (Fig. 3). Research has the function of a "radar" for future trends here.

In a futurology workshop Henkel is attempting to determine both future-relevant developments in research and technology as well as in society, the markets and in environmental conditions, and then to transfer this knowledge into Henkel-relevant impulses, ideas and topics. As well as the shortening of the innovation cycles mentioned above, trends in globalization, interdisciplinarity, miniaturization and even the use of computer simulations are of particular importance. Exploratory research stands under the higher aims of indicating the unexpected, of looking for new ideas away from the beaten track and thus of laying the foundations for technical innovation. The results of exploratory research are utilized as new options in the planning of R + D projects.

With the new instruments of innovation management the set targets are achieved even more efficiently. Successful R + D management uses different tools in the different phases of the innovation process:

– portfolio management,
– project management,
– value contribution analysis,
– innovation characteristics.

In the final analysis the systematization of the innovation process is used for lasting improvement and thus to a shortening of the often-mentioned "time to market".

Fig. 3 Overview of the innovation process

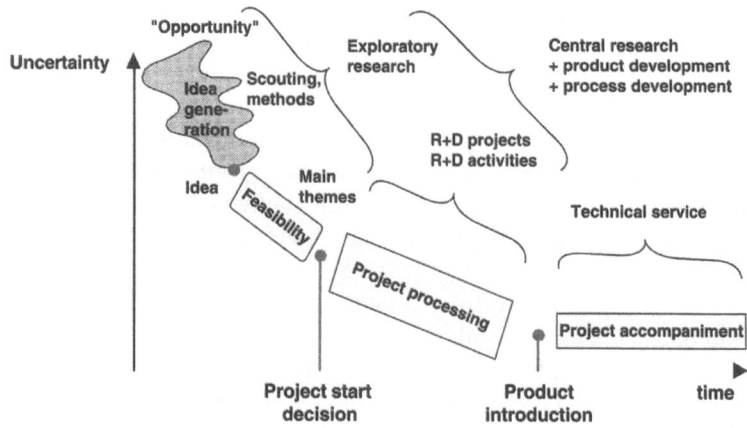

Progr Colloid Polym Sci (1998) 111:9–16
© Steinkopff Verlag 1998

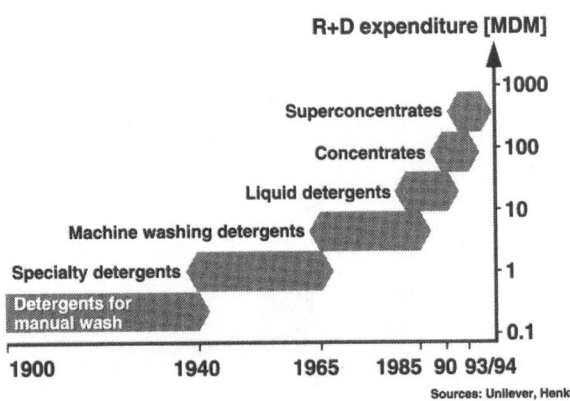

Fig. 4 Product cycles and R + D expenditure

Figure 4 shows very clearly just how important this is for the development of heavy-duty detergents. This survey shows how the product development cycles have been shortened in recent decades and how the development costs have increased even when inflation is taken into account. Although the R + D expenditure is only based on a very coarse estimate it can be seen that the development costs of a new detergent were of the order of approx. DM 100 000–200 000 at the start of this century, while today for modern superconcentrates their order of magnitude clearly exceeds 100 MDM. This means that false decisions in the innovation process, as can easily be estimated from this example, not only lead to the delayed introduction of products but also mean a considerable increase in the formerly planned research and development costs up to product launch.

Examples from industrial practice

In the past years numerous new products and methods based on colloid and interfacial chemistry research work

have been developed. The following examples from industrial practice are intended to make clear the importance of colloid and interfacial chemistry, while also pointing out the importance of efficient project control within the context of the innovation process. The topics were selected from business segments of the Henkel group, which include the following sectors:

- detergents and cleansers,
- cosmetics/body care,
- surface technologies,
- adhesives.

Detergents and cleansers

If the demands placed on modern detergents and cleansers are considered as our first example then for the consumer the following points stand in the foreground:

- optimized performance,
- concentrated products,
- improved convenience.

Today surface chemistry parameters are used for optimizing the performance, e.g. as described by Nickel et al. [4] for various surfactant mixtures.

Liquid products with a high content of active ingredients place very high demands on their pourability on use and storage stability across a wide range of temperatures. These criteria are determined by the phase behavior of the concentrated liquid formulations, with a decisive parameter being the optimization of complex surfactant–polymer–electrolyte–solid systems. There characteristics are based on simpler systems which are topics for basic research [5], but their application is additionally influenced by the complex composition. Industrial interfacial research provides the real work and know-how here.

Fig. 5 Viscosity of surfactant mixtures

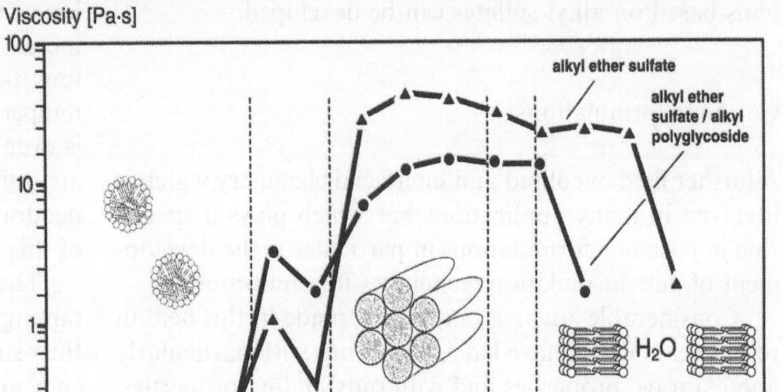

Figure 5 shows as an example the influence of surfactant mixtures on phase behavior and thus on the rheological properties which are relevant for detergents and cleansers [6]. The relationship between the viscosity and the concentration is shown for an anionic surfactant (alkyl ether sulfate) and for a mixture of this alkyl ether sulfate and a nonionic surfactant (alkyl polyglycoside). It can be clearly recognized that for the surfactant mixture the viscosity at higher concentrations decreases strongly when compared with the anionic surfactant, while in the concentration range between 20 and 30% w/w a micellar gel is formed, which in the case of the surfactant mixture leads to a significant increase in viscosity. This example shows that a knowledge of the phase behavior of surfactants, which also influences the pourability, is of paramount importance for the optimization of product properties under different conditions. For example, it is possible to manufacture liquid dish-washing concentrates which, apart from their naturally high cleansing performance, also guarantee an adequate pourability for their application while at the same time they have a consistency which the consumer finds pleasant even at different temperatures.

The dependency of the phase sequence and the phase behavior which results from the alkyl chain length was described by Hofmann et al. [7] for a different anionic surfactant (alkyl sulfate). If a move is made from dodecyl sulfate to longer alkyl chains then the essential features of the phase diagram remain unchanged; however, the concentration ranges in which the phases appear alter. The hexagonal phase shows a larger expansion with longer alkyl chains and is also shifted to lower concentrations. The solubility limit is shifted towards higher temperatures as the alkyl chain length increases. This solubility limit can then again be positively influenced by the addition of a nonionic surfactant ($C_{8/10}$ alkyl polyglycoside), so that even at lower temperature ranges suitable surfactant systems based on alkyl sulfates can be developed.

Cosmetic formulations

A further field of colloid and interfacial chemistry which is involved in many applications but which plays a special role in cosmetic formulations in particular is the development of new formulation techniques for emulsions.

Considerable advances have been made in this field in recent years which have led to emulsions with particularly good storage properties and with outstanding properties. The basis for these developments has been the fundamental work by Kahlweit [8] and Shinoda [9] on the phase behavior of ternary oil/water/surfactant systems in which pure surfactants were used.

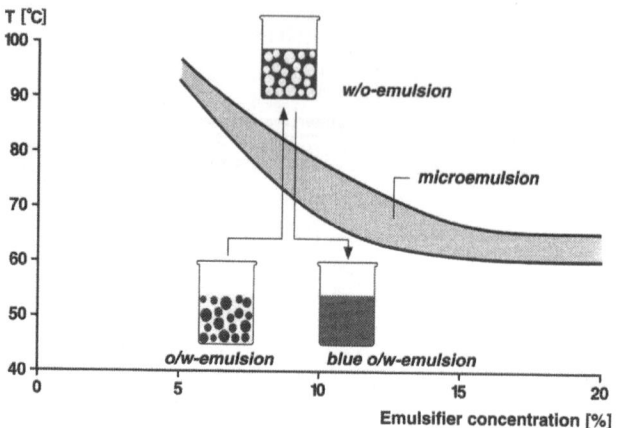

Fig. 6 Principle of the PIT method

Special mention must be made here of the so-called PIT emulsions (PIT = Phase Inversion Temperature) and microemulsions. Both are characterized by particularly fine droplets which give the emulsions an increased stability with microemulsions additionally allowing a clear formulation of oil/water systems.

The principle of the phase-inversion method (Fig. 6) is based on the ability of systems of water, oil and ethoxylated nonionic surfactants to invert from an oil/water emulsion to a water/oil system when the temperature is increased [10]. During this process a microemulsion range is passed through where the temperature range for technical systems is a function of the emulsifier concentration. If the water/oil emulsion is cooled below the inversion range, an oil/water emulsion is obtained which because of its fineness (droplet size < 200 nm) has a blue glimmer and is therefore often known as a "blue emulsion". Because of its fineness this type of emulsion provides the ideal condition for the preparation of oil/water emulsions which are stable over long periods.

It should be mentioned that this type of emulsion is kinetically stabilized and, depending on further parameters such as the type of additive, can be kept stable for long periods in regions considerably below the inversion temperature. With the aid of the phase-inversion method it is even possible to formulate oils and oil mixtures which are difficult to emulsify in a stable emulsion. New types of deodorant emulsions can be used as an example for the use of this principle.

The second approach to formulating emulsions containing fine particles are microemulsions, which because of their small particle size are characterized by their transparency and thermodynamic stability. They also have a high solubilizing capacity for oil- and water-soluble substances. Despite these outstanding advantages microemulsions have until now not been used to the extent which might have been expected. This is because of the fact that in

Progr Colloid Polym Sci (1998) 111:9–16
© Steinkopff Verlag 1998

many surfactant systems microemulsions are only stable over a limited temperature range (see Fig. 6) and therefore cannot be used in numerous applications. With new types of surfactants (such as the alkyl polyglycosides [11]) it is, however, possible to extend the stability range considerably and to formulate microemulsions with increased temperature stability. The necessary adjustment of the hydrophilic/hydrophobic balance of the emulsifier system for the microemulsion is carried out by the selection of suitable co-emulsifiers and fixing a suitable concentration ratio between emulsifier and co-emulsifier.

This procedure is illustrated in Fig. 7 using a perfume oil microemulsion as an example [12]. Conductivity and particle size of the emulsion are shown as a function of the concentration ratio between co-emulsifier and total emulsifier. Although some applications already exist today for microemulsions, we are without doubt just at the start of a further great development.

The type of emulsion is not just responsible for the consistency of formulations, but for example in the cosmetics sector it is also responsible for the interaction with the skin and therefore for the penetration of active ingredients into the skin layers. It has been shown here that not only the original formulation but, in particular, the phases passed through in sequence as it dries on the skin must be taken into account if transport into the skin is to be optimized [13]. This confirms that a knowledge of the phase behavior is of fundamental importance and therefore should be the object of further research.

Computer simulations

Many physical–chemical measurements such as the study of phase diagrams require a lot of time. This is particularly the case when complex systems, as found in practice, are to be investigated. In interfacial chemistry computer simulations or molecular modeling are increasing in importance to the extent that they can support experimental work. Simple calculation methods, for example, the HLB concept for the selection of emulsifiers [14] or the CAPICO method for calculating phase-inversion temperatures [15] already exist today, but approaches to calculating surfactant structures in aqueous systems or the association of surfactants in aqueous systems or at interfaces on a molecular basis are also on the increase [16, 17].

Figure 8 shows the simulation of a nonionic surfactant of the ethylene oxide type in water as an example [18]. It is possible to simulate, for example, the dependence of this structure on the temperature and in this way to make statements about the influence of parameters on the structure of different surfactants and the way that they will behave in formulations. With these calculations, which are being carried out today by various research groups with different approaches, we are without doubt at the start of a great development whose target could be the calculation of the behavior of surfactants or surface effects in complex processes, such as for instance the washing process.

Fig. 7 Droplet size and conductivity of a perfume oil microemulsion

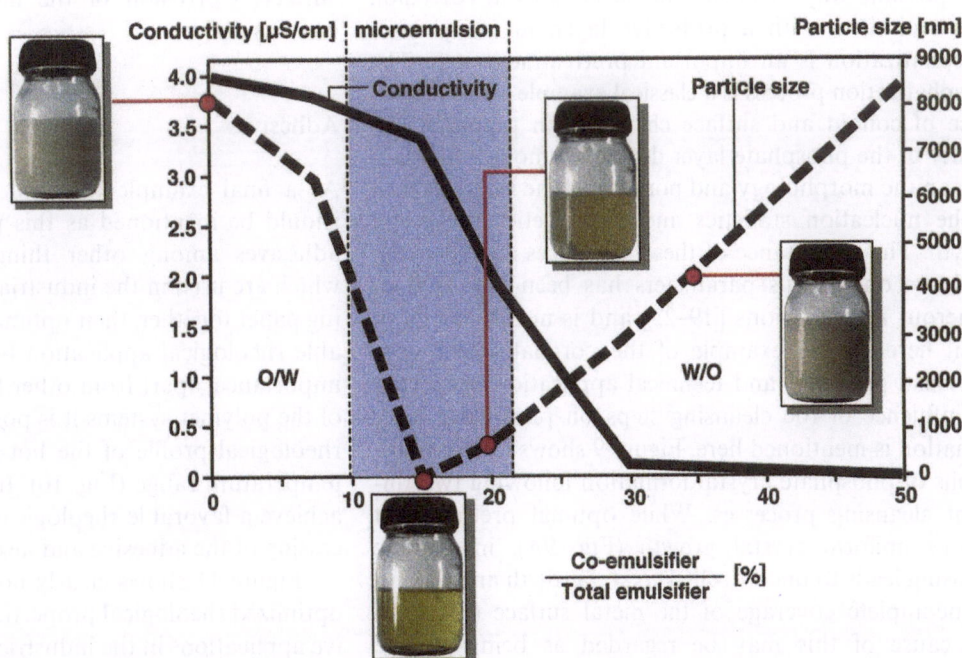

Fig. 8 Computer simulation of
a nonionic surfactant in water
(dodecyl alcohol triglycol ether)

Surface technologies

A great problem encountered when using metals is that
they are susceptible to environmental and weathering influences, particularly those of water and oxygen. This
means that corrosion protection is extremely important.
One possible way of protecting surfaces from corrosion
is to seal them with a protective layer; in many cases
phosphatization is an important pretreatment step. The
phosphatization process is a classical example of the importance of colloid and surface chemistry in practice. The
quality of the phosphate layer depends, among other factors, on the morphology and porosity of the layers as well
as the nucleation statistics and the kinetics of crystal
growth. The importance of these influences and their dependence on various parameters has been indicated in
numerous investigations [19–22] and is not discussed in
detail here. As an example of the correlation between
interfacial processes and technical application properties
the influence of the cleansing steps on phosphate layer
formation is mentioned here. Figure 9 shows microphoto-
graphs of phosphate crystal formation following two different cleansing processes. While optimal pretreatment
ensures uniform crystal growth (Fig. 9A), inadequate
cleansing leads to uncontrolled crystal growth and thus to
an incomplete coverage of the metal surface (Fig. 9B).
The cause of this may be regarded as being surface

inhomogeneities which remain if the surface cleansing process is not thorough enough. As nucleation during phosphatization begins immediately after the metal surface
is immersed in the phosphatizing bath and is already
complete after about 30 s this means that the phosphate
crystals will grow unevenly on a not thoroughly cleaned
surface. Corrosion of the metal surface is a foregone
conclusion.

Adhesives

As a final example the field of polymer characteristics
should be mentioned as this plays an important role in
adhesives among other things. In hot-melt adhesives,
which are used in the industrial adhesives sector for stick-
ing paper together, their optimal adhesion as well as favorable rheological application behavior are of outstanding
importance, apart from other factors. By skilled selection
of the polymer systems it is possible to suitably adjust the
rheological profile of the hot-melt adhesive over a wide
temperature range (Fig. 10). In this way it is possible to
achieve a favorable rheological behavior both in the processing of the adhesive and also in the use of the product.

Figure 11 shows clearly how important the setting of
optimized rheological properties is, particularly for adhesive applications in the industrial sector. If the adhesive has

Progr Colloid Polym Sci (1998) 111:9–16
© Steinkopff Verlag 1998

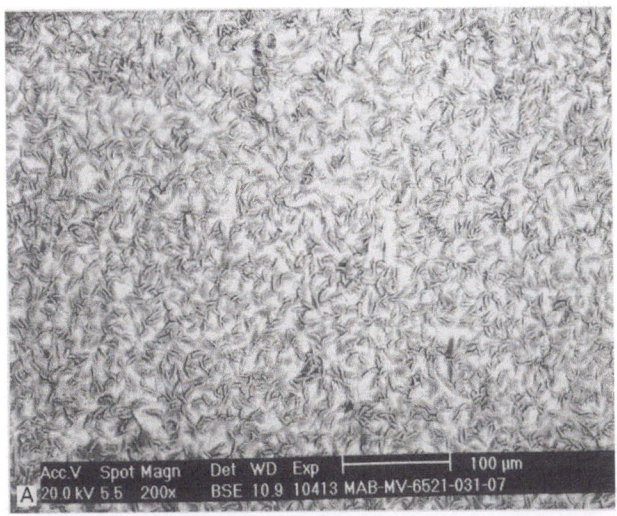

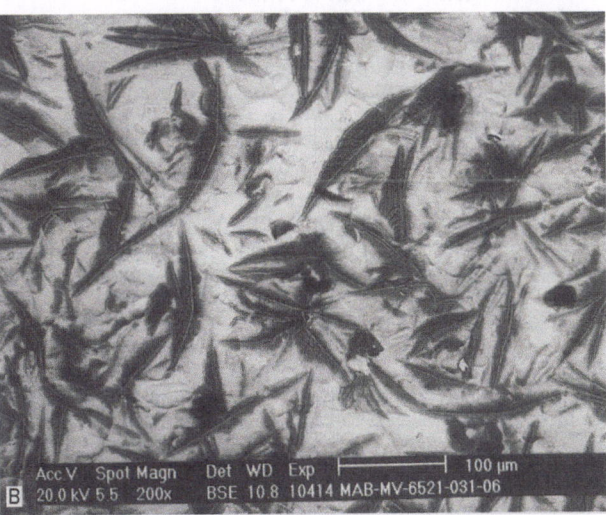

Fig. 9 Micrographs of phosphate layers: (A) optimal pretreatment; (B) inadequate cleansing

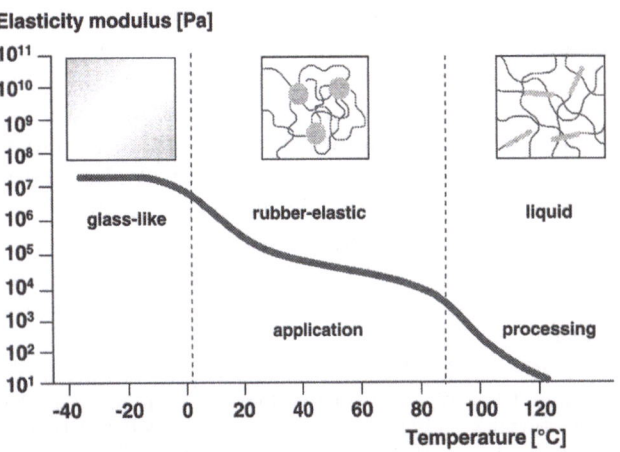

Fig. 10 Elasticity modulus of hot-melts as a function of temperature

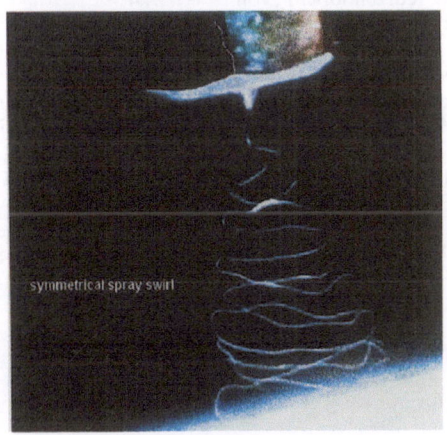

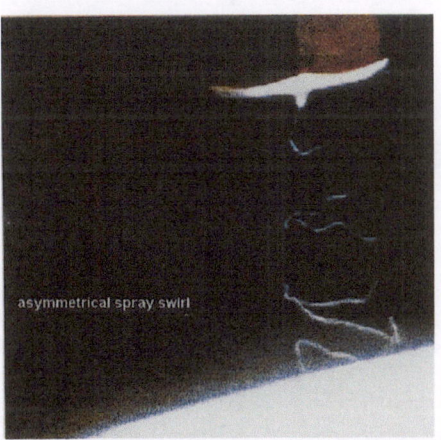

Fig. 11 Spray swirl of different adhesives: (A) favorable rheological properties; (B) unfavorable rheological properties

favorable flow properties then the spray pattern when the adhesive is sprayed is very uniform (Fig. 11A), whereas with unfavorable rheological properties an asymmetric spray cone is seen which produces an uneven spray pattern (Fig. 11B). Not only systematic investigations into the polymer properties help here, but also the use of new measuring techniques for optimizing the adhesive.

A preview

The few chosen examples make it clear that colloid and interfacial chemistry are acquiring an ever-increasing importance in the development of modern products and methods which extends far beyond the approaches already made in the past. This trend will doubtlessly become even stronger in future, as the demands placed on products become increasingly more complex.

Also new developments such as nanostructured systems and new types of coating technologies are beginning to emerge in which questions of interfacial chemistry play a central role. Apart from the technical applications, the understanding of biological systems from a colloid chemistry standpoint will increase and will also affect product sectors such as cosmetics and pharmaceuticals. As many experiments in colloid and interfacial chemistry are complicated and require a lot of time, simulations of interfaces will achieve an ever-increasing importance. First approaches to calculating the aggregation behavior of surfactants and the modification of surfaces by surface-active substances already exist, but require even greater efforts to be able to achieve a better understanding of complex systems.

References

1. Götte E, Kling W, Mahl H (1954) Melliand Textilber 35:1252
2. Kling W, Lange H (1960) J Am Oil Chem Soc 37:30
3. Lange H (1975) Tenside Detergents 12:27
4. Nickel D, Nitsch C, Kurzendörfer P, von Rybinski W (1992) Progr Colloid Polym Sci 89:249
5. Sjöblom J, Stenius P, Danielsson I (1987) Nonionic Surfactants – Physical Chemistry. In: Schick MJ (ed) Marcel Dekker, New York, p 369
6. Andree H, Hessel JF, Krings P, Meine G, Middelhauve B, Schmid K (1997) Alkyl Polyglycosides. In: Hill K, von Rybinski W, Stoll G (eds) VCH, Weinheim, p 99
7. Hofmann R, Nickel D, von Rybinski W (1994) Tenside Surf Detergents 31:63
8. Kahlweit M, Strey R (1985) Angew Chem 97:655
9. Shinoda K (1967) J Colloid Interface Sci 24:4
10. Förster T, von Rybinski W, Wadle A (1995) Adv Colloid Interface Sci 58:119
11. Hill K, von Rybinski W, Stoll G (1997) Alkyl Polyglycosides. VCH, Weinheim
12. von Rybinski W, Guckenbiehl H, Tesmann H (1998) Colloids Surf A, to be published
13. Förster T, Jackwerth B, Pittermann W, von Rybinski W, Schmitt M (1998) Cosmetics and Toiletries 112:73
14. Griffin WC (1949) J Soc Cosmetic Chemists 1:311
15. Förster T, von Rybinski W, Tesmann H, Wadle A (1994) Int J Cosmetic Sci 16:84
16. Esselink K, Hilbers PAJ, van Os NM, Smit B, Karaborni S (1994) Colloids Surf A 91:155
17. Kuhn H, Rehage H (1997) Ber Bunsenges Phys Chem 101:1493
18. Paschek D, Geiger A. Unpublished results
19. Losch A, Schultze JW, Speckmann HD (1992) Metalloberfläche 46:133
20. Roland WA, Gottwald KH (1988) Metalloberfläche 42:301
21. Tegehall PE, Vannerberg NG (1991) Corrosion Sci 32:635
22. Cheever GD (1967) J Paint Techn 39:504

Progr Colloid Polym Sci (1998) 111:17–26
© Steinkopff Verlag 1998

G. Rossmy

From polyurethane foam stabilization to amphiphilic particles – An example of industrial R & D in colloid chemistry

Plenary lecture presented at the "38. Haupt-versammlung der Kolloidgesellschaft". (29.10.–1.11.1997, University of Essen/Germany)

Dr. G. Rossmy
Im Hadkamp 18
D-45721 Haltern-Lavesum
Germany

Abstract In this article, a new function of siloxane polyether surfactants is described. It turns out that these compounds are able to stabilize the rising flexible poly-urethane foam by adsorption on the surface of precipitating urea particles. In this way, foam collapse is prevented. The influence of siloxane surfactants on the properties of suspended urea particles was investigated by TEM and AFM. Basic research in the field of polyurethane foam induced the synthesis of amphiphilic particles with designed anisotropic distributions of hydrophilic and hydrophobic domains. Particles derived from hollow glass spheres showed some principal properties of this class of amphiphiles: size and curvature of these particles allowed the stabi-lization of emulsified droplets of unusual shape and size (cauliflower, inverted cauliflower, and highly distorted structures). Several synthetic approaches for Janus-like particles are given. A very simple and efficient way is the restructuring of clusters of coated silica particles with different degrees of hydrophobicity at a water/oil interface. These amphi-philic particles are an excellent example for "physically coupled surfactants".

Key words Polyurethane foam – siloxane surfactant – Janus bead – amphiphilic particle – physically coupled surfactants

From polyurethane foam stabilization to amphiphilic particles

In flexible polyurethane foam, which is often used for typical slabstock formulations, polysiloxane polyether copolymers are an essential compound for increasing foam stabilization. Originally, this phenomenon was just traced-back to the rather unique capability of the siloxane deriva-tives to reduce the surface tension of the main component of the foam, a polyether-triol. Optimization was thought to be realized by adjusting the dynamical parameters of the system.

In former investigations we noticed [1] that the col-loidal aspects of stabilizing flexible polyurethane foams are, indeed, more complicated. The surface tension mea-sured by addition of polysiloxane polyether copolymers gives no information on the quality of suitable surfactants even if one includes the time dependence of the surface tension reduction. By close inspection of the mechanisms of the reactions of toluene diisocyanate with water and polyol, it becomes clear that the main function of the surfactant is related to the precipitation of polyurea par-ticles in the foam mixture. (In this approach we have neglected the important but not critical assistance of the surface active compound in the nucleation of gas bubbles.) Precipitation of polyurea particles takes place in a late phase of the foam rise. Without a suitable surfactant, urea precipitation is immediately followed by a foam collapse. The quality of the surfactant can easily be demonstrated by the constant growth of the foam after urea precipitation. But here a delicate compromise has to be achieved. First of all, the blowing gas should be utilized as long as possible to reach maximum volume of the foam. On the other hand, the cell structure should open as much as possible: a flexible foam with closed cells cannot be used in technical applications because it will shrink. As a consequence, air flow through the flexible foam should be as large as possible.

The rising foam – as soon as its starts to form a polyhedral system – represents a special case of a dissipative system, which has interesting features. As was shown already by Plateau, a pressure difference between the curved parts of the cell struts and the flat membranes forming the windows of the polyhedral cells exists. The cell struts, which can be defined as Plateau borders, induce suction on the liquid film of the membranes. This has to be counteracted by the pressure increase of the blowing gas. In this way the premature opening of the foam is effectively prevented. It is interesting to note that expansion of the foam is a stabilizing factor in itself. We have developed speculative ideas about the roughening of the film surfaces induced by the polymeric surfactant. In this way, the energy transfer of the blowing gas into the liquid phase is optimized. In this simple model the surfactant has somehow the function of a clutch. A late cell opening can then be attributed to a tight coupling of the expanding gas and the fluid phase.

We have not been able to prove this simple idea. Another function of the surfactant can, however, easily be demonstrated. It is well known that after about three-quarters of the normal rise time a foam without any surfactant tends to collapse. In the time before, the rise profile does not differ too much from that of a foam stabilized by adding suitable surfactants [1]. We have shown that a few seconds before the foam collapses urea particles precipitate out of the foaming liquid. The foam destroying property of fine particles is well known and technically used in many defoamer formulations.

The siloxane surfactant in the polyurethane foam mixture does not prevent the basic process of urea precipitation. But, obviously, it seems to modify those particles in a way which makes them tolerable for the foam system. The polydimethylsiloxane groups are the basis for the low surface energy, which makes this class of surfactants fairly unique for organic systems. The polyether part consists of two different parts, ethylene oxide and propylene oxide. The ethylene oxide groups do not only lead to the solubility of the surfactant in water but also provide acceptor functions for hydrogen bonds. Of course, a well-defined balance between the size of those blocks has to be adjusted. This is similar to selecting a certain HLB value of surfactants in aqueous solutions. It is well known that the

foam stabilizing surfactant must have a branched siloxane structure – linear block copolymers are not suitable at all. The potency of the surfactant increases with the degree of branching. The limitation is only set by the need for sufficient cell opening. Also, the length of the polyether branches is critical: for most practical requirements the molecular weight of the polyether, which is used to form the copolymer, exceeds 1600. The polyether block has also to contain a certain amount of ethylene oxide. Additionally, one has to take into account the requirement that ethylene oxide and propylene oxide have to be arranged in the polyether block in an approximately statistical manner. For successful applications one has to avoid ethylene oxide blocks exceeding five groups. These numbers are given for standard slabstock formulations. For special purposes, for example, with the usage of highly reactive polyols, the structural requirements may be significantly different. The most effective surfactants with pending polyether branches are not taken into consideration here. Although ethylene oxide groups tend to promote stabilization, the useful range is fairly limited. This problem is caused by incomplete cell opening which usually occurs at higher ethylene oxide contents. Additionally the molecular weight distribution has an important influence on the activity of the surfactant.

Taking all this into consideration and keeping in mind that special foam formulations require specific compromises, one may understand that still today the design of an optimized surfactant is more an art than really a science.

From the above-mentioned structure effects it is possible to develop a simple model from which additional guidelines can be derived. It seems to be essential that the surfactant adsorbs at the surface and especially in the cell windows, which provide large surface areas. As discussed before, the surfactant has to modify the polyurea particles, so that the whole system remains stable. It is certainly a plausible hypothesis that this occurs by forming hydrogen bonds between the NH groups of polyurea and the oxygen atoms of the surfactant ethylene oxide groups. The requirement of a branched structure can best be explained by a model in which the urea particles are integrated in an elastic network. This leads to an enhanced stability, possibly even converting a defoamer into a foam stabilizing compound. Such processes may also lead to an elevated surface viscosity. The elastically controlled integration of the polyurea particles provides also an increased surface roughness.

There is a very simple but persuading proof for a close contact of surfactant and polyurea particles: if one replaces in a foaming system the main component – the polyether triol – completely by a silicone surfactant one also gets very stable foams. Of course this system is not a

Progr Colloid Polym Sci (1998) 111:17–26
© Steinkopff Verlag 1998

polyurethane foam; after curing it consists of a stiff dispersion of polyurea particles in excess surfactant. The foam can easily be pulverized and a brittle powder is obtained. If one tries to recover the polyurea by continuous extraction of the surfactant with hot ethanol, the residue consists of approximately 50% polyurea and 50% surfactant. The "surfactant foam" itself acts in a normal foam formulation like a nearly perfect stabilizer. Experiments with fine particles of different origins support the thesis that the adsorption of siloxane polyether copolymers on polyurea is the principal reason for foam stabilization.

Fumed silica is commercially available in different degrees of hydrophobicity. We added 1% of such silica (size of primary particles about 12 nm, but under normal conditions associated to much larger clusters) into a foam formulation having a surfactant content, which was just sufficient to prevent collapse of the system (0.3–0.4 parts surfactants per 100 parts of polyol). These experiments were made under carefully standardized conditions, allowing the observation of small deviations. The experiments have shown that unmodified hydrophilic fumed silica (e.g., A 200 from Degussa) and the very hydrophobic silica (HDK 2000 from Wacker) both destabilize the foam, whereas an intermediate product like R 972 (Degussa) tend to increase stabilization (Table 1).

These results are in fairly good agreement with experiments performed with glass beads and quartz particles. Extensive experiments of Johannsen and Pugh [2] and Aveyard et al. [3] revealed that particles with intermediate contact angles gave the best improvement in foam stability. More hydrophobic particles clearly destabilized the foam, and the most hydrophilic additives did not show much influence. The destabilization by the hydrophobic particles was explained by a bridging-dewetting mechanism in the films thinning by drainage, whereas the stabilization by the amphiphilic particles was attributed to their capability to adsorb at the surface and finally pack themselves at the interface of the plateau borders.

Keeping in mind that aqueous and non-aqueous foams are quite different in many aspects, the parallelism of the observed relationships is really remarkable.

We tried to investigate the interaction of urea particles with surfactants by applying more sophisticated analytical tools. This work was done in cooperation with Prof. G.L. Wilkes at the Department of Chemical Engineering and Polymer Materials and Interfaces Laboratory, Virginia Polytechnic Institute and State University.

There is a principal difficulty in comparing a foam specimen with the collapsed system that one gets in the absence of any surfactant. We cannot expect a homogeneous distribution of the surface active compound within the whole structure; in the rising foam most of the surfac-

Table 1 Effect of fumed silica on the stabilization of PU-foams; basic stabilization of [4 H$_2$O] PU foam with 0.3 wt% BF 2370

Additive [1 wt%]	Methanol number	Settling of foam [cm]
—	—	3.1
A 200	0	Collapse
R 972	40	2.0
R 805	45	3.2
A 202	58–60	9.7
HDK 2000	> 60	Collapse

tant will probably be found at the interface between the cell windows and the gas phase. The surfactant concentration in the bulk phase will be much lower. In the cell-opening phase the windows drain into the struts. But in a well-stabilized foam this flow-back is retarded by an increased viscosity and will for this reason be incomplete. Therefore, we cannot expect to have a homogeneous surfactant distribution. Also, the interaction of urea particles and surfactant will be far from equilibrium.

The first experiments with TEM, WAXS, and SAXS did not show a reliable influence of the surfactant. We examined if this was possibly the consequence of the temperature increase after the precipitation of urea has occurred. For this reason we quenched the specimen shortly after urea precipitation by putting them into liquid nitrogen. This procedure limited the temperature rise in the foam to 90–110 °C, whereas a typical 4 water PU foam reaches temperatures of 140 °C. Using this technique TEM indeed showed an important difference between samples containing no surfactant and a foam stabilized with 1.5% surfactant based on polyol. Without the addition of surfactants, the urea particles tend to form clusters, whereas in the foam containing surfactant the particles are much better separated and more condensed. Figure 1 shows the comparison of the specimen without and with surfactant. The sample was quenched after 100 s. If we make the same comparison 20 s later, the difference has become much smaller. This means that the temperature increase also favors the de-clustering of the urea particles.

The clustering of urea particles will certainly depend on hydrogen bonds between C=O and NH groups. The FTIR spectra of samples without and with surfactant (Fig. 2) show a remarkable difference in the region of the C=O absorption of urea which interacts by hydrogen bonds.

Also, the AFM experiments gave a good indication of an influence of the surfactant on urea particle properties. In these types of experiments we used the tapping mode as phase imaging. Figure 3 gives a typical scanning micrograph: dark spots correspond to softer material.

Fig. 1 TEM micrographs of
[4 H$_2$O] PU foam quenched
in liquid nitrogen: (A) no
stabilization, after 100 s;
(B) 1.5% stabilizer, after 100 s;
(C) no stabilization, after 120 s

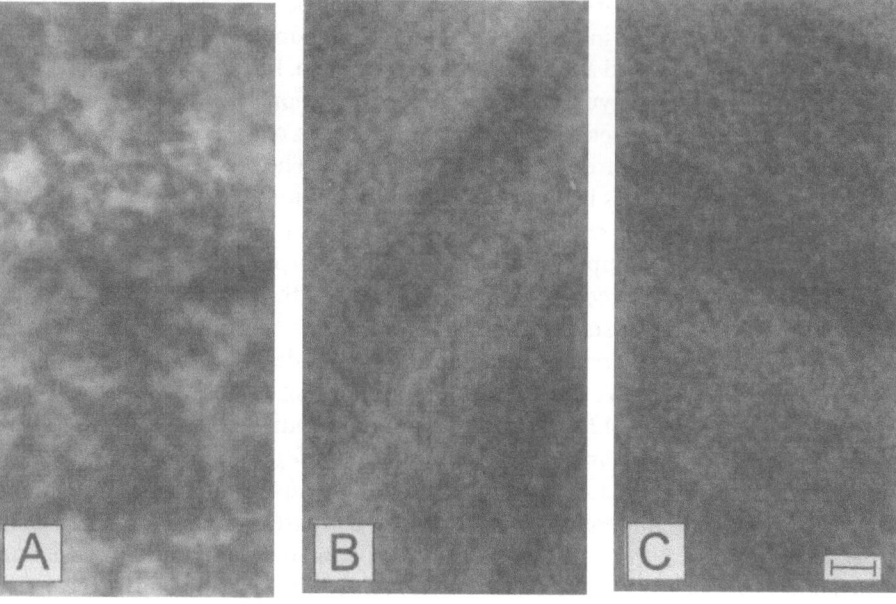

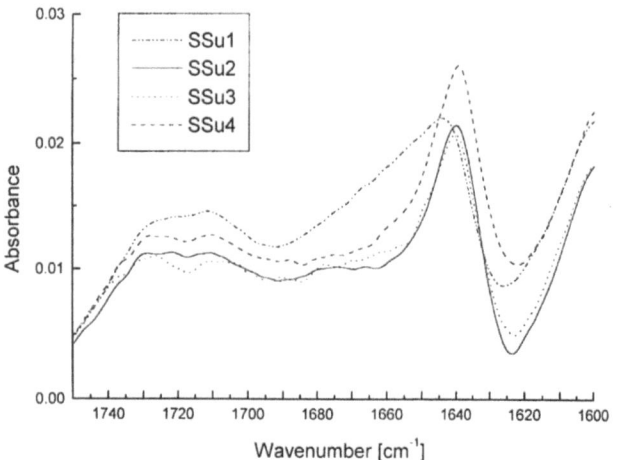

Fig. 2 FT-IR absorbance of different PU foam samples: SSu1: no
stabilization; SSu2: 0.3% stabilizer; SSu3: 0.75% stabilizer; SSu4:
1.5% stabilizer

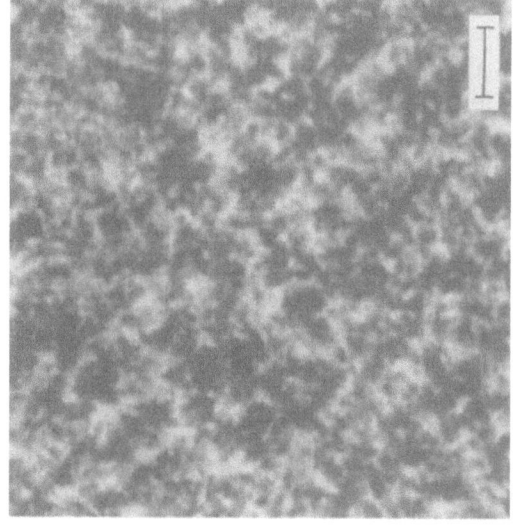

Fig. 3 AFM image of PU foam sample (tapping mode; 0.3% stabilizer; bar = 0.5 μm)

A visual inspection of the micrographs did not reveal
reliable differences. But a numerical evaluation by taking
the Fourier transform of the phase image showed an
interesting relationship: with increasing surfactant concentrations the average distance between particles becomes
larger (Fig. 4). This is in good agreement with the results
from TEM showing more compact particles upon addition
of surfactant. The numerical values for the quenched
sample are also included in this plot. The trend seems to be
the same but the absolute values are much lower.

We intend to perform experiments with much higher
surfactant concentrations (10%, 20% and 100%) to make

the discovered trend more reliable. Most interesting
should be micrographs being taken from isolated windows. This material can be obtained from a foam with an
incomplete cell opening.

Our finding of particles converted by absorption of
a surfactant from a defoamer into a foam supporting
material induced additional research on amphiphilic solid
particles. There is already a long history of using particles having hydrophilic and hydrophobic properties.
They have been first examined by Pickering. Since the

Progr Colloid Polym Sci (1998) 111:17–26
© Steinkopff Verlag 1998

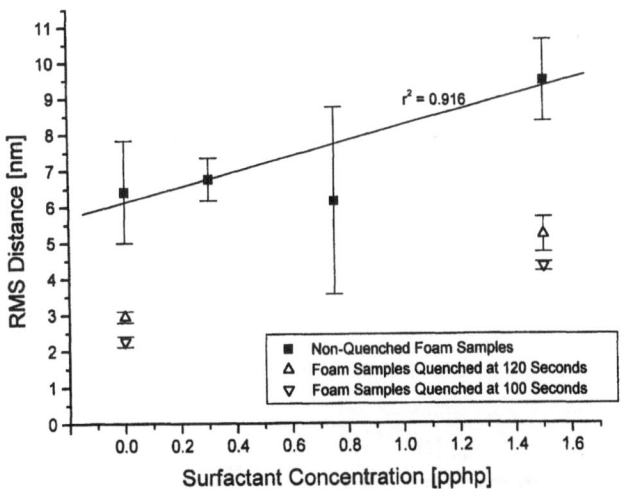

Fig. 4 Average phase offset across the sample for various PU foams

beginning of this century, many investigations have been made. One of the general rules resulting from the work of different research groups was that the character of emulsions is determined by the preferred wetting of these particles by one of the components: particles preferably wetted by oil will lead to w/o emulsions; predominantly hydrophilic particles tend to form o/w emulsions. The stability of those emulsions can be extraordinarily high. The amphiphilic particles at the interface obviously act as a rigid barrier avoiding drop coalescence.

However, all the amphiphilic particles known as emulsifying agents or defoamers exhibit a statistical distribution of hydrophilic and hydrophobic surface areas. Our goal was now to synthesize particles with a well-designed anisotropic distribution of hydrophilic and hydrophobic domains.

For this purpose we developed different strategies for chemical synthesis [4]. Some of them have a common principle: the linkage of predominantly hydrophilic and hydrophobic particles by chemical bridging reactions. For example, fumed silica was modified with hydrophobic silanes and a small amount of an amino-functional silane. After reaction with excess diisocyanate, we obtained hydrophobic silica particles with some isocyanate functions. In a similar way, we synthesized hydrophilic silica using reactions with polyether-substituted and amino-functional silanes. The reaction between the silica particles with different surface modifications then resulted in the desired anisotropic aggregate with hydrophilic and hydrophobic surface domains.

Other starting materials were based on microgels derived from divinyl benzene with OH groups at the surface, carbon black with COOH and phenylic OH groups

at the surface and starch. The pattern of making hydrophilic and hydrophobic particles and bridging them by chemical reactions was similar in all cases. A different approach consisted of modifying silica particles at the interface between a polar and an nonpolar solvent with different silanes.

In special applications, our amphiphilic particles could be used as emulsifiers, demulsifiers, foamers, defoamers, and in the process of enhanced oil recovery. Of course, the specific design of the particles had to be carefully chosen according to the desired function, as is usually the case with all surfactants, especially polymeric compounds.

One special method was extremly useful to procure well-defined particles with just two nicely separated surface domains. The resulting products demonstrated the basic properties of well-designed amphiphilic particles (Fig. 5). We used hollow glass microspheres with an average particle diameter between 20 and 180 μm and a wall thickness of about 1 μm as the starting material. The outer surface was modified by the reaction of its SiOH groups with a silane, making the glass surface hydrophobic by using octa-decyl-methyl-dimethoxy-silane or rendering it still more hydrophilic by attaching polyethylene-oxide-chains via a trialkoxy-silane. After completion of this modification step the spheres were broken and ground, giving now access to the inner surface of the former hollow spheres. This glass surface can now be modified with silanes of opposite polarity, or it could also be left untreated. Our average fragment size was chosen between 5 and 50 μm.

If we use these particles as emulsifiers for w/o emulsions, we can show their position at the interface by optical microscopy (Figs. 6–9). Size and shape of the droplets depend on the dimensions of the amphiphilic particles, on the polarity of the outer surface of the sphere, and on the shear rate. The curvature of the particles, defined by the diameter of the starting material, also had a significant influence.

Figure 6 shows a spherical droplet of water immersed in liquid paraffin. It is interesting to study how nicely the glass fragments with an hydrophobic outer surface have arranged themselves at the interface. It really reminds one of a puzzle put together by an outstanding, diligent child.

If we use smaller fragments of the same type, they imprint their curvature on smaller droplets. We get water droplets of a cauliflower shape (Fig. 7). In contrast, fragments of hollow spheres with a hydrophilic outer surface change the shape to an "inverted cauliflower" (Fig. 8). Increasing the shear rate in the emulsifying process leads to well stabilized sausage-like water droplets; this is obviously achieved by immobilizing the interface.

Fig. 5 Synthesis of Ho–Hi
particles, ((Ho) hydrophobic –
outer surface, (Hi) hydrophilic –
inner surface)

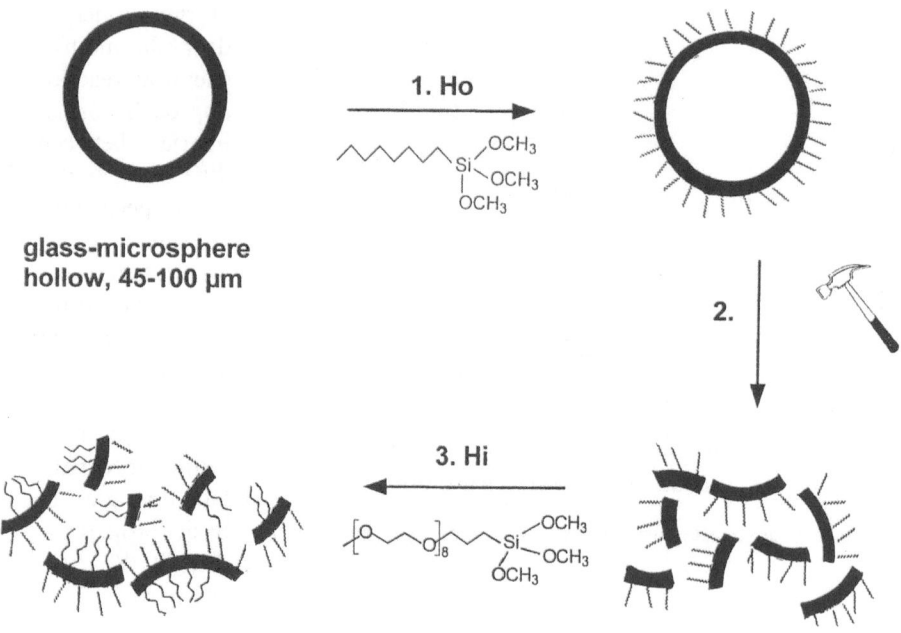

Fig. 6 Ho–Hi particles at the
interface of a water/liquid
paraffin emulsion; (A) 15 min
pulverized; (B) 2 min pulverized
(bar = 30 μm)

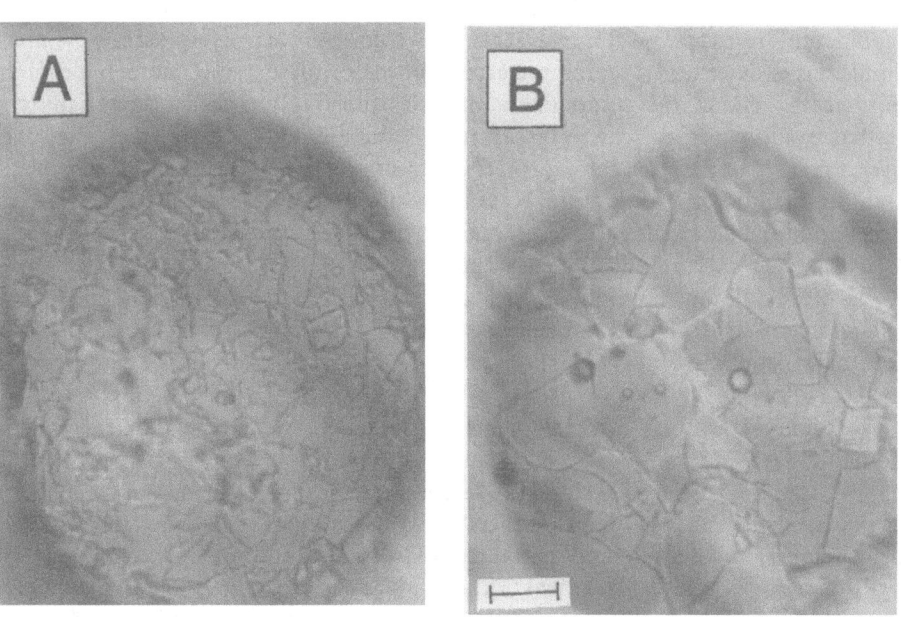

Silicates with layer structures of different chemical composition are also useful starting materials. Thus, kaolinite clay with a thickness of 50–100 nm and an average particle size of 0.5 μm could be modified with hydrophobic or hydrophilic silanes, using essentially the same procedure as we discussed for glass spheres. The hydrophobized particles tend to stabilize w/o emulsions, whereas the hydrophilically modified systems formed o/w emulsions. Also, with the modified kaolinite we observed highly distorted water droplets.

One of the most interesting aspects we learned from the study of particle-based emulsifiers is that their curvature determines the shape of the droplet. In addition, they tend to stabilize droplets with excess surface areas. This opens, for example, new opportunities for suspension polymerization techniques.

After we had finished the first stage of our synthetic development, we learned about the work of de Gennes and his co-workers [5]. They were able to make one of the ideal prototypes of amphiphilic particles by modifying

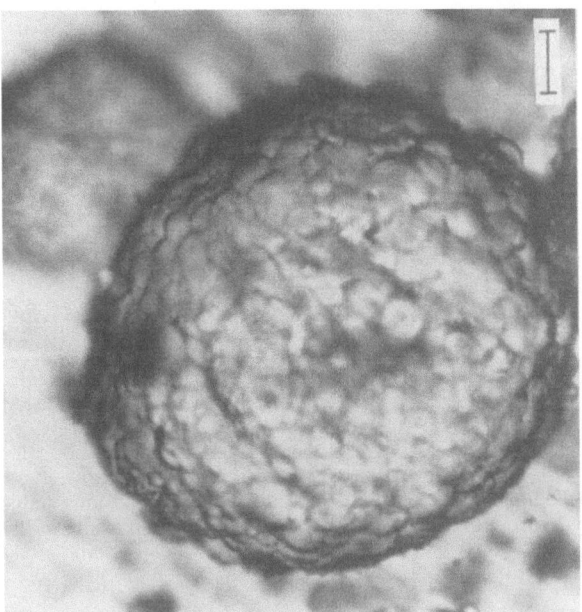

Fig. 7 Ho–Hi particles at the interface of a water/liquid paraffin emulsion: cauliflower structure (bar = 70 μm)

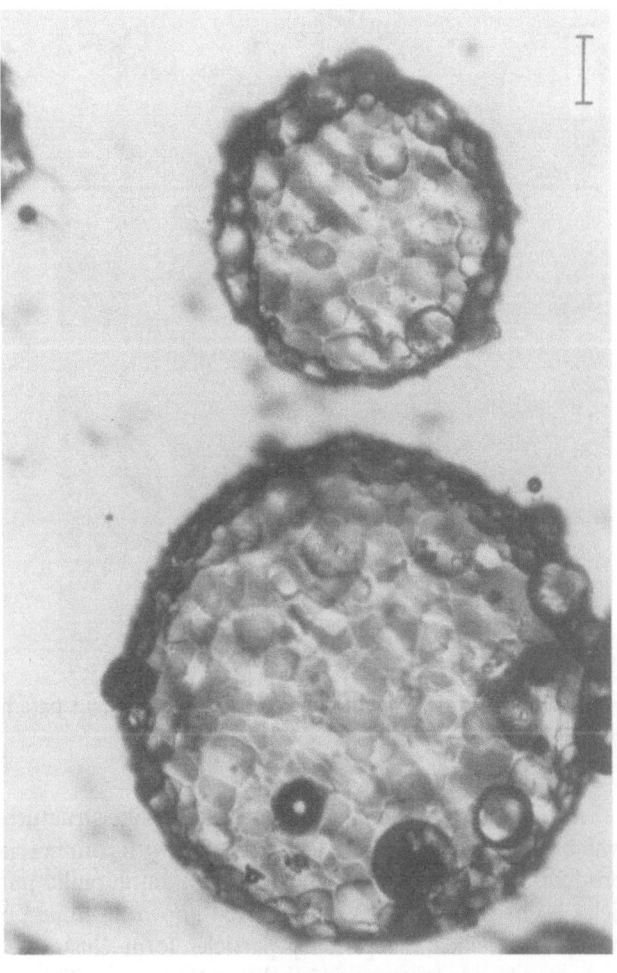

Fig. 8 Hi–Ho particles at the interface of a water/toluene emulsion: inverted cauliflower structure (bar = 70 μm)

glass spheres in a way that they show a hydrophilic and a hydrophobic hemisphere. This was a beautiful artistic approach, performed with the skills of a precision tool maker. And the name of the particles was also poetic: Janus beads. Compared to this, our amphiphilic particles may also be called Janus beads but they have distorted or smashed faces.

Of course we also had in mind technical applications. So we were mainly interested to synthesize smaller particles using simple reaction mechanisms.

Fumed silica was again the starting material for the new approach. We used commercial grade silica with a high degree of hydrophobic modification via $(CH_3)_3Si-O$ groups. All the commercial hydrophobic fumed silica still contain residual $Si-OH$ groups, which are the reason why the primary particles of a size of about 12 nm tend to form clusters in the μm range. Hydrophobic fumed silica can, therefore, also be used to stabilize w/o emulsions. It is plausible to postulate that the silica particles adsorb at the interface with the $Si-OH$ groups surrounded by water. If we use alkaline water, this gives a good chance to remove $(CH_3)_3Si-O$ groups and to replace them by $Si-O^-$ or $Si-OH$. Because the alkaline attack will occur in the neighborhood of a hydrophilic site, one can expect fairly discrete surface domains resulting from cooperative reactions. Figure 10 shows for a pH range of 7–13 the changes in emulsion characteristics which are achieved by this method. The data for the commercial grades A 972 and R 812 (Degussa) are given as references (pH = 7).

If we choose the drop size and the OH^- concentration in a droplet in a way that will finally lead to 50% of the $(CH_3)_3Si-O$ groups becoming replaced by $Si-O^-$ groups (in the equilibrium state in balance with $Si-OH$ groups), we will get primary silica particles corresponding to Janus beads. In a similar reaction, microspheres with a methyl silsesquioxane structure can be made hydrophilic at the surface. Here the position of the starting material at the interface will also be regulated by the amount of residual $Si-OH$ groups. Small siloxane beads have a rather large $Si-OH$ concentration at the surface. In this case the alkaline attack is not restricted to the surface, but leads to a removal of the basic polymeric units $[-Si(CH_3)-O_{1.5}]$ additionally. Part of the surface reaction may result in the fissure of strained siloxane rings. The resulting fragments are irregular in shape, but they also consist of sharply seperated hydrophobic and hydrophilic domains (Fig. 11).

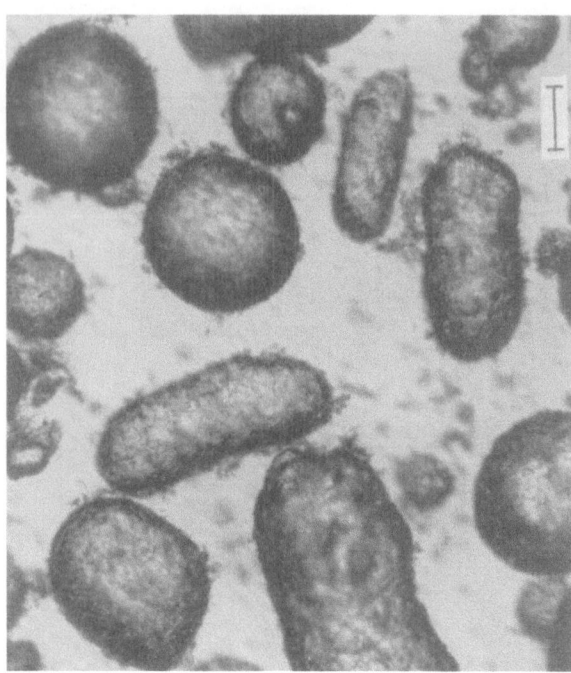

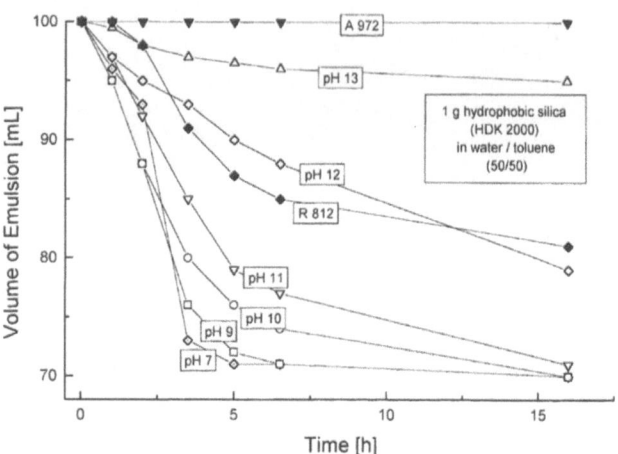

Fig. 10 Sedimentation behavior of water/toluene emulsions stabilized by hydrophobic silica (HDK 2000) as a function of pH in the water phase

Fig. 9 Ho–Hi particles at the interface of a water/liquid paraffin emulsion (bar = 70 μm)

A different approach to synthesize Janus-like structures, which is most easily achieved and can start with commercial products, consists in the self-assembly of amphiphilic particles by hydrogen bonds. As it is well known in fumed or precipitated silica the primary particles form clusters by hydrogen bonds between residual Si–OH groups. This occurs even when the surface is densely packed with hydrophobic groups. On the action of shear forces those clusters break, but they reassemble after stopping the mixing procedure.

We found that shearing a mixture of hydrophobic and hydrophilic silica finally resulted in forming clusters with intermediate surface properties. This can be shown by the methanol wetting test, but also by the behavior of these mixed particles in different colloidal systems. For example, we mentioned above that both hydrophilic and strongly hydrophobized silica act as defoamers in typical flexible polyurethane foams. The silica aggregates prepared by shearing a suspension of both types have lost this strong defoaming property.

Of course, in this way, the restructuring of silica clusters containing hydrophilic and hydrophobic primary particles will result in a statistical distribution of different types of aggregates. In order to come closer to the ideal

Fig. 11 Fluorescence micrographs of silsesquioxane beads labeled with fluoresceine thioisocyanate (FITC); (A) before pH-treatment; (B) after treatment at pH 13 (bar = 15 μm)

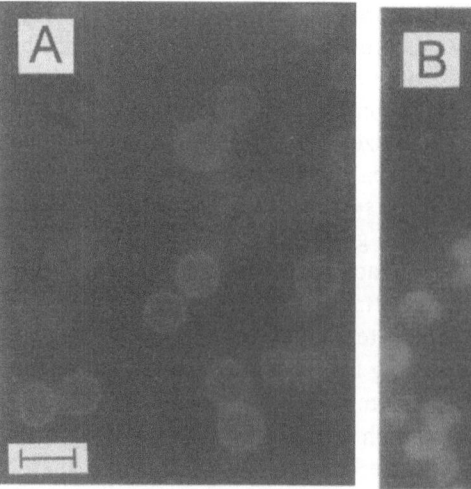
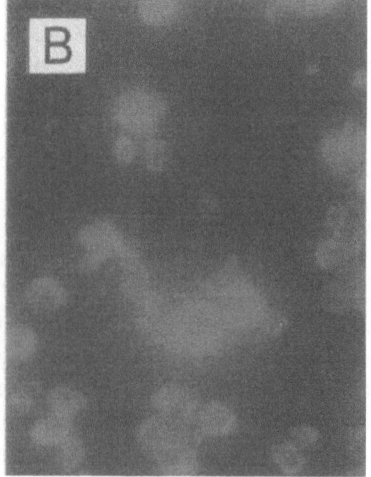

Progr Colloid Polym Sci (1998) 111:17–26
© Steinkopff Verlag 1998

structure of Janus beads, we again used a fluid interface for the restructuring. If we use, for example, toluene and water phases, the hydrophobic particles will approach the interface from the toluene side, and the hydrophilic particles are adsorbed from the water phase (Fig. 12). The coupling of both species by hydrogen bonds will result in an amphiphilic particle stabilizing the interface. This holds not only for primary particles of 12 nm but also for small clusters. In technical applications we need, of course, large interface area. This can most conveniently be achieved by shearing the toluene/water mixture and the silica particles. The formation of a stable emulsion is strongly related to aggregation processes at the interface. This special technique allows one to make reasonable amounts of product, which can be used for in situ processes or recovered by the usual methods of evaporating the fluids. Figure 13 shows how hydrophilic and hydrophobic silicas can be used by specific mixing procedures turned from defoaming into foam-stabilizing materials.

An essential condition for a sufficiently stable amphiphilic particle of this type is the formation of multibridging or sterically shielded hydrogen bonds. On the other hand, it can be a considerable advantage that these "physically coupled surfactants" may be destroyed by controlled procedures, such as temperature increase, pH changes or by the addition of molecules having acceptors for hydrogen bonds. These might be alcohols, amines or similar chemical compounds.

Of course, this principle is not restricted to silica particles. Each material with a certain number of OH, NH or COOH groups can be used as a partner in this assembling game. The above-mentioned silsequioxane beads, starch, polysaccharides, and proteins are typical examples. The self-organization of biochemically important molecules like the DNA helix and folded proteins give a good guideline for the principles of selection. Some useful particles like carbon black will have to be chemically modified (by esterification with polyols and/or ethoxylation, for example) in a first step in order to provide the critical concentration of OH groups or hydrogen bond acceptors.

Having now dealt with nanometer sized particles, the way to surfactant monolayers with a permanent rigidity is no longer very far. This is a traditional field of university research. Of course, we all know that many surfactants can be arranged in a two-dimensional crystalline or quasi-crystalline structure by compressing the surface layer (Langmuir trough). But these two-dimensional particles melt when the pressure is released. The goal is to synthesize stable particles, which have often a complex fractal pattern.

I have the privilege to show here some recent results of professor Rehage's group at the University of Essen [6].

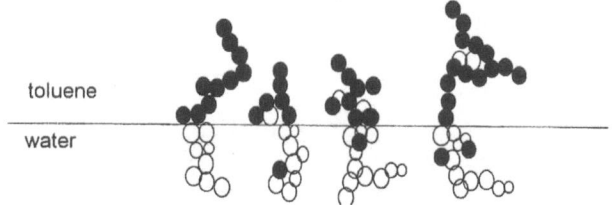

Fig. 12 Model of the coupling of hydrophilic (●) and hydrophobic (○) silica at the interface

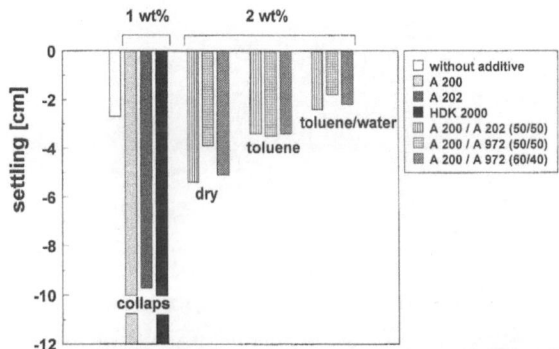

Fig. 13 Effect of fumed silica on PU foams (basic stabilization of [4H₂O] PU foam with 0.3 wt% BF 2370): pure fumed silica and combinations of hydrophilic and hydrophobic silica mixed by different methods (mixing of the dry powders, in toluene suspension and at a toluene/water interface)

From many relevant experiments I took one example which comes nearest to our approach: the quasi-crystalline network of a suitable surfactant which was crosslinked by Coulomb forces using multivalent ions in the subphase. Figure 14 gives a typical structure obtained by "Brewster Angle Microscopy". This network structure immediately forms by spreading tetracosane diacid on the surface of an aqueous solution of aluminium chloride. Cationic surfactants like quaternary ammonium bromides could also be crosslinked by adding sulphate ions in the subphase. Polymerization at the surface was a different approach [6].

In this context it is interesting to note that Ringsdorf and co-workers [7] were able to form expanded and compressable films of monocarbonic acids at the air/water interface due to the interaction with polymeric counterions (polyethylene imine). These films could be transformed onto solid substrates without damaging the typical surface morphology.

The above-mentioned experiments show that detailed knowledge about amphiphilic particles at interfaces has been considerably increased during the last few years. The anisotropic Janus beads offer the advantage of much

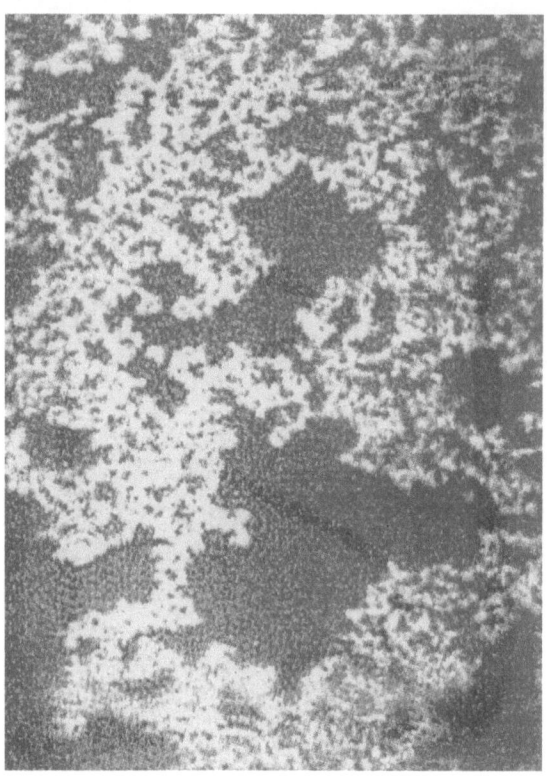

Fig. 14 Clustering of tetracosane diacid induced by AlCl₃ (Brewster angle microscopy)

clearer designed structures which can be adjusted easily to specific problems.

There are also new aspects; refined studies of the importance to control the surface roughness of membranes and the shape of droplets seem to be rewarding and are also a challenge for theoretical studies.

Janus-like particles could be used as new encapsulating materials. If it becomes possible to build windows in the rigid cages by applying well-defined mixture between amphiphilic solids and normal surfactants at the interface, a new way of regulating the release of encapsulated materials could be opened.

No doubt a lot of detailed studies will have to be done to improve the analytical insight into the complicated structure of the amphiphilic particles. It is also essential to explore in more detail the structure-efficacy relationships of these special systems.

Acknowledgements I would like to thank Dr. Peter Gimmnich, Helmut Schator, Dr. Andreas Weier, Dr. Burghard Grüning, Th. Goldschmidt AG for their contributions. I would also like to acknowledge the support from Dr. Philippe Sonntag and Prof. Heinz Rehage from the University in Essen/Germany and Bryan Kaushiva and Prof. Garth Wilkes from the Virginia Polytechnic Institute in Blacksburg/Virginia.

References

1. Rossmy G, Kollmeier HJ, Lidy W, Schator H, Wiemann M (1981) J Cellular Plastics 319–327
2. Johansson G, Pugh RJ (1992) Int J Miner Process 34:1–21
3. Aveyard R, Binks BP, Fletcher PD, Rutherford CE (1994) J Disper Sci Technol 15(3):251–271
4. Grüning B, Holtschmidt U, Koerner G, Rossmy G (1987) EP 0 156 270
5. Casagrande C, Fabre P, Raphaël E, Veyssié M (1989) Europhys Lett 9:251–255
6. Rehage H, Achenbach B, Kaplan A (1997) Ber Bunsenges Phys Chem 101:1683–1685
7. Chi LF, Johnston RR, Ringsdorf H (1991) Langmuir 7:2323–2329

Progr Colloid Polym Sci (1998) 111:27–33
© Steinkopff Verlag 1998

D. Horn
J. Klingler
W. Schorf
K. Graf

Experimental progress in the characterization of colloidal systems

D. Horn (✉) · J. Klingler
W. Schrof · K. Graf
Polymers Laboratory
Department of Polymer
and Solid State Physics ZKM
BASF Aktiengesellschaft
D-67056 Ludwigshafen
Germany

Abstract The increasing complexity of queries arising from the manufacture and application of colloidal systems demands continuously improving for their characterization. The objectives on one hand are the characterization (preferentially in-situ) of the properties of individual particles of the disperse phase like particle size distributions and, on the other hand, characterization of the manifold interactions with other components of the colloidal system like polymers and surfactants.

In this paper, some novel developments of characterization techniques for colloidal systems are highlighted on examples taken from practical applications. The methods discussed are quasielastic light scattering and its fiber optical variant, atomic force and chemical force microscopy and the optical confocal correlation techniques fluorescence correlation spectroscopy and Raman correlation spectroscopy.

Key words Particle polymer interactions – quasielastic light scattering – fiber optical quasielastic light scattering – chemical force microscopy – fluorescence correlation spectroscopy – Raman correlation spectroscopy

Introduction

In numerous applications, the technical performance of chemicals is determined by their colloidal state. Examples are pigments, polymer dispersions, fine chemicals in pharmaceutical formulations and crop protection agents. Recent advances in ecologically sensitive technologies like paper making and waste water treatment could only be achieved by a basic understanding of the mechanism of action of functional polymers applied as process chemicals controlling the specific interactions in such colloidal systems.

Obviously, rational progress in this field depends heavily on the availability of characterization techniques addressing static and dynamic effects in colloidal systems. In this respect, recent advances in laser-optical technologies and scanning force microscopies have revolutionized experimental colloidal science.

We exemplify this development on a number of experimental characterization techniques for colloidal systems, currently being in different states of development. They range from the *well established* method of quasielastic light scattering over *state of the art* visualization techniques using atomic force microscopy and the *cutting edge method* of fluorescence correlation spectroscopy to Raman correlation spectroscopy, *a recent invention*.

Quasielastic light scattering – particle size and particle size distribution

Today, quasielastic scattering (QELS, or dynamic light scattering, DLS) – first demonstrated 33 years ago [1] – has become a standard technique and commercial instrumentation is readily available. QELS and a number of its variants are used in fields ranging from macromolecular chemistry to biophysics to industrial quality control [2–4]. For the determination of particle size distribution (PSD) of submicron particles – one of its major applications in industrial practice – QELS has reached a high level of sophistication. The principal problem is the decomposition of the measured autocorrelation function (ACF) into the contributions of the individual components. Mathematically, this decomposition is unstable ("ill-posed problem") and different strategies have been developed to deal with this difficulty [5]. Figure 1A shows the state of the art of such decompositions for a model sample.

QELS experiments were performed on a mixture of three different polystyrene lattices (particle diameters 55, 126, 278 nm, volume ratios 16:4:1) on a commercial QELS setup (ALV, Langen, Germany). The resulting ACFs were analyzed with three data evaluation algorithms: CONTIN (S. Provencher, [6]), NLREG (group of J. Honerkamp [7]) and KOR (BASF AG, [8]). CONTIN and NLREG use only the ACF from one scattering angle (90° in this example) as input, while KOR uses the ACFs from several scattering angles (7 angles from 30° to 90° in this example) simultaneously. All three algorithms use a "regularization parameter", i.e. a parameter that sets the overall smoothness of the allowed solutions for the particle size distributions. The only analysis reproducing the original particle size distribution is the analysis with KOR, using the multi-angle information and setting the regularization manually to a low value, thus allowing a sharp

multimodal distribution. With higher values for the regularization parameter, KOR also produces unrealistic bimodal or monomodal distributions. Essentially, determination of a particle size distribution like the above – with diameter ratios of the individual components of less than 2.5 – by QELS requires the input of additional knowledge about the system.

In one of its practical applications, QELS is used for determining particle size distributions of carotenoid suspensions prepared by the process of carotenoid micronization reported elsewhere ([3], Fig. 1B). In short, the carotenoids are dissolved in a water-miscible solvent at high temperatures followed by a rapid quenching of the solution in water containing gelatin as a protective polymer. This process results in stable colloidal suspensions of the carotenoids which can be spray dried and resuspended in water. They are used as food colorants, anti-oxidants and vitamin supplements. The resulting particle sizes determining color shade and insuring bioavailability are of the order of 50–200 nm. Though QELS analysis of the end product shows bimodal distributions (Fig. 1C) with several decomposition methods and over a wide range of regularization parameters, certainty about the PSD could only be obtained by comparison with electron microscopy (Fig. 1D). Some current work in QELS is directed towards improving the technique, e.g. achieving faster data collection rates through simultaneous multi-angle measurements, using new scattering geometries to make increased concentration ranges accessible [9–13], using new wavelength regimes like X-rays [14] and harnessing multiple scattering as a tool instead of avoiding it [15]. Nevertheless, we feel that for the determination of PSDs in colloidal

Fig. 1 Determination of particle size distribution by QELS. (A) Particle size distributions of a mixture of polystyrene latices (diameters 55 126 278 nm, volume ratios 16:4:1), obtained from QELS data with three algorithms (CONTIN, NLREG and KOR). (B) Carotenoid micronization technique – schematic. (C) Particle size distributions of carotene micronizate. (D) EM picture of the same sample

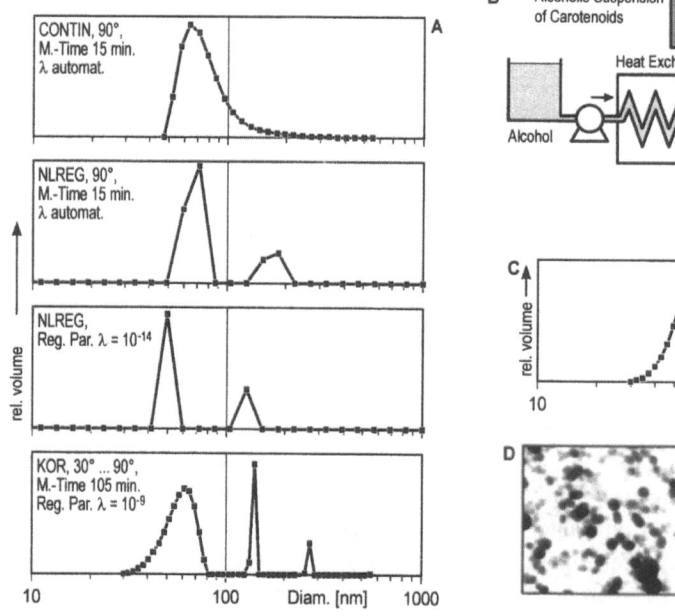

Fig. 2 Applications of fiber optical quasielastic light scattering (FOQELS). (A) FOQELS setup. (B) Measured relative diffusion coefficients vs. volume concentration for a 115 nm PS/BA latex. (C) Setup for measuring particle size development during emulsion polymerization. (D) Comparison of the development average particle sizes in emulsion polymerizations of two different types. Without seed: continuous growth of particles formed by initial nucleation. Seeded polymerization: Formation of a second generation of particles leads to intermittently decreasing average size

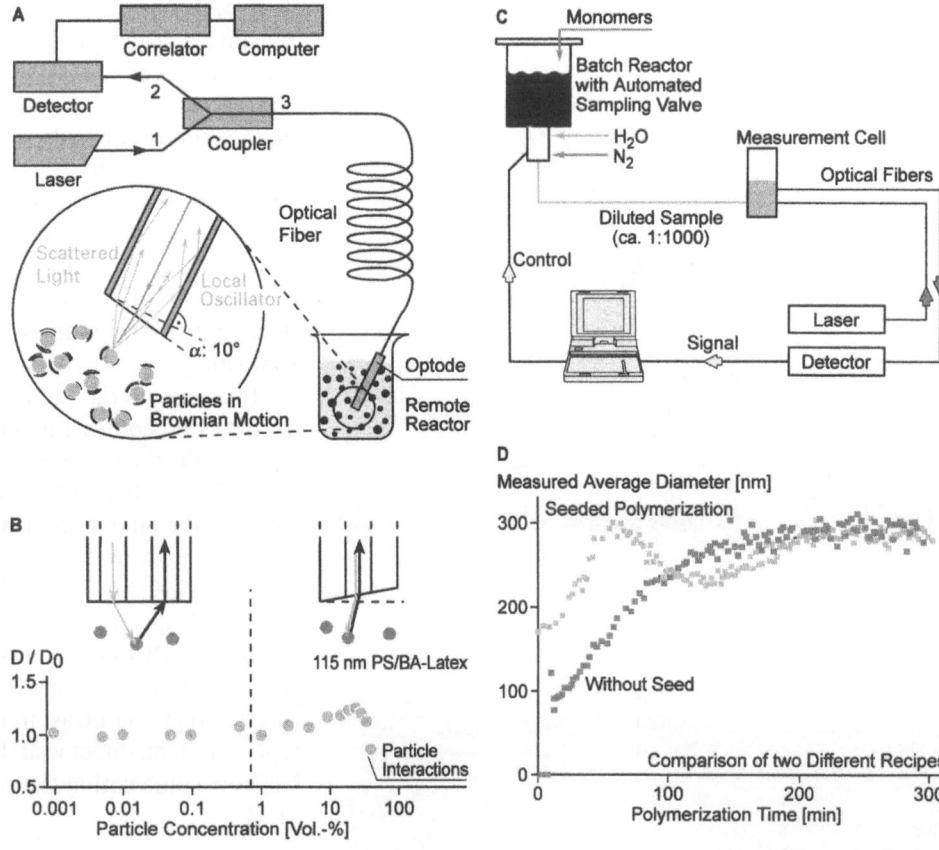

systems with QELS to progress much beyond the state of the art described above, an additional key invention would be necessary.

For practical purposes, the extension of QELS into a robust method for on-line and in situ measurements is a more important objective than refinements of the data analysis procedures. The introduction of fiber optical quasielastic light scattering (FOQELS) has made such an extension possible [3, 12]. The optical fibers serve a dual purpose: First, the FOQELS technique, where an optical fiber is placed directly into an optical dense medium and acts simultaneously as emitter of laser light and receiver of back-scattered light, has made possible QELS in optically dense samples by avoiding multiple scattering (Fig. 2A). Using an additional separate fiber as a receiver in the case of lower particle concentrations, a wide concentration range is accessible to the FOQELS technique (Fig. 2B). The second purpose of the optical fibers is to spatially separate the sensitive optical and electronic components of the setup from the location of the measurement. Thus FOQELS measurements are possible even in harsh industrial environments, as has been shown for the process of carotenoid micronization [3]. Figure 2 shows the use of QELS as an in situ measurement technique for particles formed during emulsion polymerization as another example.

Chemical force microscopy – particle/polymer interactions

Polyethylenimine (PEI) is used in the paper industry as a drainage and retention aid, i.e. PEI allows the control of flocculation in stock suspensions [16]. For the mode of action of PEI as a flocculant, experimental results pointed toward the patch charge model. This model explains the flocculation induced by PEI by a mosaic of adsorbed, planar PEI molecules on the particle. This mosaic gives rise to a bipolar interface of positively and negatively charged patches. Recently, we could visualize this pattern of charged patches directly by chemical force microscopy (CFM, [17]).

In CFM [18, 19], the tip of a standard scanning force microscope (SFM) is coated with a monolayer of functional molecules that show a specific interaction with the molecules targeted for visualization (Fig. 3A). For detection of PEI molecules adsorbed on polymer later spheres, a micromachined Si_3N_4 SFM tip was coated with gold and a chemisorbed monolayer of $SH–(CH_2)_{10}–COOH$

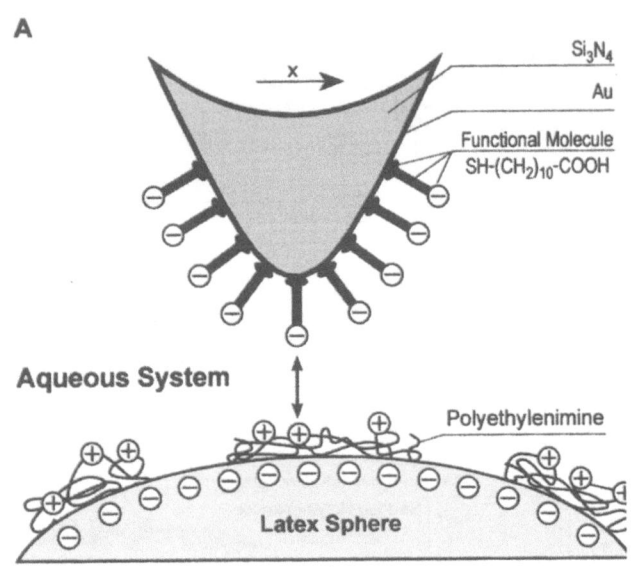

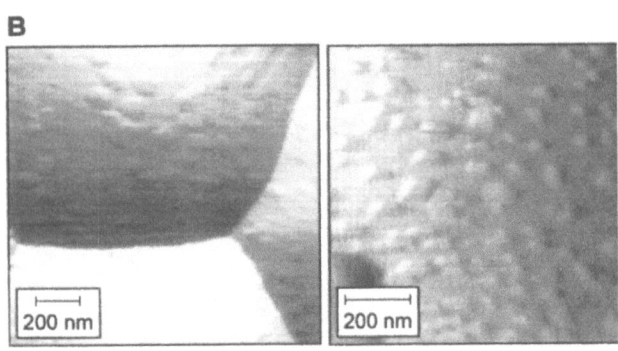

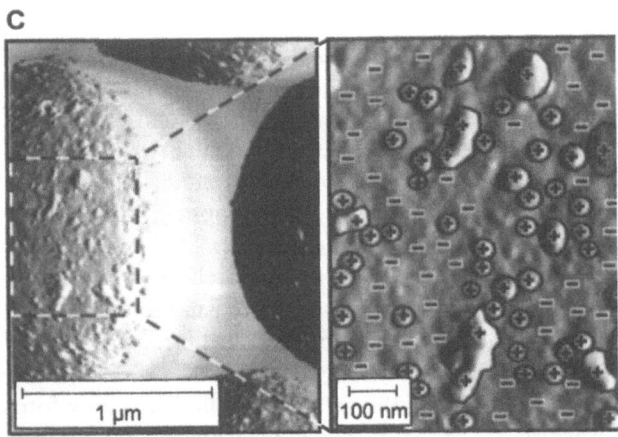

Fig. 3 Direct visualization of adsorbed polymer molecules by chemical force microscopy (CFM). (A) Principle of CFM. (B) CFM images of single PEI molecules adsorbed on polymer latices. (C) Left: Tapping mode SFM pictures of PEI on latices. Right: Resulting patch-charge pattern (schematic)

(Fig. 3A). The carboxylic groups on the tip interact with the amino groups of the adsorbed PEI molecules. Normal force-images of latex spheres with adsorbed PEI molecules show an unperturbed, smooth surface for lattices with zero and maximum PEI adsorption (not shown for brevity) and a patterned surface for lattices in the region of partial surface coverage (Fig. 3B). Size and density of the observed patches fit well with the model of single PEI molecules adsorbed on the latex surface in a flattened "pancake" configuration [17]. The experiments yield similar results using either dry samples or samples in an aqueous environment.

Imaging of the same samples with unmodified tips produces no contrast in a conventional SFM microscope. Apparently, the adsorbed single PEI molecules are too soft to produce sufficient force contrast without the additional electrostatic interaction with the tip. In a SFM microscope equipped with the tapping-mode [20], however, even unmodified tips show the individual PEI molecules in good contrast (Fig. 3C: amplitude picture). Apparently, even though the PEI molecules are too soft to give direct force contrast, they are damping the oscillating tip of the SFM enough to give contrast in the tapping mode [21]. For this type of system, direct visualization, first obtained by CFM, has now come within the scope of improved conventional SFM.

Fluorescence correlation spectroscopy – association processes

Fluorescence correlation spectroscopy (FCS) is an analytical tool that uses number fluctuations of fluorescent molecules (or particles) in a very small observation volume (the confocal volume of a setup analogous to confocal laser scanning microscopy) to obtain information on the diffusive properties of these molecules [22, 23].

The experimental setup is shown in Fig. 4A [24]. The light of a laser is coupled into a single-mode optical fiber and passes through the illumination pinhole onto a dichroic mirror. The reflected light is focused by a microscope objective. The fluorescence excited in the sample is collected by the objective and focused through the mirror and the detection pinhole onto a detector. The electric signal of this detector is fed into a hardware correlator, where an autocorrelation function (ACF) of the observed fluorescence intensity fluctuations is calculated. If the concentration of the fluorescent molecules is in the nanomolar range, the average number of molecules present in the confocal volume (ca. 10^{-18} m^3) is of the order of 1–10. The observed intensity fluctuations are caused by Brownian diffusion of the fluorescent molecules leading to number

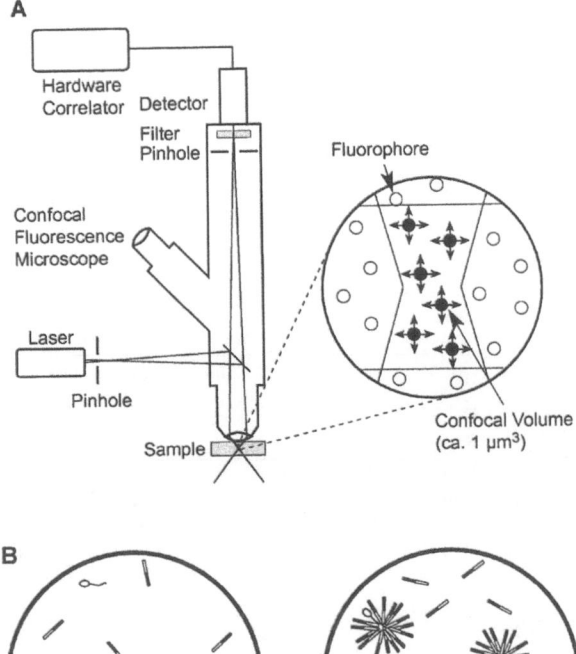

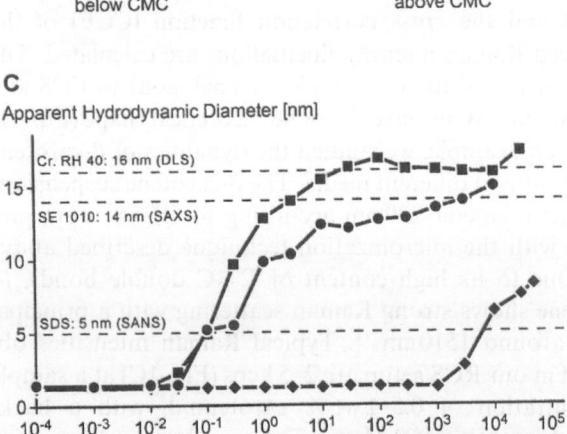

Fig. 4 (A) Experimental setup for fluorescence correlation spectroscopy (FCS, experimental details in [24]). (B) Principle of detecting micelle formation with FCS. Fluorescent probe molecules (o⌐) are incorporated in micelles formed from a surfactant (—). (C) Apparent hydrodynamic diameter of the probe molecules vs. surfactant concentration for three surfactants measured with FCS. Cr. RH40: Cremophor®RH40 (Polyoxyethyleneglycerol-trihydroxystearate, BASF AG, Germany), SE1010: Polystyrene-block-polyethyleneoxide-copolymer (each block 1000 D molecular weight, Th. Goldschmidt AG, Germany), SDS: sodium dodecylsulfate

fluctuations in the observation volume. For an idealized experiment, the resulting ACF is given by [23, 25]:

$$G(t) = 1 + \frac{1}{N}\left(\frac{1}{1 + 4Dt/r^2}\right), \qquad (1)$$

where N is the average number of fluorescent molecules in the observation volume, D their thermal diffusion coefficient and r the radial width of the observation volume. The diffusion coefficient can be converted to a hydrodynamic diameter by the Stokes–Einstein equation

$$d = \frac{kT}{3\pi\eta D}, \qquad (2)$$

where d is the hydrodynamic diameter of the molecule, k the Boltzmann constant, T the absolute temperature and η the viscosity of the medium.

The main focus of FCS so far has been the field of biophysics [25, 26]. The high sensitivity and the spectral selectivity of FCS, namely the ability to derive diffusion coefficients of a very low concentration of fluorescent molecules in a complex system, makes FCS an interesting technique for colloid chemistry. For systems that show no intrinsic fluorescence, one has to either label one of the components by covalent binding of a fluorophore or introduce a fluorescent probe into the system that specifically associates with the component of interest.

An example for the later strategy is shown in Figs. 4B and C. There, the formation of micelles of several surfactants is observed by adding a small amount (final conc. 1 nM) of a surface active fluorescent molecule (BODIPY®-hexadecanoic acid, Molecular Probes Inc., Eugene, Oregon) to the system. At low surfactant concentrations, no micelles are present and the apparent hydrodynamic diameter of the fluorescent probe molecules, calculated from the measured diffusion coefficient, is 1.1 nm (Fig. 4C). With increasing surfactant concentration, the fluorescent probe becomes incorporated into micelles and the measured size increases. At high surfactant concentrations, the plateau value for the measured size indicates the size of the micelle. With such an experiment, both micelle size and critical micellar concentration (CMC) can be measured. The advantage of this method over more established ones to obtain the same information (e.g. light scattering or surface tension measurements) is its sensitivity and directness. FCS allows the direct detection of micelle formation in the bulk phase at surfactant concentrations well below 1 ppm.

With the availability of a commercial FCS apparatus (ConfoCor, Carl Zeiss Jena GmbH, Jena, Germany), we expect FCS to make the transition from the cutting edge to a state of the art method as its use in colloidal chemistry becomes more widespread.

Fig. 5 (A) Experimental setup for Raman correlation spectroscopy (RCS). (B) Raman intensity traces (at $1510\,\mathrm{cm}^{-1}$) of carotenoid particles (β-carotene, particle diameter 220 nm, 1 wt % β-carotene in suspension): influence of medium viscosity (water/glycerol mixtures, glycerol content varied from 0–86 wt %). (C) Resulting ACFs (29). (D) Viscosity of the medium (●), calculated from the ACFs (Eqs. (1) and (2)), viscosity of water glycerol mixtures from the literature (line) (28)

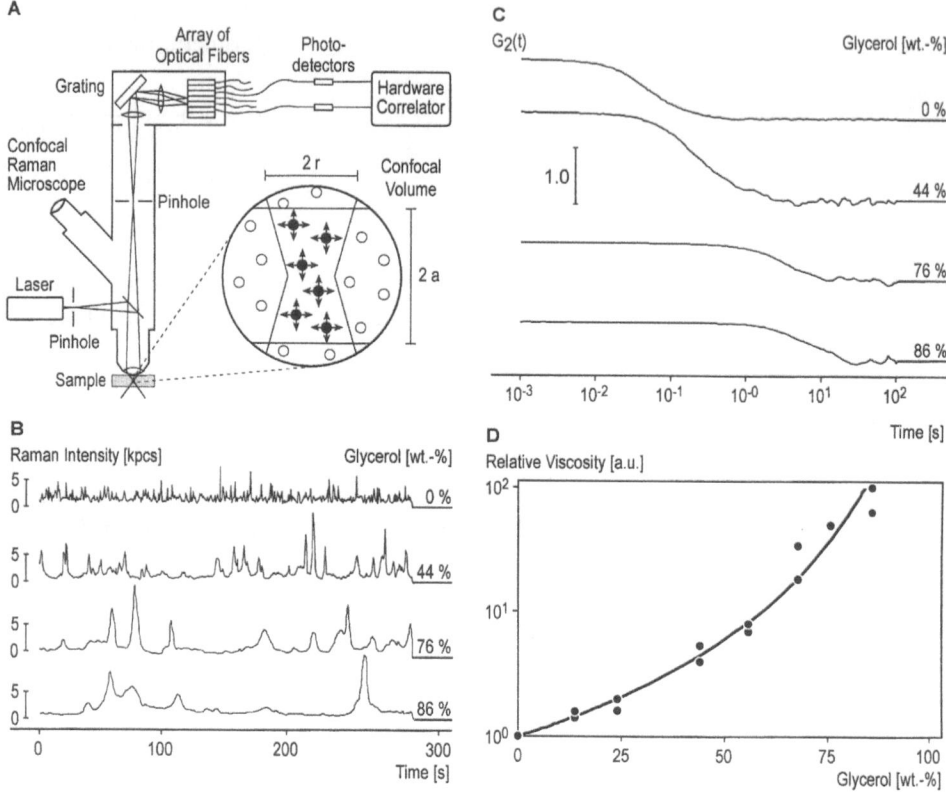

Raman correlation spectroscopy – dynamics and chemical information

The principles of fluorescence correlation spectroscopy shown above should work equally as well for any kind of spectroscopic technique. In principle, it is possible to take any kind of spectroscopic information and combine its selectivity with the dynamical information gained from measuring number fluctuations in a small observation volume. Recently, we have shown for the first time the possibility of this approach for Raman scattering. We combined the chemical information from Raman scattering with the structural and dynamical information obtained by fluctuation spectroscopy in a method we named Raman correlation spectroscopy (RCS, [27]).

The experimental setup, shown in Fig. 5A, is based on a modified confocal Raman microscope. In order to follow fast temporal fluctuations, the CCD-detector of the original microscope is replaced by a linear array of 40 optical fibers (diameter 200 μm, corresponding to a Raman shift of $30\,\mathrm{cm}^{-1}$ between adjacent fibers). Two avalanche photodiode detectors can be connected to any two fibers, thus setting the detection wavelengths to the desired bands in the Raman spectrum. The signals of the detectors are fed into a hardware correlator, where the autocorrelation functions

(ACF) and the cross correlation function (CCF) of the observed Raman intensity fluctuations are calculated. The optical setup of RCS is completely analogous to FCS and the resulting ACFs have the same theoretical shape (Eq. (1)).

As an example, we studied the dynamics of β-carotene suspensions in different media. The β-carotene suspensions (particle diameter 220 nm according to QELS) were prepared with the micronization technique described above [3]. Due to its high content of C=C double bonds, β-carotene shows strong Raman scattering with a principal band around $1510\,\mathrm{cm}^{-1}$. Typical Raman intensities observed in our RCS setup are 2–5 kcps (Fig. 1C) at a sample concentration of 0.5–1 wt % carotenoid, with a background intensity of 0.3 kcps. The samples were measured in water/glycerol mixtures of increasing glycerol content. Figure 5B shows the resulting intensity vs. time traces. The dynamics of the intensity fluctuations is slowed down as the viscosity increases with increasing glycerol content. Figure 5C shows the resulting ACFs. Assuming a constant particle size, Eqs. (1) and (2) can be used to calculate the apparent relative viscosity of the medium from the ACFs. Figure 5D shows the results for this apparent viscosity from the RCS experiments, together with the literature values [28] for the viscosity of water/glycerol mixtures. The good correspondence proves the claimed RCS-principle.

Progr Colloid Polym Sci (1998) 111:27–33
© Steinkopff Verlag 1998

With two simultaneous detection channels, the intensity fluctuations of two different Raman bands can be observed at the same time and cross correlated. A positive CCF of the fluctuations of two Raman bands of two chemically different types of particles then indicates a coordinate diffusion of the two species and therefore the formation of a complex. As an additional technique, a simultaneous detection of fluorescence and Raman scattering might be possible for some systems.

The major advantage of RCS over FCS is that RCS can distinguish the components of a colloidal system in their native state without labeling and monitoring their behavior selectively. We think that with technical improvements in sensitivity, RCS has the potential to be an important step forward in the characterization of complex collodial systems like coatings, pigments and pharmaceutical formulations.

Summary and outlook

Understanding of structure, dynamics and properties of colloidal systems depends on experimental insights and therefore on the development of experimental techniques to characterize them. Hence, the development and refinement of such techniques is a key element for a rational design of colloidal systems and for improving their performance in applications.

Despite the steady progress, illustrated by the above examples, there are still many interesting and important questions that cannot be satisfactorily addressed with the current experimental techniques. This is especially true for many colloidal systems used in practical applications, which mostly are high-concentration multicomponent systems, where the performance in applications is controlled by specific interactions of the individual components: particles, functional polymers and surfactants.

References

1. Cummins H, Knable N, Yeh Y (1964) Phys Rev Lett 12:150
2. Berne B, Pecora R (1976) Dynamic Light Scattering. Wiley, New York
3. Wiese H, Horn D (1993) Ber Bunsenges Phys Chem 97:1589
4. Stein R, Srinivasarao M (1993) J Polym Sci B 31:2003
5. Johnsen R, Brown W (1992) In: Harding S (ed) Laser Light Scattering in Biochemistry. Cambridge, UK
6. Provencher S (1979) Macromol Chem 180:201
7. Honerkamp J, Maier D, Weese J (1993) J Chem Phys 98:865
8. Wu C, Unterforsthuber K, Lilge D, Lüddecke E, Horn D (1994) Part Part Systems Charat 11:145
9. Phillies G (1981) J Chem Phys 74:260
10. Schätzel K, Drewel K, Ahrens K (1990) J Phys: Condens Matter 2:SA 393
11. Aberle L, Wiegand S, Schröder W, Staude W (1997) Progr Colloid Polym Sci 104:121
12. Auweter H, Horn D (1985) J Colloid Interface Sci 105:399
13. NIBS (non invasive backscattering) detector, ALV GmbH, Langen, Germany. Company Publication
14. Thurn-Albrecht T et al (1996) Phys Rev Lett 77:5437
15. Weitz D, Pine D (1993) In: Brown W (ed) Dynamic Light Scattering. Clarendon, Oxford, pp 652–720
16. Horn D, Linhart F (1996) In: Roberts J (ed) Paper Chemistry. Blackie Academic & Professional, London, pp 64–82
17. Akari S, Schrepp W, Horn D (1996) Langmuir 12:857
18. Akari S, Horn D, Keller H, Schrepp W (1995) Adv Mater 7:549
19. Frisbie C, Rozsnyai L, Noy A, Wrighton M, Lieber C (1994) Science 265:2071
20. Zhong Q et al (1993) Surf Sci 290: L688
21. Pfau A, Schrepp W, Horn D, in preparation
22. Magde D, Elson E, Webb W (1972) Phys Rev Lett 29:705
23. Rigler R, Widengren J, Mets Ü (1993) In: Wolfbeis O (ed) Fluorescence Spectroscopy. Springer, Berlin, pp 13–24
24. Klingler J, Friedrich T (1997) Biophys J 73:2195
25. Thompson N (1991) In: Lakowicz J (ed) Topics in Fluorescence Spectroscopy. Plenum, New York, pp 337–378
26. Rigler R (1995) J Biotechnol 41:177
27. Schrof W, Klingler J, Rozouvan S, Horn D (1998) Phys Rev E 57:2523
28. Weast R (ed) (1988) CRC Handbook of Chemistry and Physics, 69th edn. CRC Press, Boca Raton, p D-232
29. The straight part at the end of the ACFs indicates the respective baseline, i.e 1.0
30. We are indebted to H. Schuch, A. Pfau, S. Akari and S. Rozouvan for their contributions to the work presented

Progr Colloid Polym Sci (1998) 111:34–40
© Steinkopff Verlag 1998

R. Lipowsky

Colloids and interfaces: a theoretical perspective

R. Lipowsky
MPI für Kolloid- und
Grenzflächenforschung
Kantstr. 55
D-14513 Teltow-Seehof
Germany

Abstract Colloids and interfaces exhibit several levels of self-organization. As one moves through this hierarchy of levels, new phenomena emerge on each level which are characteristic for that level and cannot be anticipated on lower levels. The theoretical description of these phenomena leads to unusual concepts such as (i) morphological transitions, (ii) colloidal forces of entropic origin, (iii) elasticity of self-assembled structures, and (iv) colloidal machines. These concepts are illustrated here for (i) wetting of structured or imprinted surfaces, (ii) flexible membranes in contact with dispersed particles, (iii) fluid bilayers composed of amphiphilic molecules, and (iv) molecular motors.

Key words Wetting – membranes – dispersed particles – molecular motors

Introduction to "colloid mechanics"

Colloidal structures are large compared to atoms but small compared to macroscopic bodies. Thus, these structures cover a wide range of length scales and have a dimension or size which is roughly between 1 nm and 1 μm. Extended interfaces and membranes represent very anisotropic examples of such structures which have only one dimension, namely their thickness, within the colloidal domain.

In general, both the artificial colloids produced by the chemical industry and the natural colloids found in biological systems represent compounds which contain a variety of colloidal structures. A simple classification scheme is based on the shape of these structures. One may then distinguish three types of "elementary" colloids: (i) polymers which represent linear chains of monomers.[1] On small scales, these chains represent thin rods of variable length but of fixed diameter as determined by the size of the monomers; (ii) membranes which are surfaces composed of amphiphilic monomers. On small scales, these membranes are thin films with variable area but with a fixed thickness; (iii) particles in which the monomers are packed in a three-dimensional manner. These particles will grow in all directions if one adds monomers to them.

From the theoretical point of view, the most interesting feature of colloidal systems is the appearance of new phenomena and mechanisms which have no analogues or counterparts in the atomic or in the macroscopic world. In order to understand these phenomena, one has to develop new concepts. One such concept which is rather general but also somewhat fuzzy is self-organization. In fact, colloidal systems typically exhibit several levels of self-organization as will be discussed in the next two sections. More precise theoretical concepts which refer to different levels of self-organization are morphological transitions, colloidal forces of entropic origin, elasticity of self-assembled

---

[1] The term monomer is used here for a molecular group or a molecule which represents the basic building block of a polymer, membrane or particle.

Progr Colloid Polym Sci (1998) 111:34–40
© Steinkopff Verlag 1998

structures, and colloidal machines, see Sections 5–8 below.

The behavior of atoms and small molecules is governed by the laws of quantum mechanics whereas the properties of macroscopic bodies are well described by continuum mechanics. The unusual phenomena, which are both typical and generic for the colloidal domain, and the new concepts, which have emerged in order to understand these phenomena, may eventually lead to a new type of mechanics, namely "colloid mechanics".

Self-organization: bottom-up view

Since higher levels of self-organization on larger scales are built upon lower levels on smaller scales, I will first discuss these different levels in a bottom-up manner. As an example, let us consider amphiphilic molecules (or monomers) in aqueous solution: these molecules represent short chains in which the submolecular groups are connected by covalent bonds; the amphiphilic chains form aggregates such as bilayer membranes via self-assembly; these bilayers form closed vesicles without edges; the vesicles exhibit different shapes and may undergo transitions between these different morphologies; the vesicles may interact with other colloids; several vesicles may be crosslinked into larger clusters or "super-colloids", etc.

It is obvious that this hierarchy of structural organization involves several characteristic length scales. In the case of a single vesicle, one has (i) the size of the amphiphilic molecules which determines the thickness l_{me} of the bilayer membrane and (ii) the lateral size of the membrane which determines the overall size R_{ve} of the vesicle. On length scales which are small compared to l_{me}, we see the atoms and molecular groups of the amphiphilic molecules which interact via various intermolecular forces. On length scale which are comparable to l_{me}, the membrane resembles a thin film which is bounded by two membrane/water interfaces. On length scales large compared to l_{me} but small compared to R_{ve}, the membrane behaves as a flexible sheet which undergoes thermally excited shape fluctuations. Finally, on length scales which are comparable to R_{ve}, the vesicle exhibits a certain characteristic shape.

A similar separation of length scales as discussed here for membranes also applies to polymers and particles, i.e. to the two other "elementary" colloids mentioned in the introduction. Likewise, such a separation of length scales is also useful if one considers the next level of complexity consisting of binary solutions of different types of "elementary" colloids. Classical examples for such binary systems are provided by dispersed particles sterically stabilized by polymers. Binary systems consisting of membranes +

polymers and of membranes + particles will be briefly discussed in Section 6 below.

The different length scales involved in these colloidal systems provide a simple and rather natural way in order to identify the different levels of self-organization. What I would like to emphasize here, is that these different levels are characterized by new phenomena which emerge on that level and which cannot be anticipated on the levels lying underneath it. Therefore, colloids and interfaces provide many examples for the epistemological principle of "emergence".

Self-organization: theoretical challenges

For the theoretician, the hierarchy of self-organization in colloidal systems provides several challenges. Indeed, for each level of this hierarchy, one needs to develop a separate theoretical description which identifies the useful concepts and the relevant parameters for that level. This is a challenge which is not always appreciated. In fact, one often hears statements of the sort that, in principle, any system consisting of many atoms is governed by a "huge" Schrödinger equation which describes the quantum-mechanical motion of the atomic nuclei and of the electrons. From this point of view, the whole problem is a purely technical one, namely how to solve this equation, and one may hope that such a solution becomes feasible as soon as one has sufficiently powerful computers.

However, even if we were able to obtain a complete solution for the quantum-mechanical motion of all the nuclei and electrons, we would not obtain the concepts which are useful for the description of a certain level of self-organization. Neither would we be able, by looking at this solution, to identify the parameters relevant for that level. Indeed, these concepts and parameters would still represent additional elements of the theory which were necessary in order to reduce the enormous amount of mostly useless information as contained in the complete solution of the "huge" Schrödinger equation.

The most important criterion for the usefulness of a theoretical description is its predictive power. This necessarily implies that such a theory depends only on a relatively *small* number of parameters. Indeed, the predictive power of theories which involve *many* parameters is very limited: The parameter space has many dimensions and cannot be explored in a systematic way; it will always be possible to choose these parameters in such a way that one can fit a certain set of experimental data; this choice is, however, not unique and, thus, the whole modelling is, to a large extent, arbitrary.

Another important criterion for theories of colloidal systems is their consistency. Of course, any reasonable theory has to be internally consistent. Here, we also have to worry about external consistency: since the different levels of self-organization are built up in a hierarchical fashion, the theoretical descriptions for these different levels must be consistent with each other. This implies that those parameters, which are relevant for a certain level characterized by a certain length scale (or a range of such length scales), depend on the properties of the underlying levels on smaller scales. This dependence provides another challenge: once we understand which parameters are relevant for a certain level of description, we would like to express those parameters in terms of the more microscopic parameters as used for the lower levels.

Self-organization: top-down view

In order to emphasize the fact, that each level of structural organization exhibits its own characteristic phenomena and requires its own theoretical description, I will proceed in the following discussion from large to small scales (rather than in the opposite direction). This top-down approach will be illustrated for the behavior of interfaces and membranes. The top level from which I start is the micrometer range which lies at the upper end of the colloidal domain and which can be observed in the optical microscope. On this length scale, the behavior of interfaces and membranes can be understood, to a large extent, in terms of three relevant parameters: interfacial tension, "spontaneous" curvature,[2] and bending rigidity.

These parameters will be discussed in the context of the following systems and phenomena: (i) wetting on structured or imprinted surfaces. The behavior of these systems can be understood in terms of interfacial tensions and contact angles; (ii) flexible membranes in contact with dispersed particles. In this case, the membrane is viewed as an ultrathin film which is a few nanometers thick and is bounded by two interfaces. The corresponding interfacial tensions have a contribution which arises from the translational entropy of the dispersed particles. The competition of these tensions with the bending rigidity of the membrane leads to a "spontaneous" membrane curvature; and (iii) bilayer membranes arising from the self-assembly of model amphiphiles. When prepared in a tensionless state, these bilayers undergo thermally excited bending undulations from which one can extract a value for the bending rigidity of the self-assembled bilayer.

The three phenomena just described correspond to systems in (or close to) thermal equilibrium. In the last section, molecular motors which operate far from equilib-

rium will be addressed. The theoretical description of these latter systems is still in its infancy and, thus, restricted to a single coarse-grained level.

Morphological transitions

Colloids can exhibit a large variety of different shapes. In some cases, this polymorphism can be controlled by external parameters such as temperature or osmotic conditions. One example is provided by the amazing polymorphism of lipid vesicles which has been studied in some detail; for reviews, see [1–3]. Here, I want to illustrate this concept for a different system, namely for wetting layers on structured or imprinted surfaces [4].

Several experimental methods are available by which one can create structured or imprinted surfaces with domain sizes of a few micrometers. Three examples are: (i) microcontact printing based on elastomer stamps by which one can create patterns of hydrophobic alkanethiol on metal surfaces [5]; (ii) vapor deposition through grids which cover part of the surface [6]; and (iii) photolithography of amphiphilic monolayers which contain photosensitive molecular groups [7].

The morphology of wetting layers on such surfaces reflects the geometry of the underlying pattern of surface domains. Even for a single surface domain, one must distinguish different droplet regimes depending on the lyophilicity and lyophobicity of the surface domain and the surrounding surface matrix. If the surface domain and the surface matrix are strongly lyophilic and strongly lyophobic, respectively, the contact angle θ of the droplets does not satisfy the well-known Young's equation but can attain any value within the interval $0 < \theta < \pi$ [4].

For a surface pattern consisting of N identical surface domains, the wetting layer can exhibit several distinct morphologies, see Fig. 1 [4]: (i) a homogeneous droplet pattern (A) where all droplets have the same size and, thus, the same contact angle θ; (ii) a heterogeneous droplet pattern (B) characterized by one large and many small droplets. This pattern is N-fold degenerate since the large droplet can be located on any of the N surface domains; and (iii) a film state (C) for which the wetting layer covers both the hydrophilic and the hydrophobic surface regions. The relevant parameters which determine the corresponding phase diagram are (i) the water volume v per hydrophilic domain and (ii) the area fraction X of the hydrophilic domains, see Fig. 1. As one moves across the phase boundaries, the wetting layer undergoes transitions

---

[2] Here and below, "spontaneous" appears in quotes since this curvature arises from the interactions of the surface with its surroundings.

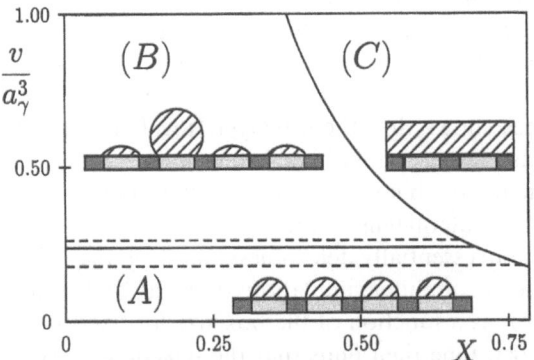

Fig. 1 Droplet morphologies for circular hydrophilic domains with radius a_γ as determined by the reduced volume v/a_γ^3 of the liquid and the area fraction X of the hydrophilic domains. The data given in this figure correspond to the case of four domains

between these different morphologies. These transitions exhibit spontaneous symmetry breaking.

The transition from the homogeneous to the heterogeneous droplet pattern is continuous for $N = 2$ but discontinuous for $N > 2$. For $N = 2$, the critical contact angle $\theta_{AB}^*(N = 2) = \pi/2$. For $N > 2$, one has a hysteresis loop for the contact angle range $\theta_{AB}^{min}(N) < \theta < \theta_{AB}^{max}(N)$ where $\theta_{AB}^{max}(N)$ has the universal value [4]

$$\theta_{AB}^{max}(N) = \pi/2 \quad \text{for all } N. \tag{1}$$

Colloidal forces of entropic origin

It has been realized for some time that colloidal forces can be of enthalpic or entropic origin. In fact, the concept of forces arising from entropy is rather unique to the colloidal domain and has no counterpart for macroscopic bodies. Entropic forces arise if one subsystem is confined or constrained by another subsystem.

Entropic forces play an important role for the behavior of flexible membranes such as lipid bilayers. If such a membrane interacts with another surface, it loses configurational entropy which leads to a fluctuation-induced repulsive force [8] and, in general, to the renormalization of the interaction potentials [9]. Likewise, polymers which are anchored to flexible membranes exert bending moments onto these membranes arising from the configurational entropy of the polymer [10–13]. In the following, another example provided by membranes in contact with dispersed nanoparticles or colloids which lose translational entropy by the presence of the membrane will be discussed [14, 15].

Thus, consider dispersed particles which (i) are rigid and (ii) cannot permeate the membrane on the experi-

mentally relevant time scales. Since water can permeate the membrane easily, the dispersed particles have an osmotic effect, and the two water compartments, which are separated by the membrane, will adapt their volume until the system is osmotically balanced.

In addition to this osmotic effect, the dispersed particles also change the curvature of the membrane. This can be understood intuitively if one envisages the membrane as a thin film which is bounded by two surfaces. These two surfaces exert attractive or repulsive forces on the dispersed particles. The resulting adsorption or depletion layers contribute to the interfacial tensions of the two surfaces. If these interfacial tensions are different, they will induce a "spontaneous" curvature of the membrane. Such a picture was originally discussed, in a qualitative way, for monolayers in emulsions [16]. In the context of lipid bilayers, the "spontaneous" curvature was first introduced by Helfrich [17] as a phenomenological parameter.

In the absence of a "spontaneous" curvature, the curvature energy of a membrane segment is scale invariant, i.e., it does not depend on the overall size of this segment but only on its shape. The "spontaneous" curvature represents an inverse length scale which breaks this scale invariance.

The "spontaneous" curvature induced by the dispersed particles depends on the particle size R_{pa}. One must distinguish several cases. If all particle species are non-adhesive, the membrane curves towards the larger particles, see Fig. 2(a). In this case, the "spontaneous" mean curvature M_{sp} increases with increasing size R_{pa} of the particles. For a binary solution with two particle species which have size R_1 and R_2, respectively, one obtains [14]

$$M_{sp} = (T/4\kappa) \, \Delta N_1 (R_2 - R_1) \, (l_{me} + R_1 + R_2), \tag{2}$$

where T is the temperature (in energy units), κ the bending rigidity of the membrane, $\Delta N_1 \equiv N_1^{ex} - N_1^{in}$ the difference in the particle number density of species 1 across the membrane, and l_{me} the membrane thickness as before.

Fig. 2 "Spontaneous" curvature of a membrane segment which is in contact with dispersed particles: (a) two non-adhesive particle species which differ in their size; (b) one adhesive and one non-adhesive species; and (c) Two adhesive species which differ in the adsorption strength

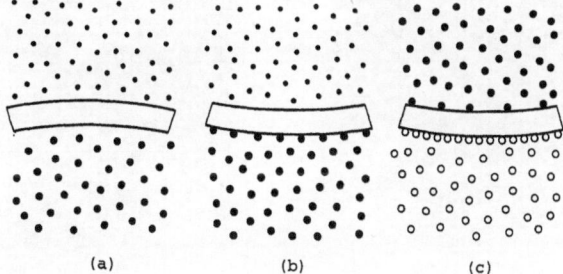

If all the particle species are adhesive and small compared to the membrane thickness l_{me}, the membrane curves away from the more strongly adsorbed particles, see Fig. 2(c), and M_{sp} decreases with increasing R_{pa}. Finally, if the solution contains one non-adhesive species and one adhesive species which is small compared to l_{me}, the membrane curves away from the adhesive particles and towards the non-adhesive ones as shown in Fig. 2(b).

Flexible membranes in contact with large adhesive particles behave differently: the membrane typically encapsulates the whole particle as soon as the particle size R_{pa} exceeds a certain threshold value $R_* \equiv [2\kappa/|W|]^{1/2}$ where W is the adhesion energy per unit area [18].

Elasticity of self-assembled structures

The behavior of flexible membranes is determined, to a large extent, by their bending rigidity κ. For thin films, this parameter can be expressed in terms of the elastic moduli of the material [19]. For a fluid layer with a vanishing two-dimensional shear modulus, this leads to the bending rigidity $\kappa_1 = K_1 l_1^2/12$ where K_1 and l_1 are the area compressibility modulus and the thickness of the layer, respectively. If this is applied to both monolayers of the bilayer membrane, the bending rigidity of the bilayer is given by $\kappa = K_A l_{me}^2/48$ where $K_A = 2K_1$ and $l_{me} = 2l_1$ are now the area compressibility modulus and the thickness of the bilayer membrane.

Real bilayers are formed by the self-assembly of lipid molecules and other amphiphiles, see, e.g., [20]. Thus, one would like to determine the elastic parameters directly for self-assembled bilayers. This is possible if one studies binary model systems consisting of "solvent" and "surfactant" molecules via molecular dynamics simulations [21–23]. In the context of such a model, one may first observe the self-assembly of the surfactant molecules into micelles and bilayer membranes, see Fig. 3. Direct observation of the motion of the surfactants within the bilayer membranes shows that these bilayers are fluid since the surfactants undergo rapid lateral diffusion.

In general, the bilayer which spans the simulation box is subject to a lateral tension arising from this box. In order to attain an essentially tensionless state of the bilayer membrane in computer simulations, one has to determine this tension as a function of the box size for fixed surfactant number. One then finds that the tension changes its sign at a certain box size and, thus, at a certain projected area per molecule, A_{s0}. For molecular areas A_s close to A_{s0}, the tension Σ behaves as $\Sigma \approx K_A(A_s - A_{s0})/A_{s0}$. Thus, by measuring the tension as a function of the molecular area, one can determine the area compressibility modulus directly [21].

It is also possible to deduce the bending rigidity κ of the bilayer directly from the analysis of its shape fluctuations [22, 23]. The values for K_A and κ which are obtained in this way are found to satisfy the relation $\kappa = K_A l_{me}^2/48$ as obtained from the continuum description within the error bars.

Colloidal machines

The colloidal systems which have been discussed in the preceding sections are in (or close to) thermal equilibrium. Biocolloids, on the other hand, are characterized by energy fluxes arising from chemical reactions. This leads to the concept of colloidal machines which transform chemical energy into mechanical work on the nanometer scale. Examples are provided by molecular motors in the living

Fig. 3 Aggregates formed by flexible amphiphiles consisting of a single chain with one headgroup particle (white) and four chain particles (black) in water (transparent) for several concentrations c_s: (a) cylindrical micelle for $c_s = 0.278$, (b) flattened cylindrical micelle for $c_s = 0.347$ and (c) bilayer membrane for $c_s = 0.382$

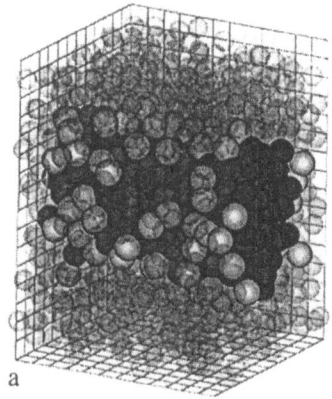
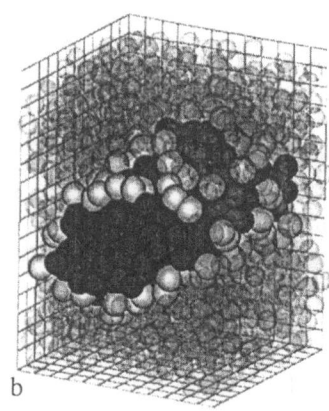
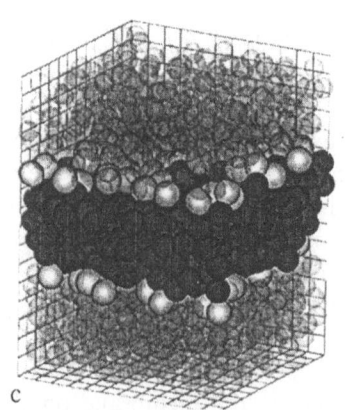

a   b   c

cell, such as the linear motors kinesin and dynein [24], which consume chemical energy obtained from ATP hydrolysis and perform directed motion along filaments such as microtubuli. Other examples are rotary motors, which move the flagellar structure of bacteria, and DNA polymerase molecules, which move along DNA strands as part of the replication machinery.

It is generally believed that the work cycle of such a motor corresponds to a sequence of conformational changes of the motor protein. What is not clear, however, is to what extent this cycle is deterministic or stochastic. One can distinguish two extreme cases: (i) caterpillar motors, often termed power-stroke motors, which undergo a deterministic sequence of conformational changes much like a macroscopic engine; and (ii) motors based on Brownian ratchets which act as rectifiers for thermal fluctuations.

The distinction between these two types of motors can be illustrated in the context of "two-state motors", which are presumably the simplest theoretical models for molecular motors. In these models, the motor can be in two different states which we call the "ground state" and the "excited state", respectively. The ATP consumption induces transitions between these two states. Furthermore, in order to move in a certain direction, the motor must be on a polar rail which generates an asymmetric force potential such as a sawtooth potential [25–28]. Thus, depending on its internal state, it feels the ground-state potential or the excited-state potential.

Now, consider a ground-state potential which has the form of a sawtooth. If the corresponding potential barriers are large compared to the thermal energy T, thermally excited diffusive motion over these barriers is strongly suppressed. Depending on the form of the excited-state potential, the two-state motor may now act like a caterpillar or like a thermal ratchet. It acts like a caterpillar if the excited-state potential has the same shape as the ground-state potential but is shifted with respect to this latter potential. On the other hand, the motor acts like a thermal ratchet if the excited-state potential is flat and the motor undergoes free Brownian motion in its excited-state.

Both types of motor mechanisms have their advantage [29]: (i) for an ordered rail, the caterpillar motor is more effective than the ratchet since all steps proceed in the same direction; but (ii) the ratchet motor is more robust with respect to perturbations arising from rail defects or other forms of rail disorder.

The behavior of a two-state motor acting as a ratchet is shown in Fig. 4. The top figure displays the stochastic motion of the ratchet for completely ordered rails, the bottom figure for a rail with defects. It is obvious from this figure that the motor is slowed down by rail defects but it is still able to undergo directed motion. For defects which

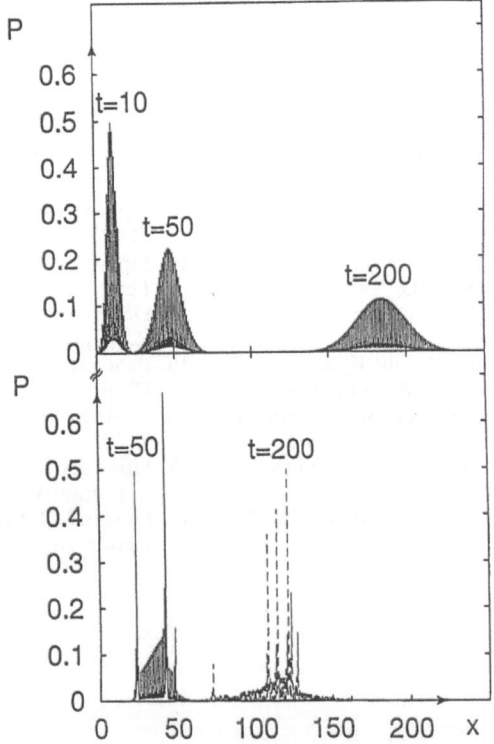

Fig. 4 Snapshots of the position probability P of a ratchet motor as a function of the rail coordinate x at time t. The x-coordinate is measured in units of the sawtooth potential period. (Top) Ordered rail: the position probability behaves as a Gaussian wave packet which drifts to the right and which is modulated on the scale of the potential period. (Bottom) Disordered rail with defect concentration $q = \frac{1}{20}$. At time $t = 10$, the position probability for the disordered rail is identical with the one displayed for the ordered case. For $t = 50$, the motor has encountered three defects; for $t = 200$, it has overcome the latter defects but is now slowed down by new ones. Each defect leads to a sharp peak in the position probability

correspond to reversed sawteeth, the corresponding reduction of the motor velocity v can be estimated via an effective hopping model. For defect concentration q, one then finds the velocity [27]

$$v(q) = v_0[q_v - q]/[q_v + q(1 - 2q_v)] \qquad (3)$$

with $q_v \equiv 1/[\exp(av_0/D_0) + 1]$ where v_0 and D_0 are the velocity and the diffusion coefficient in the absence of defects, and a is the potential period.

In contrast to stochastic ratchets, deterministic caterpillars are much more sensitive to such rail defects and can be blocked even by a single defect. Because of this distinction, it would be highly desirable to perform experiments with real molecular motors which move on rails with a controlled amount of disorder or defects.

Acknowledgements I thank Hans-Günther Döbereiner, Rüdiger Götz, Gerhard Gompper, Thomas Harms, and Peter Lenz for enjoyable collaborations.

References

1. Lipowsky R, Sackmann E (eds) (1995) Structure and dynamics of membranes, Vol. 1 of Handbook of biological physics. Elsevier, Amsterdam
2. Lipowsky R (1995) Current Opinion Struct Biol 5:531
3. Seifert U (1997) Adv Phys 46:13
4. Lenz P, Lipowsky R (1998) Phys Rev Lett 80:1920
5. Drelich J, Miller JD, Kumar A, Whitesides GM (1994) Colloids Surfaces A 93:1
6. Herminghaus S, Jacobs K, Schlagowski S, Gau H, Mönch W, to be published
7. Möller G, Harke H, Motschmann H. Langmuir (in press)
8. Helfrich W (1978) Z Naturforsch A 33:305
9. Lipowsky R, Leibler S (1986) Phys Rev Lett 56:2541
10. Lipowsky R (1995) Europhys Lett 30:197
11. Hiergeist C, Indrani V, Lipowsky R (1996) Europhys Lett 36:491
12. Hiergeist C (1997) PhD thesis, Universität Potsdam
13. Lipowsky R (1997) Colloids Surfaces A 128:255
14. Lipowsky R, Döbereiner HG. Europhys Lett (in press)
15. Döbereiner HG, Selchow O, Lipowsky R, preprint
16. Bancroft W, Tucker C (1927) J Phys Chem 31:1681
17. Helfrich W (1973) Z Naturforsch C 28:693
18. Lipowsky R, Döbereiner HG, Hiergeist C, Indrani V (1998) Physica A 249:536
19. Landau L, Lifshitz E (1989) Elastizitätstheorie. Akademie-Verlag, Berlin
20. Tanford C (1991) The Hydrophobic Effect: Formation of Micelles and Biological Membranes, 2nd ed. Krieger Publ. Comp., Malabar, FL
21. Goetz R, Lipowsky R (1998) J Chem Phys 108:7397
22. Götz R (1997) PhD thesis, Universität Potsdam
23. Götz R, Gompper G, Lipowsky R, preprint
24. Howard J (1996) Ann Rev Physiol 58:703
25. Astumian R, Bier M (1994) Phys Rev Lett 72:1766
26. Prost J, Chauwin J-F, Peliti L, Ajdari A (1994) Phys Rev Lett 72:2652
27. Harms T, Lipowsky R (1997) Phys Rev Lett 79:2895
28. Harms T (1998) PhD thesis, Universität Potsdam
29. Harms T, Lipowsky R, in preparation

Progr Colloid Polym Sci (1998) 111:41–47
© Steinkopff Verlag 1998

Z. Adamczyk
B. Siwek
P. Weronski
M. Zembala

Adsorption of colloid particle mixtures at interfaces

Z. Adamczyk (✉) · B. Siwek
P. Weronski · M. Zembala
Institute of Catalysis and Surface Chemistry
Polish Academy of Sciences
30-239 Cracow, Niezapominajek 1
Poland

Abstract Adsorption of polydisperse colloid mixtures, including the limiting case of adsorption at surfaces precovered with smaller sized particles, was studied theoretically and experimentally. The theoretical analysis of these phenomena was carried out using the generalized random sequential adsorption (RSA) model suitable for irreversible systems. In order to determine the range of applicability of the RSA model for reversible systems, numerical simulations were compared with the analytical results stemming from the equilibrium scaled particle theory. Some theoretical predictions concerning adsorption at precovered surfaces and adsorption of bimodal mixtures were discussed, i.e., the blocking functions, the kinetics and the jamming coverages. These theoretical predictions were compared with experimental data obtained for model latex suspensions using the direct microscope observation method combined with the impinging jet technique. Adsorption kinetics at the mica surface precovered with smaller particles was studied as well as adsorption from bimodal mixtures of particles differing widely in size. The characteristic features of the RSA models were quantitatively confirmed in these experiments which supported the hypothesis that small colloid particles, polymer or surfactants present in trace amounts may significantly reduce adsorption rates of larger particles.

Key words Adsorption of colloids – colloid adsorption – kinetics of colloid adsorption – structure of adsorbed colloids

Introduction

Adsorption of colloid and bioparticles is of a large practical significance for polymer and colloid science, biophysics and medicine enabling a better control of colloid, protein and cell separation processes (e.g., by filtration, chromatography), enzyme immobilization, thrombosis, biofouling of transplants and artificial organs, etc. Often in these processes, especially in filtration, polydisperse suspensions or mixtures occur, e.g., colloid/polymer, colloid/macroscopic

particle or protein/surfactant. Due to their higher diffusivity and number concentration the smaller particles or molecules will first adsorb at the interface forming a layer which may prevent adsorption of larger particles. Thus, the competition from the mobile component will result in "poisoning" of interfaces before an appreciable accumulation of larger particles takes place as reported often in the literature [1–3]. Similar problems appear in model experiments concerning particle or protein adsorption when the usual cleaning procedure may produce a layer of contamination at the substrate surface difficult to detect. This is expected to affect adsorption kinetics in the proper experiments. Hence the problem of polydisperse mixture adsorption seems important from a practical viewpoint considering the fact that few systematic studies were carried out on this subject, either theoretical or experimental.

Thus, the goal of our present work was to develop theoretical models for a quantitative analysis of kinetic and structural aspects of polydisperse mixture adsorption. In view of the complexity of the problem we confined our experiments to bimodal suspensions composed of smaller sized and larger sized monodisperse polystyrene latex suspensions mixed at various concentration ratios. As a reference state we also consider the practically important situation of larger particle adsorption at surfaces precovered with a given amount of smaller particles.

The theoretical approaches

The RSA simulation method

Polydisperse mixture adsorption, including the limiting case of adsorption at precovered surfaces, was analyzed theoretically in terms of the random sequential adsorption (RSA) method which is one of the simplest and most efficient approach to analyze sequences of irreversible events [4–7]. In this process particles are placed randomly, one upon a time, over a plane of isotropic properties. Once an empty space element is found the particle becomes irreversibly adsorbed (with no consecutive motion allowed). Otherwise it is rejected and a new addition attempt is undertaken totally uncorrelated with previous attempts. The process usually starts from an empty plane and continues until the jamming state when no additional object can be introduced to the volume.

The simulation algorithm used in our work was similar to that used for polydisperse sphere adsorption described in [8]. The simulations were carried out over a square of unit length with the usual periodic boundary conditions at its perimeter. A subsidiary matrix was introduced in order to facilitate the overlapping test. Particle positions and size distributions (e.g. Gaussian, uniform or bimodal) were simulated by using the high-quality pseudo-random number generator as described in [6, 8].

Adsorption of larger particles at surfaces precovered with smaller sized particles was simulated using the algorithm consisting of the two main stages:

(i) first the simulation plane was covered with smaller sized particles to a prescribed dimensionless surface concentration (coverage) $\theta_s = \pi a_s^2 N_s$ (where a_s is the radius of these particles and N_s is the surface concentration of smaller particles); during this stage the usual RSA simulation algorithm was used [8],

(ii) then, the larger spheres having the radius a_L were adsorbed by choosing at random their position within the simulation area; the overlapping test between larger/larger and smaller/larger particle pairs was carried out by considering the true three-dimensional distances between the sphere centers.

In order to simulate the kinetic runs the dimensionless adsorption time τ was defined as

$$\tau = \frac{N_{att}}{N_{ch}},\qquad(1)$$

Where N_{att} is the overall number of attempts to place particles and N_{ch} is the characteristic surface concentration.

The surface blocking parameter B (also called the available surface function ASF) for larger particles was calculated according to the method of Schaaf and Talbot [5] by exploiting the definition

$$B = \frac{p(\theta_s, \theta_l)}{p_0} = \frac{N_{succ}}{N_{att}},\qquad(2)$$

where p is the probability of adsorbing the larger particle at the surface characterized by the coverages θ_s, $\theta_l = \pi a_l^2 N_l$ (N_l is the surface concentration of larger particles) and p_0 is the probability of adsorption at uncovered surface (assumed equal to one without loss of generality) and N_{succ} is the number of successful adsorption events performed at fixed θ_s, θ_l. In practice, N_{att} was about 10^6 in order to attain a sufficient accuracy of B.

The analytical approximations

Adsorption kinetics of a bimodal mixture of spherical particles is governed by the system of two coupled surface mass balance equations [9]

$$\frac{d\theta_s}{dt} = \pi a_s^2 \bar{j}_s^0 n_s B_s(\theta_l, \theta_s),$$

$$\frac{d\theta_l}{dt} = \pi a_l^2 \bar{j}_l^0 n_l B_l(\theta_l, \theta_s),\qquad(3)$$

where $\bar{j}_l^0$, $\bar{j}_s^0$ are the reduced initial fluxes (for uncovered surfaces) of larger and smaller particles, respectively and B_l, B_s are the surface blocking parameters for larger and smaller particles, respectively.

Equation (3) can be expressed in a more concise, dimensionless form as

$$\frac{d\theta_s}{d\tau} = \frac{K_s}{K_s + \lambda^2} B_s(\theta_l, \theta_s),$$

$$\frac{d\theta_l}{d\tau} = \frac{\lambda^2}{K_s + \lambda^2} B_l(\theta_l, \theta_s),\qquad(4)$$

where $\tau = \pi a_1^2 \bar{j}_1^0 n_1 (1 + (K_s/\lambda^2)) t$ is the dimensionless time, $\lambda = a_1/a_s$ is the particle size ratio and $K_s = \bar{j}_s^0 n_s / \bar{j}_1^0 n_1$ is the dimensionless adsorption constant of smaller particles.

Equation (4) can be integrated by standard methods when analytical expressions for B_s and B_1 are known. As mentioned above, in the case of irreversible adsorption governed by RSA model these blocking functions can only be derived from numerical simulations. However, useful analytical approximations for B_s and B_1 can be obtained by exploiting the scaled particle theory (SPT) formulated in [10] for disks and disk mixtures [11, 12] and extended later on for sphere adsorption [14]. These functions are explicitly given by the expressions

$$B_s = (1 - \theta)\, \mathrm{e}^{-\left[\frac{3\theta_s + (\frac{4}{\lambda} - 1)\theta_1}{1 - \theta} + \left(\frac{\theta_s + (\frac{2}{\sqrt{\lambda}} - 1)\theta_1}{1 - \theta}\right)^2\right]},$$

$$B_1 = (1 - \theta)\, \mathrm{e}^{-\left[\frac{3\theta_1 + (4\lambda - 1)\theta_1}{1 - \theta} + \left(\frac{\theta_1 + (2\sqrt{\lambda} - 1)\theta_s}{1 - \theta}\right)^2\right]}, \qquad (5)$$

where $\theta = \theta_s + \theta_1$.

It is interesting to note that the low coverage expansion of B_1 in the case of absorption at precovered surfaces assumes the form

$$B_1 = 1 - C_1(\theta_s)\theta_1 = 1 - \frac{\theta_1}{\theta_{mx}}. \qquad (6)$$

Hence, a quasi-Langmuirian behavior is predicted with θ_{mx} given by the expression [14]

$$\theta_{mx} \cong \frac{1}{4 + (4\lambda + 4\sqrt{\lambda} - 7)\theta_s}. \qquad (7)$$

In the general case the θ_1 or θ_s vs. τ dependencies were obtained by integrating Eq. (4) by the Runge–Kutta method.

Experimental procedure

The experimental setup described in detail in our previous works [3, 8, 13] was based on the circular impinging-jet principle and direct in situ microscope observations of particle adsorption. Since the size of the particle was close to a micrometer we were able to determine in real time the surface concentration (number of particles adsorbed over a given surface) of both small and large particles. We used the freshly cleaved mica sheets as the substrate. In the case where negatively charged colloid mixtures were used the negative surface charge of mica was converted into a positive one due to the irreversible adsorption of Al^{3+} salts in a procedure similar to that previously described [13].

The latex suspensions were produced in a surfactant-free polymerization procedure with the persulfate initiator in case of negative particles and a special azonitrile ini-

tiator for the positive ones [3]. In this study we used two monodisperse latex samples (i) characterized by the averaged size determined by the Coulter–Counter to be $2a_1 = 1.48 \pm 0.1\ \mu m$ (hereafter referred to as larger particles) and (ii) $2a_s = 0.68 \pm 0.05\ \mu m$ (referred to as smaller particles). Their size ratio a_1/a_s was, therefore, equal to 2.2. We also used a similar combination of positively charged latices having the size $2a_1 = 1.12 \pm 0.1$ and $2a_s = 0.55 \pm 0.05\ \mu m$, respectively, characterized by $\lambda = 2.0$.

The ionic strength in these experiments was kept at $10^{-4}\ M$ (by addition of recrystallized KCl solution) and the pH was about 6.0. The volume velocity of the suspension Q was equal to $0.009\ cm^3/s$, which corresponded to Reynolds number of 4, or $0.018\ cm^3/s$ for $Re = 8$.

The experimental procedure described in detail in our previous works [3, 13] consisted in provoking the suspension motion through the cell at a fixed rate (due to hydrostatic pressure difference between the suspension container and the cell outlet tube). Then, the number of particles adsorbing over areas close to the center of the cell (in order to minimize the hydrodynamic scattering effects) was recorded in real time using image processing. The initial fluxes of monodisperse suspensions were determined in separate experiments by differentiating a polynomial which fitted best the initial kinetic date, i.e., the coverage vs. time dependencies.

In the case of adsorption at precovered surfaces the experimental procedure was slightly different: first the smaller particles were adsorbed at the mica surface under the usual procedure [3, 13] until a given coverage θ_s was attained. In our experiments θ_s varied between 0 (uncovered surface) to 30%. Then the small particle suspension was replaced in situ by the larger particle suspension characterized by the bulk number concentration n_b. A short transition time (of the order of 1 min) was allowed before the proper adsorption experiments of larger particles were recorded.

Results and discussion

The theoretical predictions

Using the RSA numerical algorithm described above one can determine not only the blocking functions of small and large particles and their adsorption kinetics but also the jamming coverages and the structure of the adsorbed layers at transient and jammed states.

The quantity of considerable practical interest is the $B_1^0(\theta_s)$ function which represents the averaged probability of adsorbing a particle over a surface precovered by smaller particles (coverage θ_s) normalized to the probability of adsorbing the particle over an uncovered surface. Thus, by

Z. Adamczyk et al.
Adsorption of colloids

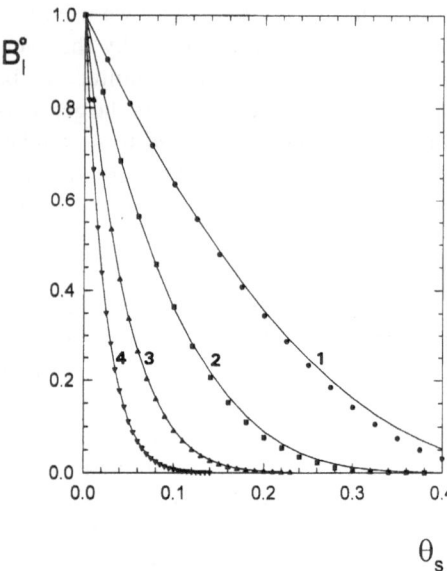

Fig. 1 The dependence of B_1^0 on the surface concentration of smaller particles θ_s. The points denote numerical simulations performed for: (1) $\lambda = 1$ (reference, monodisperse system), (2) $\lambda = 2.2$, (3) $\lambda = 5$, (4) $\lambda = 10$, the continuous lines denote the equilibrium SPT results calculated from Eq. (5)

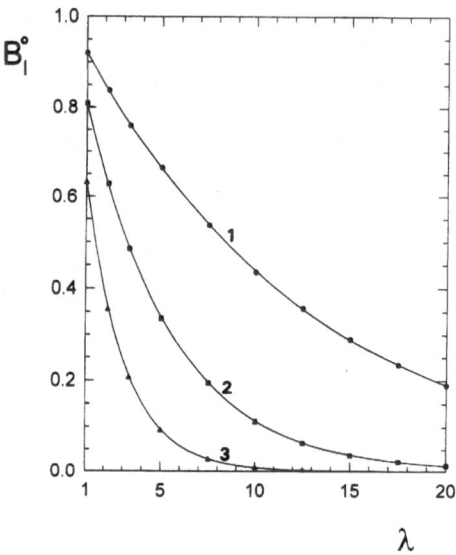

Fig. 2 The dependence of B_1^0 function of larger particles (normalized initial flux) on $\lambda = a_l/a_s$ determined numerically for: (1) $\theta_s = 2\%$, (2) $\theta_s = 5\%$, (3) $\theta_s = 10\%$, the continuous lines denote the equilibrium SPT results calculated from Eq. (5)

knowing B_1^0 one can calculate the initial flux (and hence adsorption kinetics for initial stages) at covered surfaces from the simple relationship

$$\bar{j} = \bar{j}_1^0 B_1^0(\theta_s) \ . \tag{8}$$

In Fig. 1 the dependence of B_1^0 on θ_s is plotted for $\lambda = 1$ (reference curve for monodisperse spheres), 2.2, 5 and 10, respectively. As can be noticed, adsorption probability of larger particles is considerably decreased by the presence of adsorbed small particles, especially for higher λ values. It is interesting to note that the analytical SPT results represented by Eq. (5) (when the value $\theta_1 = 0$ is substituted) reflect well the characteristic features of the B_1^0 function, especially its fast, quasi-exponential decrease for higher coverages. The agreement between the RSA simulations and these analytical predictions seems quantitative when $B_1^0 > 0.1$. As expected the deviation between equilibrium and RSA results are increased for higher θ_s values when B_1^0 became considerably smaller than 0.1

It seems therefore, in view of the limited experimental accuracy, especially for higher coverages of smaller particles, that the SPT results can be used for practical purposes as a good estimate of B_1^0 (initial flux) on precovered surfaces.

The results shown in Fig. 1 also suggest that the presence of small amounts of smaller particles, e.g., of the colloid type should result in a significant decrease in adsorption of larger particles. This "surface poisoning" effect

is further illustrated by the data shown in Fig. 2. As one can see the value of B_1^0 decreased abruptly, for a fixed θ_s as the size of the preadsorbed particles is decreased.

The kinetic curves, i.e., the θ_1 vs. τ dependencies derived from the numerical RSA-type simulations for $\lambda = 10$ and various surface concentrations of smaller particles θ_s are shown in Fig. 3. For comparison, the analytical results calculated by integrating Eq. (4) with the blocking function described by the quasi-Langmuirian model, Eq. (6), are also shown. As can be seen, as this model gives for θ_1 the explicit expression

$$\theta_1 = \theta_{mx}[1 - e^{-B_1^0 \tau/\theta_{mx}}] \ , \tag{9}$$

it can be used as a reasonable estimate of adsorption kinetics on precovered surfaces, especially for higher coverages θ_s.

It should be noted, however, that for a very long adsorption time, $\tau \gg 4$, considerable deviations from the quasi-Langmuirian model appear (see Fig. 3) and the adsorption regime is changed. It can be then better described by the power law approach at the saturation (jamming) concentration which is characteristic for many RSA processes [3, 7]

$$\theta_1^\infty - \theta_1 \sim \tau^{-1/2} \ , \tag{10}$$

where $\theta_1^\infty(\theta_s)$ is the jamming coverage of larger particles at surfaces precovered with a given amount of smaller particles.

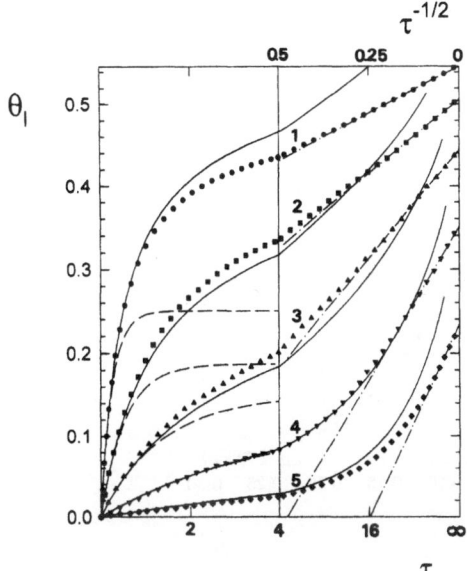

Fig. 3 Kinetics of larger particle adsorption at surfaces precovered with smaller particles (numerical RSA simulations) expressed as θ_1 vs. τ dependence, $\lambda = 10$: (1) $\theta_s = 0\%$, (2) $\theta_s = 2.5\%$, (3) $\theta_s = 5\%$, (4) $\theta_s = 7\%$, (5) $\theta_s = 10\%$, the continuous lines denote the SPT equilibrium results, the broken lines show the analytical results calculated from Eq. (9) and the $(\cdot - \cdot -\cdot)$ lines denoted the linear fits

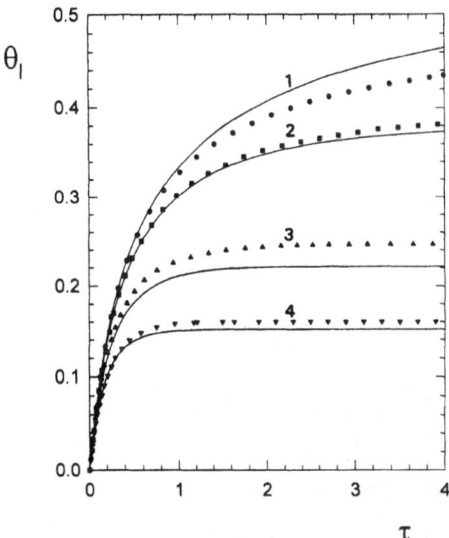

Fig. 4 Kinetics of bimodal sphere adsorption derived from RSA simulations for $\lambda = 10$: (1) $K_s = 0$ (monodisperse reference system), (2) $K_s = 1$, (3) $K_s = 5$, (4) $K_s = 10$, the continuous lines denote the analytical results obtained by numerical integration of Eq. (4)

It can be demonstrated [5, 14] that the θ_1 vs. $\tau^{-1/2}$ dependence implies that the ASF function for this asymptotic regime should assume the form

$$B_1 \sim [\theta_1^\infty(\theta_s) - \theta_1]^3 , \tag{11}$$

where the jamming coverages θ_1^∞ are dependent on the initial coverage of smaller particles.

The dependencies of these jamming concentrations θ_1^∞ on θ_s obtained for various λ values are reported elsewhere [14]. Generally speaking, the jamming coverages of large particles are considerably reduced by the presence of particles whose size is considerably smaller, $\lambda > 5$ (e.g., colloids).

Similar effects as these discussed above for precovered surfaces also occur for bimodal mixture adsorption when the dimensionless adsorption rate K_s of the smaller component becomes larger than unity. This is illustrated well in Fig. 4 where the kinetic runs are shown for $\lambda = 10$. As can be observed fast adsorption of the smaller component from the mixture blocked adsorption of larger particles whose surface coverage was saturated at a very low value. However, in contrast to the case of preadsorbed surfaces, the θ_1 values drift little with time (for $\lambda > 5$) so the θ_1^∞ values were consequently much smaller. One may draw, therefore, the conclusion that the surface poisoning effect becomes more pronounced for mixtures in comparison with precovered surfaces.

Adsorption at precovered surfaces or from bimodal mixtures differs also significantly in respect to structural aspects. For lack of space these interesting problems are discussed elsewhere [14].

Experimental evidences

The experimentally determined adsorption kinetics of large particles at surfaces precovered with a given amount of smaller particles is shown in Fig. 5. As one can notice, the increase in θ_s resulted in a considerable decrease in adsorption rate of larger particles so the surface coverages θ_1 attained after a long time in the presence of preadsorbed smaller particles became much lower than for uncovered surfaces.

The experimental kinetic runs were compared with the RSA simulations performed according to the procedure described above (solid lines in Fig. 5). As one can notice, the experimental results are well accounted for by the RSA model with electrostatic interactions neglected (hard particle limit). This can be explained by the relatively high ionic strength and rather low surface concentration of larger particles, so the electrostatic interactions became effectively screened [15].

The analytical predictions stemming from the quasi-Langmuir model (Eq. (6)) are also plotted in Fig. 5 (dotted lines). As can be seen they are in a good agreement with the numerical simulations and experimental results for the

Z. Adamczyk et al.
Adsorption of colloids

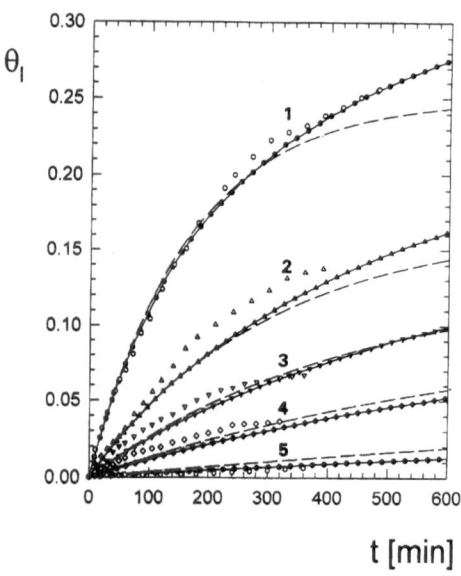

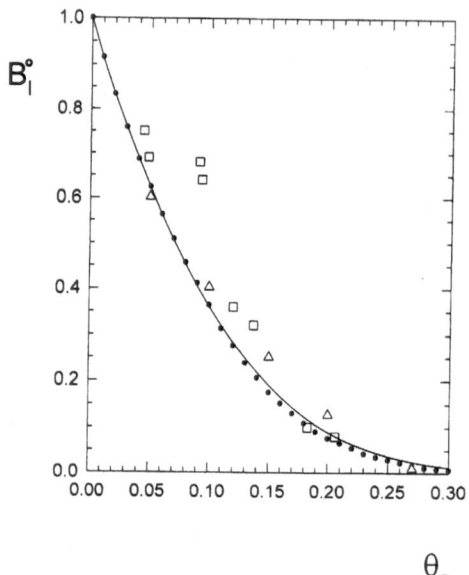

Fig. 5 Kinetics of polystyrene particle adsorption ($2a_1 = 1.48 \mu m$) at mica surface precovered by various amounts of smaller particles ($2a_s = 0.68 \mu m$): (1) $\theta_s = 0$ (uncovered surface), (2) $\theta_s = 10\%$, (3) $\theta_s = 15\%$, (4) $\theta_s = 20.5\%$, (5) $\theta_s = 27\%$, the empty points denote the experimental results, the full symbols denote the RSA simulations, the broken lines represent the quasi-Langmuir model Eq. (9)

Fig. 6 The dependence of the reduced initial flux B_1^0 of larger particles at the mica surface covered by smaller particles; the triangles and squares denote the experimental results obtained for negative and positive particles, respectively, the circles denote the theoretical results derived from the RSA simulations and the continuous line respresents the SPT results

entire range of θ, especially for shorter adsorption times ($t < 300$ min).

It should also be noted in Fig. 5 that for longer adsorption times ($t > 300$ min) the kinetic curves seem to approach stationary (saturation) values of θ. This is, however, an apparent saturation effect because from the numerical predictions discussed above one can deduce that the true limiting values of θ are much higher. However, the physical adsorption time needed to approach the true jamming limits becomes excessively long, especially for higher θ_s values (of the order of days for the bulk suspension concentration equal 4×10^8 cm^{-3}). For such long-lasting experiments, however, the chances of contaminating the system are increased which makes them less reliable.

Therefore, the kinetic runs as these shown in Fig. 5 are not suitable for determining the maximum surface coverages. They can be used, however, with a good accuracy to derive the initial flux at covered surfaces.

The results are shown in Fig. 6 in the form of the dependence of the reduced flux $\bar{j}_1/\bar{j}_1^0 = B_1^0$ on θ_s (where the initial flux at uncovered surfaces $\bar{j}_1^0$ was determined in separate experiments). This quantity can be treated as the blocking parameter of larger particles at surfaces precovered by small particles. As can be seen in Fig. 6 the experimental results (obtained both for negatively and positively charged particles) can well be described by the

numerical RSA simulations. Also, the analytical expressions stemming from the equilibrium SPT theory are in good agreement with our experimental results which demonstrated that the presence of smaller particles exerted a profound effect on adsorption of larger particles. The initial flux for θ_s as low as 10% was found almost three times smaller than for clean surfaces. For θ_s equal to 20%, the initial flux at precovered surfaces is reduced by more than 10 times.

It should be mentioned that the effects measured in our work which are stemming from the surface exclusions effects exceed by orders of magnitude the effects predicted theoretically by Dabroś and van de Ven [16] originating from hydrodynamic corrections due to the presence of adsorbed particles.

Similar effects were observed in the case of polydisperse mixture adsorption as shown in Figs. 7 and 8. One can see that the presence of small particles in the mixture significantly decreases the adsorption kinetics of larger particles whose maximum surface coverage saturated at very small values, especially for $K_s > 1$ (cf. Fig. 8). Similarly, for precovered surfaces the appearance of the poisoning effect of smaller particles was, therefore, spectacularly demonstrated. It should be noted that the experimental results shown in Figs. 7 and 8 can well be interpreted in terms of the RSA simulations and the analytical SPT results (derived by numerical integration of Eq. (4)).

Progr Colloid Polym Sci (1998) 111:41–47
© Steinkopff Verlag 1998

47

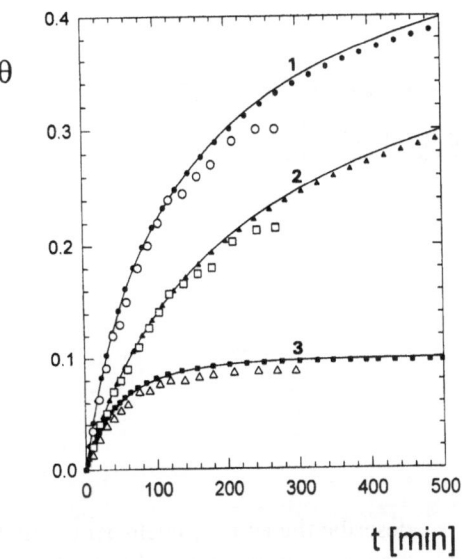

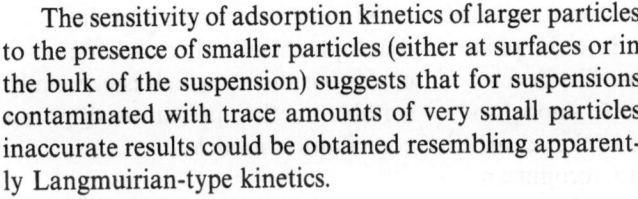

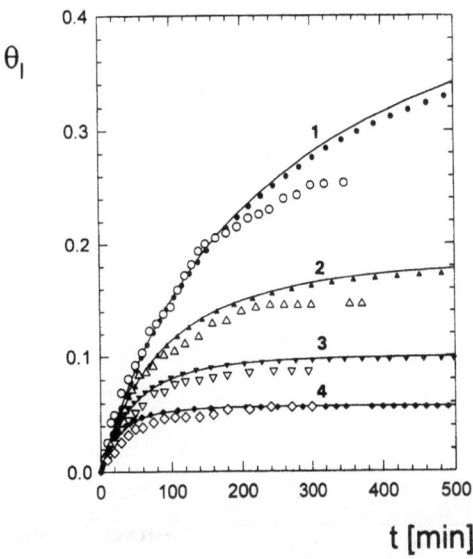

Fig. 7 Kinetics of adsorption from bimodal mixtures, $\lambda = 2.2$, $K_s = 5$; the empty symbols show the experimental results obtained for latex particles, the full symbols show the RSA numerical results and the continuous lines represent the SPT results: (1) $\theta_l + \theta_s$, (2) θ_s, (3) θ_l

Fig. 8 Kinetics of larger particle adsorption from bimodal mixtures, $\lambda = 2.2$; the empty symbols show the experimental results obtained for: (1) $K_s = 0$ (monodisperse, reference system), (2) $K_s = 2$, (3) $K_s = 5$, (4) $K_s = 10$, the full symbols show the RSA results and the continuous lines represent the SPT results

The sensitivity of adsorption kinetics of larger particles to the presence of smaller particles (either at surfaces or in the bulk of the suspension) suggests that for suspensions contaminated with trace amounts of very small particles inaccurate results could be obtained resembling apparently Langmuirian-type kinetics.

lated SPT theory with the blocking functions given by Eq. (5).

These theoretical predictions and experimental results obtained for model suspensions suggest that both the adsorption kinetics (initial flux) and the jamming coverages of larger particles are very sensitive to the presence of trace amounts of smaller sized particles (colloids, polymers) at the interfaces, often difficult to detect by conventional analytical means. These theoretical predictions may explain the persisting difficulties in obtaining reliable kinetic data and monolayer densities in colloid protein adsorption processes.

Concluding remarks

It has been found that the numerical RSA simulations performed for bimodal sphere adsorption can well be approximated in the limit of low densities by the extrapo-

Acknowledgements This work was partially supported by the KBN Grant: No 3T09 A08310.

References

1. Boluk MY, van de Ven TGM (1990) Colloids Surf 46:157–175
2. van de Ven TGM, Kelmen SJ (1996) J Colloid Interface Sci 181:118–123
3. Adamczyk Z, Siwek B, Zembala M, Belouschek P (1994) Adv Colloid Interface Sci 48:151–280
4. Hinrichsen EL, Feder J, Jossang T (1986) J Stat Phys 44:73–827
5. Schaaf P, Tablbot J (1989) J Chem Phys 91:4401–4408
6. Adamczyk Z, Weroński P (1996) J Chem Phys 105:5562–5573
7. Evans JW (1993) Rev Mod Phys 65:1281–1329
8. Adamczyk Z, Siwek B, Zembala M, Weroński P (1997) J Colloid Interface Sci 185:236–244
9. Talbot J, Schaaf P (1989) Phys Rev A 40:422–427
10. Reiss H, Frisch HL, Lebowitz JL (1959) J Chem Phys 31:369–380
11. Lebowitz JL, Helfand E, Preastgaard E (1965) J Chem Phys 43:774–779
12. Talbot J, Jin X, Wang NHL (1994) Langmuir 10:1663–1666
13. Adamczyk Z, Siwek B, Weroński P (1997) J Colloid Interface Sci 195:261–263
14. Adamczyk Z, Weroński P (1997) J Chem Phys 108:9851–9858
15. Adamczyk Z, Warszyński P (1996) Adv Colloid Interface Sci 63:41–150
16. Dabroś T, van de Ven TGM (1993) Colloids Surf A 75:95–104

Progr Colloid Polym Sci (1998) 111:48–51
© Steinkopff Verlag 1998

P. Ulbig
C. Berti
S. Schulz

Correlation and prediction of liquid phase adsorption equilibria

Dr. P. Ulbig (✉) · C. Berti · S. Schulz
Institute for Thermodynamics
University of Dortmund
Emil-Figge-Str. 70
D-44221 Dortmund
Germany

Abstract A thermodynamic framework is presented which allows the correlation and prediction of liquid-phase adsorption equilibria. In contrast to the classical formulation of phase equilibrium, the adsorbent is considered to be an additional component, thus incorporating the adsorbent's properties like functional groups of the surface into the phase equilibrium equations. The development of a modified G^E-model to describe the behavior of the adsorbed phase has to take into account an activity coefficient for each component with the limit value of pure component adsorption. In a first step the Wilson equation was used to describe the activity coefficients. First results show the good ability of the model to correlate surface excess isotherms of different types and to predict their temperature dependence. The thermodynamic approach facilitates the inclusion of enthalpies of wetting for the parameter estimation, thus leading to the development of a thermodynamic consistent theory which is especially suitable for the application of group contribution models.

Key words Liquid phase adsorption – correlation – prediction – equilibria – Wilson equation

Introduction

The adsorption from the liquid phase onto solid adsorbents represents a growing field in chemical engineering. The modelling of these processes is difficult, due to the multiplicity of variables concerning the nature of the adsorbent. Especially technical purposes require a prediction of phase equilibrium in order to compute liquid-phase adsorption processes. In the past, mainly the correlation of adsorption isotherms has been investigated by several authors [1, 2] for different adsorbents. A first approach that divides the liquids into functional groups to predict the phase equilbrium for aqueous mixtures with highly diluted organic solvents onto active carbon has been presented [3]. Considering the functional groups of the solid surface being independent of a specific adsorbent, a

preliminary formulation of the Universal Group Contribution Model for liquid phase Adsorption (UGCMA [4]) has been introduced. The aim of this paper is the description of a thermodynamic consistent framework in order to form a basis for the further development of the UGCMA-method. In a first step the well-known Wilson equation [5] was used to calculate the Gibbs excess energy G^E since this model facilitates the description of multicomponent mixtures (in this case two liquids and one solid) by using only the two respective binary parameters of each combination.

Theory

A thermodynamic description of liquid-phase adsorption equilibria depends on the definition of a reference system.

Progr Colloid Polym Sci (1998) 111:48–51
© Steinkopff Verlag 1998

Fig. 1 Definition of reference system

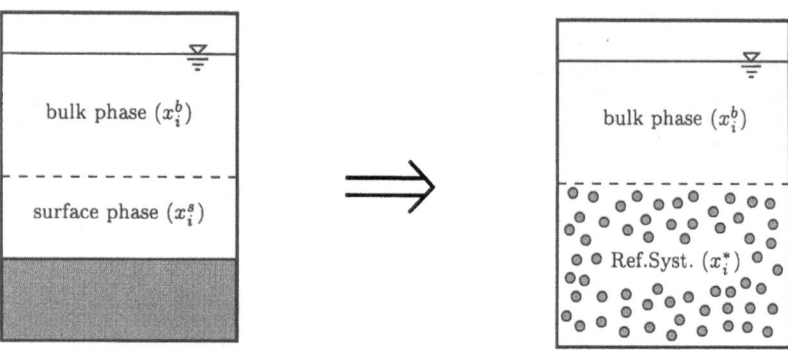

Usually, two different approaches may be distinguished: (1) the concept of a surface phase, which is influenced by the solid surface and a bulk phase, and (2) the definition of excess properties. The phase equilibrium between the surface phase and bulk phase for a component i [6] can be formulated as

$$x_i^b \gamma_i^b = x_i^s \gamma_i^s \exp\left(\frac{\phi - \phi_{0i}}{RT\,\Gamma_{mi}^s}\right), \tag{1}$$

where ϕ is the free energy of immersion of the adsorbed solution. The respective activity coefficient in the surface phase γ_i^s has to be estimated from Eq. (1) by the use of experimental data. The resulting functions $\gamma_i^s = f(x_i^s)$ show a complex variation with composition, e.g. extreme values, inflection points or non-monotonic behaviour, thus questioning the physical meaning of these activity coefficients [7–9]. Moreover, for the pure component adsorption the respective activity coefficient equals 1 though the measurable enthalpy of immersion suggests a consideration of the interactions between adsorbent and the pure liquid which, on the other hand, is described by γ_i^s. These difficulties led to the definition of an "adsorbate-solid solution" as a different reference system (see Fig. 1).

This reference system takes into account the solid adsorbent as an additional component in the adsorbed phase, thus leading to a new formulation of the phase equilibrium according to

$$x_i^b \gamma_i^b = x_i^* \gamma_i^* \exp\left(\frac{\phi^* - \phi_{0i}^*}{RT\,\Gamma_{mi}^s}\right) \tag{2}$$

The difference between the chemical potentials of the solid in contact with a liquid mixture (ϕ^*) and a pure liquid (ϕ_{0i}^*), respectively, equals the difference of the free enthalpy of adsorption ($g^{ad} - g_{0i}^{ad}$) (note that $g^{ad} \neq \phi^*$). On the other hand, this difference may be expressed in terms of the Gibbs excess energy of the adsorbate–solid solution G^{E*}:

$$\phi^* - \phi_{0i}^* = (g^{ad} - g_{0i}^{ad}) = \frac{1}{m_a}(G^{E*} - G^{Es} - G_{0i}^{E*}), \tag{3}$$

where G^{Es} is the Gibbs excess energy of the solid-free liquid mixture with the concentration x_i^s:

$$\frac{G^{Es}}{RT} = \sum_{i=1}^{k} n_i^s \ln \gamma_i(x_i^s). \tag{4}$$

Now activity coefficients for both the solid and the respective liquid components may be introduced:

$$\frac{G^{E*}}{RT} = \sum_{i=1}^{k} n_i^* \ln \gamma_i^* + n_a \ln \gamma_a^* \tag{5}$$

with the limit value of the Gibbs excess energy for the pure liquid adsorption (G_{0i}^{E*}),

$$\frac{G_{0i}^{E*}}{RT} = n_i^* \ln \gamma_{0i}^* + n_a \ln \gamma_{a,\,0i}^* \tag{6}$$

The Gibbs–Helmholtz equation establishes the relationship with the measurable adsorption enthalpy:

$$\frac{\partial(G^{E*}/T)}{\partial T} = -\frac{H^{E*}}{T^2}. \tag{7}$$

$$\Delta H_{exp}^{ad} = n^* h^{E*} - n^s h^{Eb} + n^0 (h^{Eb} - h^{E0}) \tag{8}$$

The last equation is derived from an enthalpy balance. These equations make it possible to include measured enthalpies of adsorption for the parameter estimation.

The limit of the pure component adsorption has to be incorporated in any G^{E*}-model which is suitable to describe the adsorbate–solid solution. This is achieved by splitting the activity coefficient of the respective fluid into two parts:

$$\ln \gamma_i^* = \ln \gamma_{0i}^* + \ln \gamma_{GE,i}^* \tag{9}$$

where γ_{0i} is the limit value and $\ln \gamma_{GE,i}^*$ a concentration dependent part of the activity coefficient that is given by the chosen G^E-model. $\ln \gamma_{GE,i}^*$ has to equal zero at the concentration border:

$$\ln \gamma_{GE,i}^* = \ln \gamma_{GE,i}^*(x_i^*) - \ln \gamma_{GE,i}^*(x_{0i}^*). \tag{10}$$

50

P. Ulbig et al.
Correlation and prediction of liquid phase adsorption equilibria

Fig. 2 Correlation and prediction of the surface excess of benzene (1)/cyclohexane (2) on silica gel (correlation: 273.15 K, prediction: 303.15 K, 333.15 K)

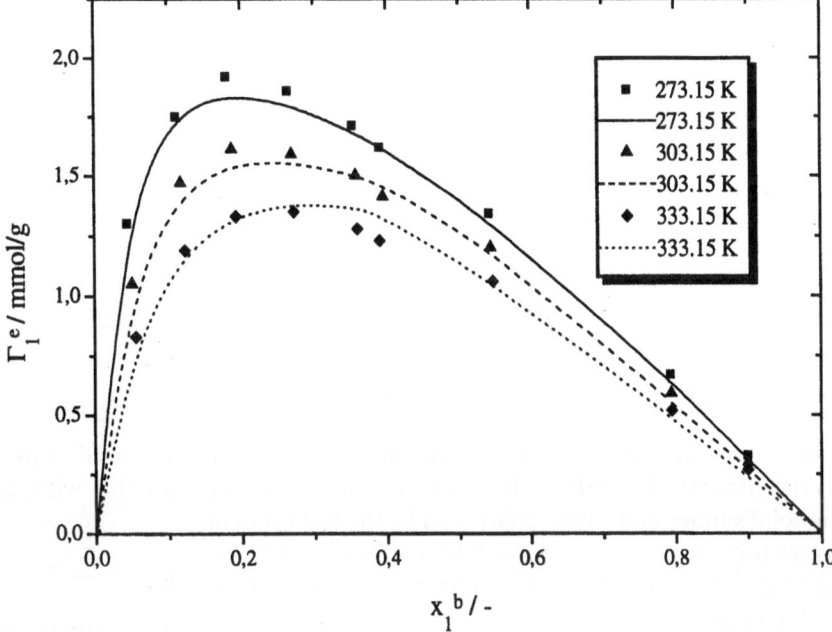

Fig. 3 Local compositions in liquid mixture (a) and in surface phase (b)

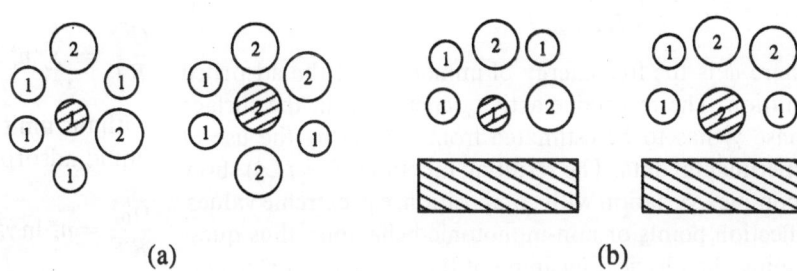

(a)   (b)

$\gamma^*_{GE,i}(x^*_{0i})$ represents the activity coefficient of a binary adsorbate–solid solution, i.e. the pure (adsorbed) liquid and the solid adsorbent. The introduction of a reference point does not have any effect on the qualitative variation of the activity coefficients with concentration.

For evaluation purposes the Wilson equation was first used as G^E-model since the required parameters for the description of the Gibbs excess energy for the liquid mixture are known from literature. Thus, only the parameters for the interactions between the liquids and the solid adsorbent have to be fitted. For a ternary mixture (two liquids, one solid) one obtains

$$\ln \gamma^*_{GE,i} = -\ln\left(\sum_{j=0}^{2} x^*_j \Lambda_{ij} \right) + 1 - \sum_{k=0}^{2} \frac{x^*_k \Lambda_{ki}}{\sum_{j=0}^{2} x^*_j \Lambda_{kj}} . \quad (11)$$

Results

In a first study surface excess data were taken from literature for the system benzene (1)–cyclohexane (2) on silica gel

(0) at different temperatures [10]. Wilson parameters were fitted at 273.15 K by using the Simplex-algorithm of Nelder and Mead [11]. Since enthalpies of immersion are not available for the examined system, only surface excess data could be consulted to determine the interaction parameters. With the calculated Wilson parameters the surface excess isotherms at 303.15 and 333.15 K were predicted. The results are shown in Fig. 2.

The mean relative deviations between the experimental and calculated data of 2.7% (273.15 K), 3.0% (303.15 K) and 4.1% (333.15 K) prove that the derived equations are suitable to correlate and predict surface excess data at different temperatures.

Conclusion

Further developement of a G^{E*}-model based on group contributions for the adsorbate-solid solution (UGCMA) will focus on the application of models that fall back upon

Progr Colloid Polym Sci (1998) 111:48–51
© Steinkopff Verlag 1998

the local composition concept like the formulation of Wilson, e.g. UNIQUAC [12] or EBLCM [13]. This is necessary as a prior stage for the use of group contribution models such as UNIFAC [14] or EBGCM [15]. The local structure of the adsorbate differs from the one of liquids due to the interactions of the center molecule with the neighbour molecules and the solid surface (see Fig. 3). This fact has to be taken into consideration in the last stage of development of UGCMA.

Acknowledgement The authors thank the German Research Foundation for financial support.

References

1. Radke CJ, Prausnitz JM (1972) AIChE J 18:761–768
2. Messow U, Bräuer P, Heuchel M, Pysz M (1992) Chem Tech 44:56–59
3. Chitra SP, Govind R (1986) AIChE J 32:167–169
4. Friese T, Ulbig P, Schulz S (1996) Ind Eng Chem Res 35:2032–2038
5. Wilson GM (1964) J Am Chem Soc 86:127–131
6. Larionov OG, Myers AL (1971) Chem Eng Sci 26:1025–1031
7. Li MH, Hslao HC, Yih SM (1991) 36:244–248
8. Minka C, Myers AL (1973) AIChE J 19:453–459
9. Schay G (1969) Surf and Colloid Sci 2:155–211
10. Valenzuela DP, Myers AL (1989) Adsorption Equilibrium Data Handbook. Prentice Hall, New Jersey
11. Nelder JA, Mead R (1965) Computer J 7:308–313
12. Abrams DS, Prausnitz JM (1975) AIChE J 21:116–128
13. Rowley RL, Battler JR (1984) Fluid Phase Equil 18:111–130
14. Fredenslund A, Jones RL, Prausnitz JM (1975) AIChE J 21:1086–1099
15. Ulbig P (1996) PhD Thesis, University of Dortmund

Progr Colloid Polym Sci (1998) 111:52–57
© Steinkopff Verlag 1998

SUSPENSIONS AND MICROCAPSULES

Large metal clusters and colloids – Metals in the embryonic state

G. Schmid

Prof. Dr. Schmid (✉)
Institut für Anorganische Chemie
Universität Essen
Universitätsstr. 5-7
D-45117 Essen
Germany
E-mail: guenter.schmid@uni-essen.de

Abstract Ligand stabilized transition metal clusters and colloids in the size range of 1–15 nm show size dependent quantization phenomena. Quantum size effects become the more evident the smaller the particle is. Whereas particles >2 nm behave like quantum dots only at low temperatures, the 1.4 nm Au_{55} cluster follows quantum mechanical rules even at room temperature. First steps to organize clusters and colloids three- (3D), two- (2D) and one- (1D) dimensionally have been performed. 3D arrays are reached by using spacer molecules to link the clusters and to enlarge the distances between them. 2D assemblies are realized in cluster and colloid monolayers on chemically modified surfaces. One-dimensional cluster wires become available by using nanoporous aluminum oxide membranes as templates.

Key words Metal clusters – colloids – quantum size effect

Introduction

The study of the size-dependent physical and chemical properties of a material is of immense importance with respect to the development of solid-state behavior. Metals are well suited to follow up the question, how many atoms of a metal are needed to make a metal. The answer depends on several conditions such as the property we look for, the method of investigation, the temperature, etc. However, it seems reasonable to assume that an assembly of only a few atoms will not provoke typical metallic behavior.

The reduction of a metal particle to a size comparable to the de Broglie-wavelength of an electron leads to the formation of stationary electronic waves with discrete energy levels. Particles of that size are called quantum dots. Owing to their special electronic characteristics, quantum dots of metal atoms are considered to play a decisive role in future nanoelectronics [1–5]. However, the application of clusters and colloids requires not only routine fabrication but also organization in three-, two- or even one-dimensional arrangements. To avoid contacts between single clusters and colloids they must be separated from each other. From a chemical point of view this is best realized by covering the particle surfaces by a shell of ligand molecules. That is why we exclusively report on so-called ligand stabilized clusters and colloids. The ligand shell not only protects the particles from aggregation but in addition causes solubility in chemically related solvents [1, 6].

Synthetic aspects of clusters and colloids will not be considered here as they have been described in detail [7–15]. This paper will focus on some physical properties regarding the special electronic situation and further it will deal with some aspects of possible structurization for future applications.

Properties

As has been discussed in the introduction, the size reduction of a piece of metal must somewhere result in a beginning quantization of electronic states. This has recently

Progr Colloid Polym Sci (1998) 111:52–57
© Steinkopff Verlag 1998

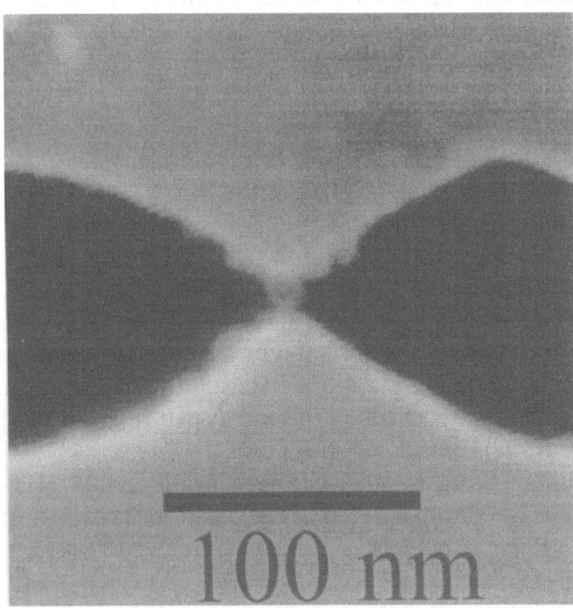

Fig. 1 A ~17 nm Pd colloid fixed between two Pt tips

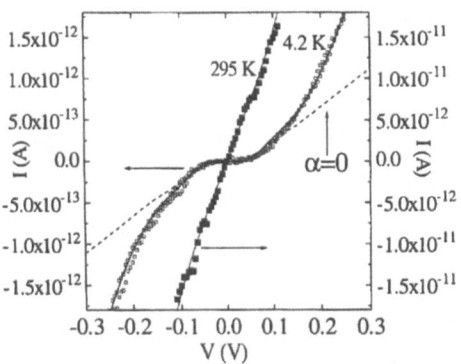
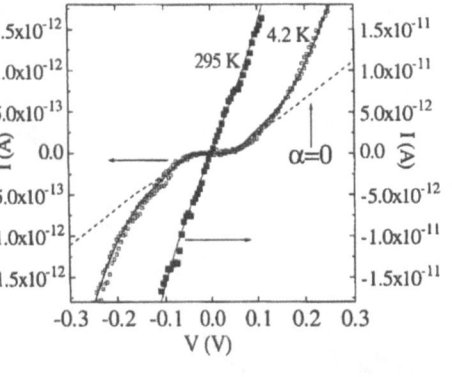

Conditions:

1. $\boxed{\dfrac{e^2}{2C} \gg k_b T}$

   C = Capacity of the tunnel contact

   $C = \varepsilon\varepsilon_o \dfrac{A}{d}$

   ε = dielectric constant
   ε_o = eletric field constant
   A = surface of the electrode
   d= distance of the electrodes

2. $\boxed{R_t \gg R_Q}$

   R_t = Tunneling Resistance
   R_Q = Quantum-Hall-Resistance

   $R_Q = \dfrac{h}{4e^2}$

Fig. 2 I–V curves of the 17 nm Pd particle at 295 K (solid squares) and at 4.2 K (open circles). The solid lines denote the fits

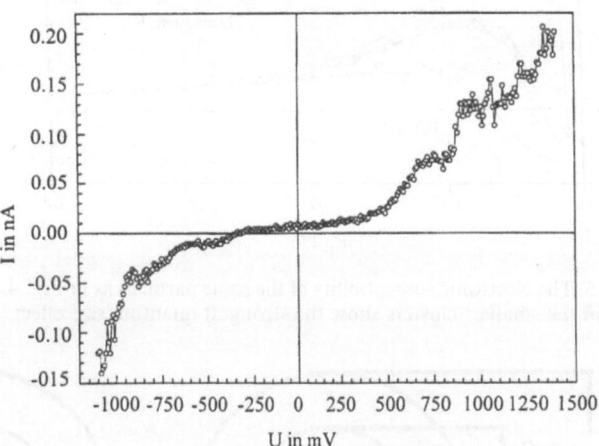

Fig. 3 I–V curve of $Au_{55}(Ph_2PC_6H_4SO_3H)_{12}Cl_6$ clusters in a monolayer on a gold surface at room temperature

impressively been demonstrated by investigating a single 17 nm Pd colloid, fixed between two platinum tips, as is shown in Fig. 1 [16].

This is still a relatively large particle consisting of hundred thousands of atoms, however, quantum size effects at low temperature can already be observed by measuring the current/voltage (I/V) characteristic (Fig. 2). At room temperature the particle behaves metallic. At 4.2 K, when the electrostatic energy is larger than the thermal energy of the electron, there is a pronounced Coulomb gap to be observed indicating energy quantization. The smaller the particle the higher the working temperature can arise. Indeed, the I/V curve of a 1.4 nm Au_{55} cluster, enveloped by a shell of PPh_3 ligands, shows a Coulomb blockade already at room temperature [17] (Fig. 3). This is one of the conditions if metal clusters shall be considered as future building blocks in nanoelectronic devices.

However, quantum size effects can also be observed by others than current/voltage measurements. Palladium particles of four different sizes have been used to study the electronic specific heat C_{el} and the electronic susceptibility χ_{el} at very low temperature [18]. Figures 4 and 5 inform on the results.

The five- (Pd5), seven- (Pd7), and eight-shell (Pd8) clusters $Pd_{561}phen_{36}O_{200}$, $Pd_{1415}phen_{54}O_{1100}$ and $Pd_{2052}phen_{78}O_{1600}$ with nucleus diameters of 2.2, 3.0 and 3.6 nm have been studied as well as a 15 nm Pd colloid, stabilized by sodium sulfanilate below 1 K. As can be seen form Fig. 4 the 2.2 nm cluster shows the strongest

deviation from the bulk curve. With increasing particle size the electronic specific heat resembles more and more to the bulk ore. This unprecedented thermodynamic quantum size behavior has been confirmed by susceptibility measurements of the same particles, also at very low temperatures. The formation of maxima in the susceptibility curves becomes more expressed the smaller the particle is. This predicted effect is explained by spin–spin interactions between single clusters depending sensitively

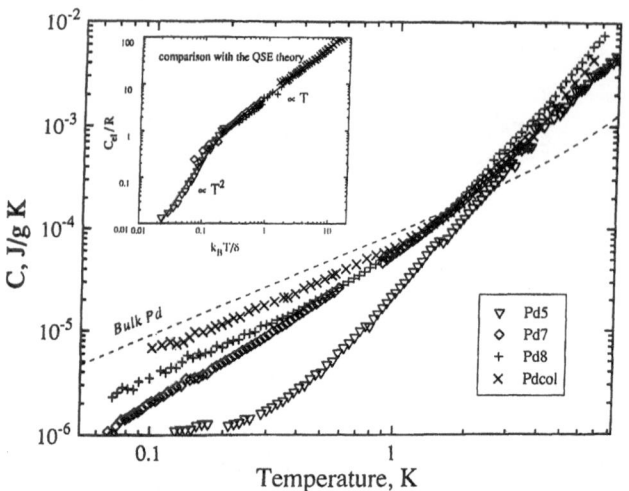

Fig. 4 The specific electronic heat of different Pd particles at very low temperatures. The smaller the particle the more pronounced the deviation from bulk behavior

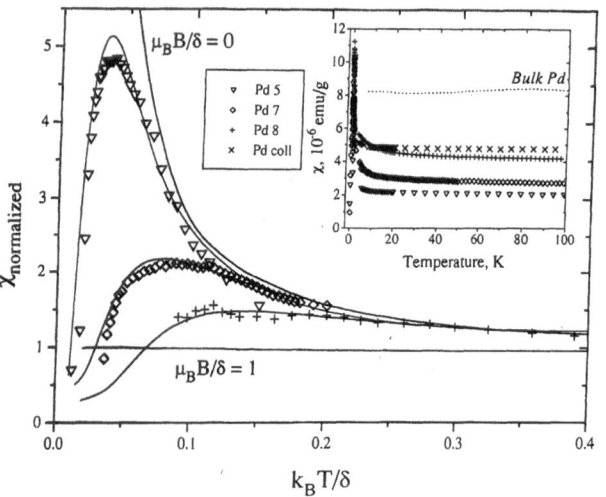

Fig. 5 The electronic susceptibility of the same particles as in Fig. 4. Again the smallest clusters show the strongest quantum size effect

on whether the clusters have an odd or an even number of electrons [19].

The inclusion of electrons in a quantum dot which is isolated from others by a nonconducting material (ligand shell) is ultimately realizable if the particle diameter corresponds with $\lambda/2$, if λ is the de Broglie wavelength. This situation is realized in ligand protected Au_{55} clusters and is shown in Fig. 6a. This state can be compared with an "s orbital" of a very big gold atom which can be occupied by two electrons in the ground state. Detailed impedance measurements of dense packed pellets of $Au_{55}(PPh_3)_{12}Cl_6$ clusters have indeed shown that a first excited state can be reached by applying a 60–100 kHz alternating voltage, corresponding with an activation energy of 0.16 V [20, 21]. The situation is elucidated in Fig. 6b.

Doubling of the frequency in the clusters enables electron tunneling through neighboring ligand shells of two or more clusters. This event is nothing but a single electron tunneling (SET) process between at least two clusters and so stands for the ultimate miniaturization of an electronic switch. Collectively spoken, this "metallizing" of a sample of Au_{55} clusters represents the behavior of a semiconductor the band gap of which can be bridged by the energy of 0.16 V. As we will see later, enlarging of the cluster distance necessarily leads to an increase of the activation energy.

These only briefly discussed physical properties and several others, which will not be described here, indicate clearly that metal particles in the size range between 1 and 15 nm behave in many respects as materials placed between bulk and molecule or as "embryonic metals".

No wonder that scientists around the world try to test small metal particles as electronic building blocks. However, before being able to realize such applications, the giant problem of organizing clusters or colloids in a distinct manner is unrenounceable. The next chapter will therefore deal with our efforts to find appropriate routes to reach this goal.

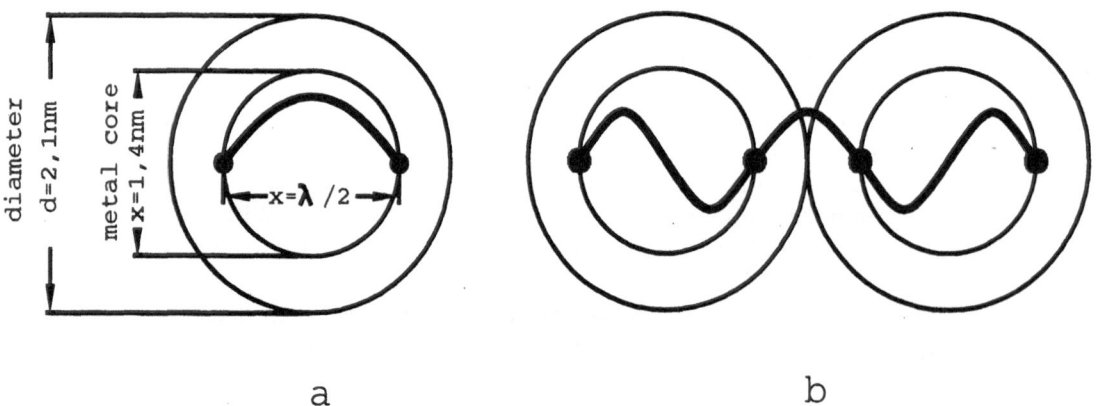

Fig. 6 A sketch of the electronic situation in a ligand stabilized cluster in the ground state (a) and between two clusters in the 1. Excited state (b), indicating electronic tunneling through the ligand shells

Progr Colloid Polym Sci (1998) 111:52–57
© Steinkopff Verlag 1998

Organization of clusters and colloids

Three-dimensionally organized ligand stabilized metal clusters and colloids would be best realized in crystals. As particles of that size can only hardly be crystallized, densely packed pellets may act as substitutes as has been done for impedance measurements. However, as could be demonstrated, the capacity between two Au_{55} clusters, separated by a ligand shell of PPh_3 or similarly sized other molecules, is too small to restrict SET processes between two or any number of clusters. Extension of the capacity should be possible by introducing spacers between the clusters as is schematically shown in Fig. 7.

Indeed, the activation energy increases with the length of the spacer as can be followed from the diagram in Fig. 8.

The interaction between cluster and spacer happens via the SO_3H and the NH_2 functions.

Two-dimensional colloid and cluster arrangements can be created by using chemically modified smooth surfaces. Silicon or glass surfaces, both characterized by layers of OH groups, react with silylated thiols such as $(MeO)_3Si(CH_2)_3SH$ to form very stable Si–O–Si bonds with elimination of MeOH. The SH functions cover the surface and react easily with all kinds of gold particles, e.g. with gold colloids from an aqueous solution. The density of the colloid layer depends just on the time of reaction. Figure 9 shows an Atomic Force Microscopic (AFM) image of isolated, well fixed 13 nm gold particles [23].

The acidified gold cluster $Au_{55}(Ph_2PC_6H_4SO_3H)_{12}Cl_6$ can also easily be fixed on various surfaces, e.g. on atomically flat gold surfaces which have before been treated with

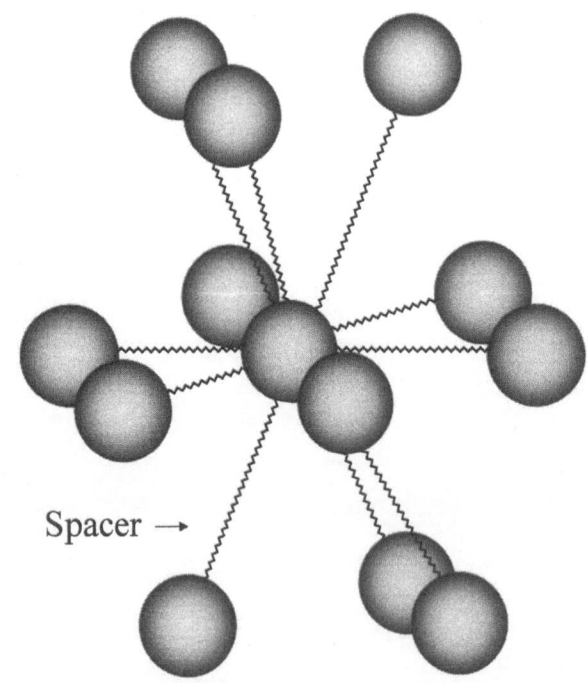

Fig. 7 A model of a "densely packed" arrangement of clusters linked by spacer molecules

an amino thiol, $H_2N(CH_2)_xSH$. In this case the thiol function first interacts with the gold surface, in a second step the clusters are coordinated via the NH_2 groups due to their sulfonic acid functions [23].

An additional route to generate cluster monolayers on supports has just recently been described by us [17].

Fig. 8 The influence of spacer lengths on the Coulomb gap of linked Au_{55} clusters

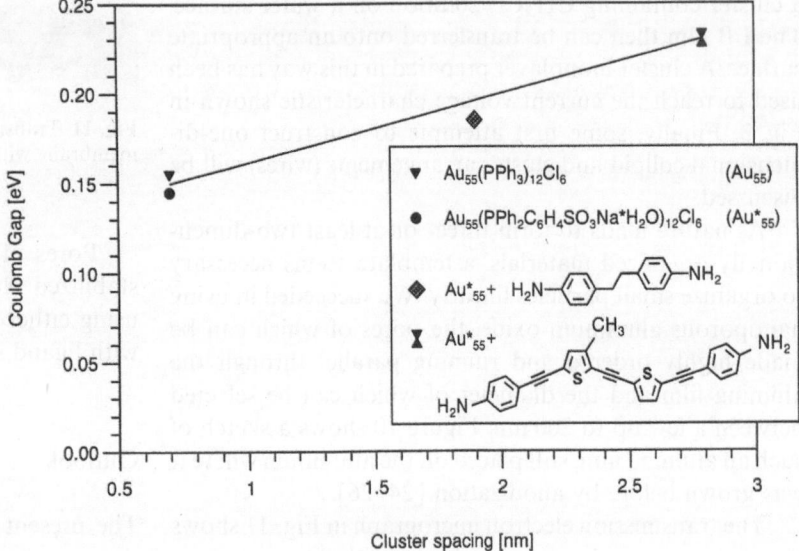

Fig. 9 AFM image of chemically fixed 13 nm gold colloids on silicon the surface of which is covered by O–Si(CH$_2$)$_3$SH functions

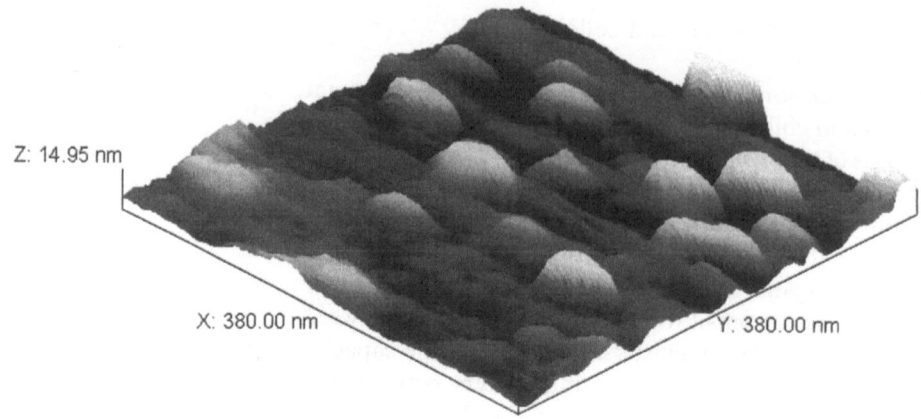

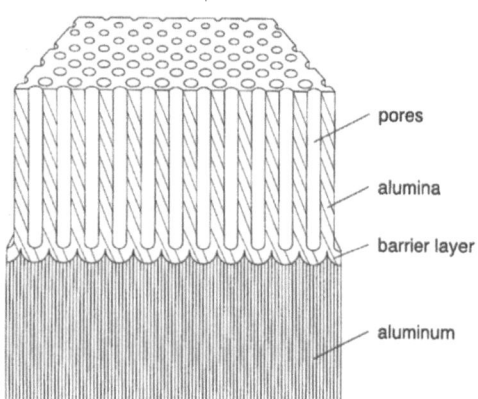

Fig. 10 Idealized model of a nanoporous alumina membrane on aluminum

Water-insoluble gold clusters such as Au$_{55}$(PPh$_3$)$_{12}$Cl$_6$ are used to form a Langmuir–Blodgett (LB) film by spreading a cluster containing CH$_2$Cl$_2$ solution on a water surface. The LB film then can be transferred onto an appropriate surface. A cluster monolayer prepared in this way has been used to reach the current voltage characteristic shown in Fig. 3. Finally, some first attempts to construct one-dimensional colloid and cluster arrangements (wires) will be discussed.

As nature tends to form three- or at least two-dimensionally organized materials, a template seems necessary to organize small particles linearly. We succeeded in using nanoporous aluminum oxide, the pores of which can be made highly ordered and running parallel through the alumina film and the diameter of which can be selected between a few up to 200 nm. Figure 10 shows a sketch of such an alumina film, still placed on the aluminum where it was grown before by anodization [24–26].

The transmission electron micrograph in Fig. 11 shows the surface of an alumina membrane with 80 nm pores.

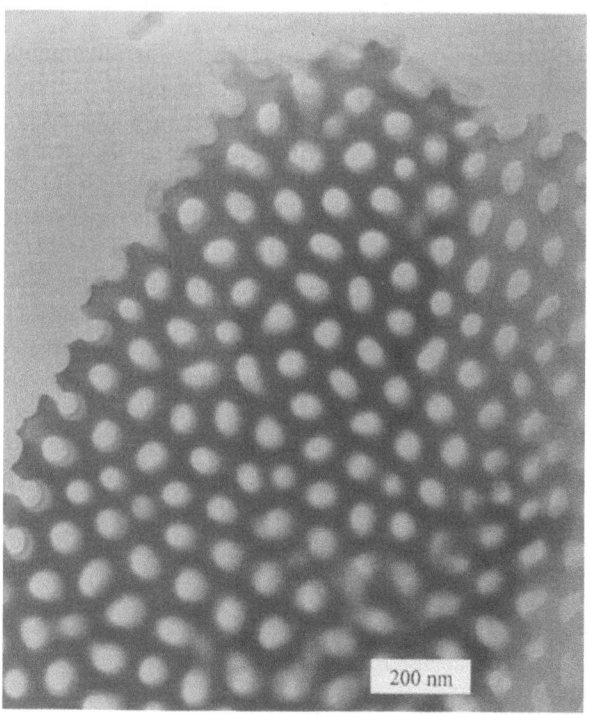

Fig. 11 Transmission electron microscopic image of an alumina membrane with 50 nm pores

Pores of approximately the same size as the ligand stabilized clusters can be filled by these quantum dots using either vacuum or electrophoresis [27]. Pores filled with ligand stabilized Au$_{55}$ clusters are shown in Fig. 12.

Outlook

The present state to organize very small metal particles three-, two- or one-dimensionally can be described as very

Progr Colloid Polym Sci (1998) 111:52–57
© Steinkopff Verlag 1998

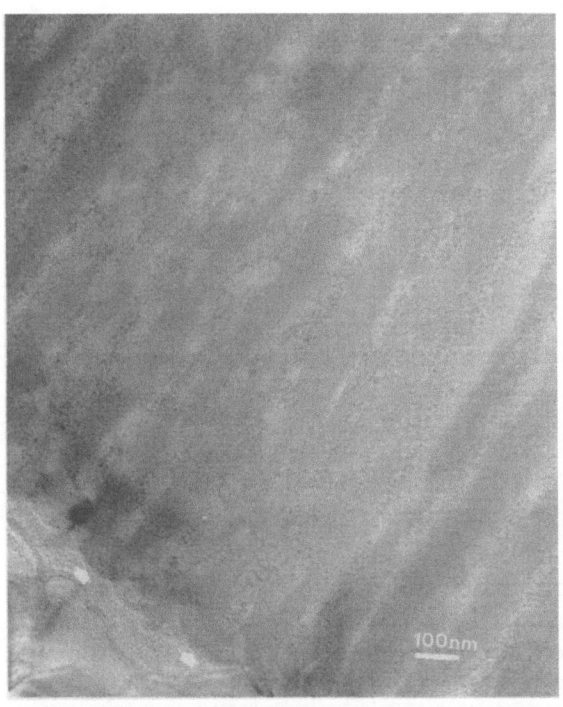

Fig. 12 ~20 nm pores filled with 1.4 nm Au$_{55}$ clusters, stabilized by a 1.4 nm thick shell of thiol ligands. The arrows indicate the barrier layer between the membrane and the aluminum metal (see Fig. 10)

promising. First successful steps have been achieved and principles become visible. However, there is still a long way to go until perfectly organized architectures will be reached.

We have first indications that the capacitance between single clusters can be tuned by the thickness of the ligand shell and the length of spacer molecules, respectively. This knowledge comes from 3D systems in pellets. We have also good knowledge about the electronics in single clusters. Larger ones show quantum size effects at low temperatures, smaller ones even at room temperature. However, we have still no information about the communication between clusters in 2D or 1D arrays. This will be the aim of the next few years.

References

1. Schmid G (1992) Chem Rev 92: 1709–1727
2. Corcoran E (1991) Spektr D Wiss 1:76–86
3. Fendler FH, Meldrum FC (1995) Adv Mater 7:607–632
4. Enderlein R (ed) (1993) Mikroelektronik. Spektrum Akademischer Verlag, Heidelberg
5. Licharev KK, Claeson T (1985) Spektr D Wiss 10:62–68
6. Schmid G (ed) (1994) Clusters and Colloids. From Theory to Applications. VCH, Weinheim
7. Schmid G, Pfeil R, Boese R, Bandermann F, Meyer S, Calis GHM, van der Velden JWA (1981) Chem Ber 114: 3634–3642
8. Schmid G (1985) Structure Bonding 62: 51–85
9. Schmid G, Klein N, Korste L, Kreibig U, Schönauer D (1988) Polyhedron 7: 605–608
10. Schmid G (1990) Endeavour, New Series 14:172–178
11. Schmid G, Morun B, Malm JO (1989) Angew Chem Int Ed Engl 28:778–780
12. Schmid G, Harms M, Malm JO, Bovin JO, van Ruitenbeek J, Zandbergen HW, Fu WT (1993) J Amer Chem Soc 115: 2046–2048
13. Schmid G, Lehnert A (1989) Angew Chem Int Ed Engl 28:780–781
14. Schmid G, Lehnert A, Kreibig U, Adamczyk Z, Belouschek P (1990) Z Naturforsch 45b:989–994
15. Schmid G, Lehnert A, Malm JO, Bovin JO (1991) Angew Chem Int Ed Engl 30:874–876
16. Bezryadin A, Dekker C, Schmid G (1997) Appl Phys Lett 71:1273–1275
17. Chi LF, Hartig M, Drechsler T, Schwaack Th, Seidel C, Fuchs H, Schmid G (1998) Appl Phys A 66:187–190
18. Volokitin Y, Sinzig J, de Jongh LJ, Schmid G, Vargaftik MN, Moiseev II (1996) Nature 384:621–623
19. Kubo R (1962) J Phys Soc 17:975
20. Schön G, Simon U (1995) Coll Polym Sci 273:101–117
21. Schön G, Simon U (1995) Coll Polym Sci 273:202–218
22. Flesch R (1997) Thesis University of Essen
23. Sawitowski Th, Schmid G, Peschel St (1997) Z Anorg Allg Chem 623:719–723
24. Masuda H, Fukuda F (1995) Science 268:1466–1468
25. Martin C (1994) Science 266:1961–1966
26. Hornyak G, Kröll M, Pugin R, Sawitowski Th, Schmid G, Bovin JO, Karsson G, Hofmeister H, Hopfe S (1997) Chemistry Europ J 3:1951–1956

Progr Colloid Polym Sci (1998) 111:58–64
© Steinkopff Verlag 1998

D. Barthes-Biesel

Mechanics of encapsulated droplets

Prof. D. Barthes-Biesel
Génie Biologique
UMR 6600
Université de Compiègne
BP 20529, 60205 Compiègne
France
E-mail: dbb@utc.fr

Abstract A capsule consists of an internal medium (pure or complex liquid), enclosed by a deformable membrane that is usually semi-permeable. Capsules are frequently met in nature (living cells) or in industrial processes (pharmaceutical, cosmetic or food industry). The motion, deformation and burst of a capsule under stress depend on three types of independent intrinsic physical properties: initial geometry, internal liquid viscosity and interface constitutive behavior. The direct determination of the latter is difficult owing to the thinness and fragility of the interface. Over the years different techniques have been used (compression between two flat plates, micropipette aspiration, etc.). Recently, a new technique has been proposed, that consists in placing a capsule on the axis of a spinning drop tensiometer. The deformation under increasing rotation rates is measured, and a mechanical model allows to compute the elastic modulus of the membrane.

Another line of approach consists in suspending a capsule in another liquid subjected to flow and in measuring the resulting deformation. In order to relate the overall capsule deformation to the particle intrinsic physical parameters, a fairly complete model of the mechanics of the particle is necessary. Analytical models have been obtained for initially spherical capsules, subjected to small deformations (less than 10–15%), with simple membrane constitutive laws (rubber elasticity, viscoelasticity, area incompressibility). They allow to infer the membrane apparent elastic modulus from deformation vs. shear rate curves obtained for artificial capsules. For large deformations and non-spherical initial shapes, numerical models have been proposed. For example, break-up of a capsule under an elongational shear is predicted. It is also possible to model the flow of a large capsule through a small short pore or through a long cylindrical tube that has a smaller diameter than the capsule.

Key words Capsule – interface mechanics – interfacial polymerization – deformation

Introduction

A capsule consists of an internal medium (pure or complex liquid), enclosed by a deformable membrane that is usually semi-permeable. Capsules are frequently met in nature (living cells) or in industrial processes (pharmaceutical, cosmetic or food industry), and are found in a variety of sizes from a few microns (living cells) to a few millimetres (artificial capsules). They take different rest shapes

Progr Colloid Polym Sci (1998) 111:58–64
© Steinkopff Verlag 1998

(e.g., discoidal geometry of a red blood cell), and the mechanical properties of the internal medium and of the membrane vary widely. It is important to be able to predict the motion and deformation of a capsule under stress and the eventual occurrence of break-up. This is a non-trivial problem since capsule deformability depends on three types of independent intrinsic physical properties: initial geometry, internal liquid viscosity and interface constitutive behaviour.

The direct determination of the membrane mechanical properties is difficult owing to its thinness and fragility. Over the years different techniques have been proposed. They consist in measuring locally the membrane (e.g., micropipette aspiration [1, 2], shear deformation of a flat sample [3]) or in determining the overall deformation of the capsule under static conditions (e.g., compression between two flat plates [4–7], deformation in a spinning rheometer [8]). The results usually provide only partial information on the mechanical properties of the membrane. Specifically, it is difficult to obtain both the shear modulus and the compression modulus of the membrane with only one experiment.

Since a capsule is often designed to be suspended into another liquid subject to flow, it is equally important to study its motion and deformation under those conditions. This is done by suspending a capsule in another flowing liquid and in measuring the resulting particle deformation under the influence of viscous shear forces [6, 7, 9]. However, it is difficult to relate the overall capsule deformation to the particle intrinsic physical parameters, unless a fairly complete model of the mechanics of the capsule is available. Such models involve the solution of two-flow problems (motion of the internal and of the external liquid) coupled to a solid mechanics problem (deformation of the interface). Analytical solutions [10, 11] have been obtained by means of perturbation methods, for initially spherical capsules. For large deformations and non-spherical initial shapes, it is necessary to resort to numerical models.

In Section 2, membrane mechanics are presented together with some experimental results. The motion of a capsule in an unbounded shear flow is presented in Section 3, and the analytical model predictions are compared to experiments performed on Nylon capsules. Some numerical simulations are then discussed in Section 4.

Determination of membrane mechanical properties

Interface mechanics

The membrane is usually modelled as a two-dimensional shell with negligible thickness, subjected to large deforma-

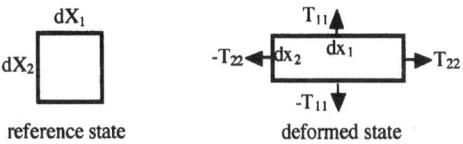

Fig. 1 Deformation of a membrane element

tions. The deformation in the membrane plane is defined by comparing a line element $\mathbf{dX}$ of the membrane in the stress-free reference configuration to its instantaneous position $\mathbf{dx}(\mathbf{X}, t)$ at time t (Fig. 1). In the reference frame (Ox_1, Ox_2) of the two principal directions of deformation and stress in the deformed state, the principal stretch ratios λ_1, λ_2 are defined by

$$\lambda_1 = dx_1/dX_1 , \qquad \lambda_2 = dx_2/dX_2 . \tag{1}$$

Owing to the thinness of the interface, stresses are replaced by tensions $\mathbf{T}$, i.e., forces per unit length of the deformed membrane. In the principal axes, those have components T_{11} and T_{22}, which are related to the local deformations by means of a rheological constitutive equation. A few classical laws are available to give a phenomenological description of typical mechanical behaviors. Some constitutive equations are presented below (for a more detailed discussion, see [9]). The expression for the principal component T_{11} is given explicitly and the corresponding expression for T_{22} can be deduced by inverting the roles of indices 1 and 2. The surface Young's modulus is denoted E_s.

(i) *Linear elasticity.* This law corresponds to Hooke's law adapted to a two-dimensional continuum, and is restricted to small deformations. The Poisson ratio v_s is unity in case of surface area incompressibility.

$$T_{11} = E_s[\lambda_1^2 - 1 + v_s(\lambda_2^2 - 1)]/2(1 - v_s^2) . \tag{2}$$

(ii) *Rubber elasticity:* The membrane is treated as an infinitely thin layer of a homogeneous, isotropic, three dimensional incompressible elastomer obeying a Mooney–Rivlin law:

$$T_{11} = \frac{E_s}{3\lambda_1\lambda_2}\left\{\Psi\left[\lambda_1^2 - \frac{1}{(\lambda_1\lambda_2)^2}\right] + \Psi'\left[(\lambda_1\lambda_2)^2 - \frac{1}{\lambda_1^2}\right]\right\} . \tag{3}$$

The coefficients Ψ and Ψ' are such that $\Psi + \Psi' = 1$ and depend on the strain invariants. The case $\Psi' = 0$ corresponds to a neo-Hookean solid.

(iii) *Red blood cell elasticity.* Skalak et al. [12] have proposed a constitutive law for the red blood cell membrane that takes into account the shearing ability and the large resistance to surface area changes. The modulus

A_s corresponds to area changes and is such that $A_s/E_s \gg 1$ (A_s is of order $10^5 E_s$).

$$T_{11} = E_s \frac{\lambda_1}{4\lambda_2}(\lambda_1^2 - 1) + A_s \lambda_1 \lambda_2 [(\lambda_1 \lambda_2)^2 - 1] \,. \tag{4}$$

The corresponding shear modulus is, respectively, $E_s/2(1 + v_s)$, $E_s/3$ and $E_s/4$ for Eqs. (2)–(4).

(iv) *Viscoelasticity*. Viscoelastic effects are usually modelled by adding a viscous contribution $\mathbf{T}^v$ to any of the above elastic laws:

$$T^v_{11} = 2\mu_s \frac{\partial \lambda_1/\partial t}{\lambda_1} \,, \tag{5}$$

where $\partial \lambda_1/\partial t$ denotes a time derivative, and where μ_s is the surface viscosity.

Finally, to close the problem, the shell equilibrium equations must be expressed:

$$\nabla_s \cdot \mathbf{T} = \mathbf{q} \,, \tag{6}$$

where ∇_s represents a gradient taken along the membrane surface, and where $\mathbf{q}$ is the load per unit area exerted on the membrane. This equation can be quite complicated for complex geometry, but simplifies in the case of an axisymmetric problem. When bending effects are included, bending moments and transerse forces must also be considered [9]. The constitutive Eqs. (2)–(4) and the equilibrium condition (6) must be completed with bending terms.

Experiments

For small capsules (with a typical size scale of the order of 10 μm or less), research has been mostly conducted on biological cells, and specifically on blood cells. The bilayer thickness is of order 10^{-8} m, so that the membrane can indeed be treated as a two-dimensional solid. The mechanical properties are measured with a micropipette with a 1 to 2 μm internal diameter. A small part of the membrane is sucked into the pipette and the height of aspiration is measured as a function of the applied depression. Such measurements, as well as relaxation experiments after suction, are used to evaluate the membrane shear elastic modulus and surface viscosity [1, 2].

The measurement of the mechanical properties of large capsules is a relatively new field of research, and as of now there are only a few published results. The overall average membrane properties may be assessed by means of experiments where the capsule is squeezed between two plates and its overall deformation is recorded as a function of the squeezing force [4–7]. Another method consists in creating a flat film of the membrane material and in measuring directly the shear modulus and surface viscosity of this film

by means of a surface torsion rheometer [3, 8]. When feasible, this technique is quite powerful since it gives access to eventual complicated material properties of the membrane such as non-linear stress–strain laws and viscoelasticity.

Recently, a new approach for the measurement of the mechanical properties of a capsule membrane has been proposed [8]. It consists in using a spinning drop apparatus originally designed to measure surface tension between two liquids. Instead of a drop, an initially spherical capsule is introduced in the device, and its deformation under increasing rotation rates is measured. A theoretical analysis of the mechanics of an initially spherical elastic shell (radius A) subjected to centrifugal forces, shows that, to a first order, the capsule deforms into an ellipsoid with radius B and length $2L$ (measured along the tube axis). The deformation D is then:

$$D = \frac{L - B}{L + B} = -\frac{\Delta\rho\omega^2 A^3}{16E_s}(5 + v_s) \,, \tag{7}$$

where $\Delta\rho$ represents the density difference between the suspending liquid and the capsule internal medium, and where the membrane constitutive behavior is described by (2). Measurements have been made on capsules that consist of an oil droplet enclosed by an ultrathin crosslinked membrane obtained by radical polymerization of surface-active aminomethacrylates. The suspending medium is water. The comparison between the measured and predicted values of deformation (Fig. 2) allows to infer the value of $E_s/(5 + v_s)$. From independent measurements of

Fig. 2 Deformation D as a function of $-\Delta\rho\omega^2 A^3$ for capsules with the same aminomethacrylate membrane but with three different radii. A (mm) = 0.63 (●), 0.73 (▲), 0.77 (■). The line corresponds to Eq. (7). From [8]

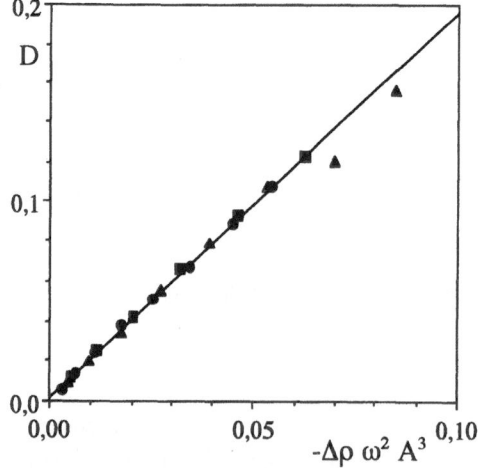

Progr Colloid Polym Sci (1998) 111:58–64
© Steinkopff Verlag 1998

the shear elastic modulus of a flat sheet of polymerized aminomethacrylates, it is deduced that v_s is nearly zero and that E_s varies between 0.1 and 0.2 N/m, depending on the degree of polymerization. These values are quite larger than the initial surface tension between oil and water, i.e. 0.05 N/m. This measurement technique is quite promising as it involves no contact between a solid surface and the deformed capsule.

The main challenge for future work is to relate the macroscopic mechanical properties of an interface to its molecular structure.

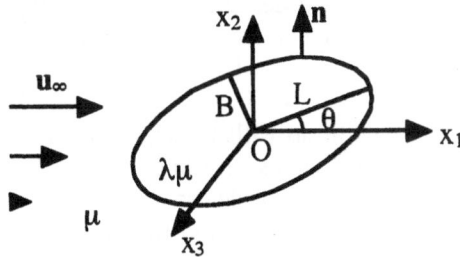

Fig. 3 Schematics of a deformed capsule freely suspended in a shear flow

Motion of a capsule in shear flow

General problem statement

A capsule with given initial unstressed geometry and size A (e.g. the radius of the sphere with the same volume) is freely suspended in a Newtonian incompressible liquid with viscosity μ. The capsule is enclosed by an infinitely thin, impermeable, elastic membrane (surface Young's modulus E_s) and filled with a Newtonian, incompressible liquid (viscosity $\lambda\mu$). Buoyancy effects are ignored and the particle Reynolds number of the flow is assumed to be very small, owing to the smallness of the capsule. The equations of motion are written in a reference frame (O, x_1, x_2, x_3) centered on the particle center of mass O and translating with it (Fig. 3). The external liquid is subjected far from the particle to an undisturbed flow field with shear rate G, velocity $\mathbf{u}_\infty$ and pressure p_∞. Under the influence of the viscous stresses, the cell deforms and the equation of its interface is given by

$$r = (x_1^2 + x_2^2 + x_3^2)^{1/2} = f(x_1, x_2, x_3) . \tag{8}$$

The function f is itself unknown and must be determined as part of the problem solution. In absence of inertia effects, the internal and external velocity and stress fields, respectively, $\mathbf{u}^*$, $\mathbf{u}$, $\boldsymbol{\sigma}^*$ and $\boldsymbol{\sigma}$, are governed by the Stokes equations:

$$\nabla\cdot\mathbf{u} = 0 , \qquad \mu\nabla^2\mathbf{u} - \nabla p = 0 \quad \text{for } r \geq f , \tag{9}$$

$$\nabla\cdot\mathbf{u}^* = 0 , \qquad \lambda\mu\nabla^2\mathbf{u}^* - \nabla p^* = 0 \quad \text{for } r \leq f , \tag{10}$$

with associated boundary conditions:

- no flow disturbance far from the cell:

$$\mathbf{u} \to \mathbf{u}_\infty \quad \text{as } r \to \infty , \tag{11}$$

- continuity of velocities at the interface:

$$\mathbf{u}^* = \mathbf{u} = \mathbf{u}_s = \partial\mathbf{x}/\partial t \quad \text{at } r = f , \tag{12}$$

- dynamic equilibrium of the interface:

$$(\boldsymbol{\sigma} - \boldsymbol{\sigma}^*)\cdot\mathbf{n} = \mathbf{q} \quad \text{at } r = f , \tag{13}$$

where $\mathbf{n}$ is the outer unit normal vector to the interface, and where $\mathbf{u}_s$ represents the velocity of the membrane, as measured by the time derivative of the position $\mathbf{x}$ of the interface material points. Condition (13) states that the viscous forces exerted by the flow of the internal and external liquids are equal to the load $\mathbf{q}$ exerted on the deformed membrane. To close the problem, $\mathbf{q}$ must be related to the surface deformation as shown in Section 2. The problem solution then depends on the following dimensionless numbers:

- the capillary number $C = \mu G A/E_s = \mu|\mathbf{u}_\infty|/E_s$, that measures the ratio between viscous and elastic forces,
- the viscosity ratio λ between the internal and external phases,
- ratio of material parameters of the membrane (e.g. A_s/E_s, $\mu_s G/E_s$).

The solution of Eqs. (8)–(13), with any constitutive Eqs. (2)–(5) represents a very difficult problem of continuum mechanics. It is of the free surface type and requires the solution of two flow problems for the internal and external liquids. Furthermore, the treatment of the interface may involve the theory of large deformations of membranes, which is also quite complex.

Spherical capsule model, small deformation

In the case of an initially spherical capsule (radius A), undergoing small deformations ($C \ll 1$), the problem is solved by a perturbation method [10, 11]. Then, to first order in C, the equation of the deformed membrane depends only on a symmetric and traceless second-order tensor $\mathbf{J}$:

$$r = 1 + C\,^T\mathbf{x}\cdot\mathbf{J}\cdot\mathbf{x} + O(C^2) . \tag{14}$$

The analytical expression for **J** depends on time, the type of external shear flow, λ and other membrane material properties such as surface viscosity [11]. For example, in the case of a simple shear flow in the 1–2 plane, the deformed profile is an ellipsoid with principal diameters L and B, making an angle θ with streamlines (Fig. 3). For a purely elastic membrane (3), the deformation D in the shear plane is a linear function of C and the orientation remains constant at 45°:

$$D = \frac{L - B}{L + B} = \frac{25C}{4} = \frac{25\mu GA}{4E_s}, \quad \theta = 45°.$$

For a membrane with linear viscoelastic law given by (3) and (5), D increases with shear rate, up to a high shear asymptotic value, $D_\infty = 5\mu A/2\mu_s$. The orientation angle θ decreases from 45° to 0° with increasing shear rates (Fig. 4). This analysis is valid provided that the membrane viscosity is large, i.e., $\mu A/\mu_s \ll 1$.

Those predictions are in qualitative agreement with experimental observations of initially spherical capsules enclosed by a Nylon membrane and suspended either in a simple shear flow or in an hyperbolic flow [6, 7]. By comparing the experimental D vs. G curves to the model predictions, it is possible to determine the value of the membrane material properties. Indeed, both the low and high shear limits (15) have been used to infer the Nylon interface surface elastic modulus and viscosity from the experimental deformation curves. The surface viscosity values are consistent with data on capsule relaxation after cessation of flow. The values obtained for E_s sometimes differ from those computed from the squeezing experiments. However, this discrepancy might be attributed to an inhomogeneous membrane that influences the squeezing measurements but that are averaged out in a simple shear flow.

In conclusion, the spherical capsule model can be useful to interpret experimental observations of the motion and deformation of capsules subjected to flow, in terms of the interfacial mechanical properties. Other similar data are also pertinent, such as membrane rotation rate or relaxation behaviour under transient conditions. Unfortunately, break-up cannot be predicted since the models are based on small deviations from sphericity.

Numerical models

The present trend is to develop numerical models, that allow for large deformations and for capsules with an initial arbitrary geometry. The Stokes equations (9) and (10) are recast as boundary integrals, and the problem is usually solved by means of a numerical collocation method. However, such models are very complicated to design and demand large computing times. Consequently, most of the presently available models deal with situations where the flow and the capsule are axisymmetric and where the viscosity ratio λ is unity. They thus only provide some qualitative insight on the mechanics of capsules suspended in actual straining motions.

A capsule in unbounded shear flow

The case of spheroidal capsules, with a Mooney–Rivlin membrane (3), suspended in a pure straining flow has been considered [13]. It is found that the deformation increases with capillary number, until a critical value of C is reached, past which there exists no steady-state solution to the equations of motion (Fig. 5). For values of C larger than the critical level, the capsule elongates without bound until obviously a failure criterion for the membrane is eventually reached. Burst of the particle is thus predicted according to what is found experimentally for a polylysine capsule [9]. The critical value of C depends on the parameters Ψ and Ψ', and also on the initial shape of the capsule. In the case of a spherical capsule with a neo-Hookean membrane, this critical value is of order 0.085. Results have also been obtained for initially spherical capsules with a strain-hardening membrane of the Hart–Smith type [9] such that

$$T_{11} = \frac{E_{s0} \exp\{K_s[\lambda_1^2 + \lambda_2^2 + (\lambda_1\lambda_2)^{-2} - 3]\}}{3\lambda_1\lambda_2}$$

$$\times \left[\lambda_1^2 - \frac{1}{(\lambda_1\lambda_2)^2}\right]. \tag{15}$$

Fig. 4 Deformation D and orientation θ of an initially spherical capsule with a viscoelastic membrane

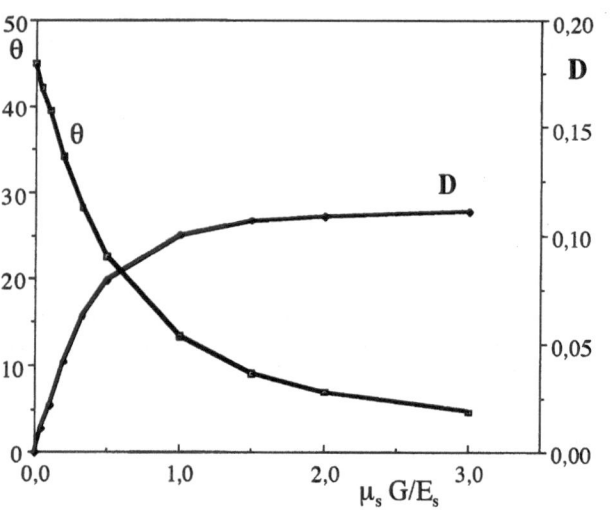

Progr Colloid Polym Sci (1998) 111:58-64
© Steinkopff Verlag 1998

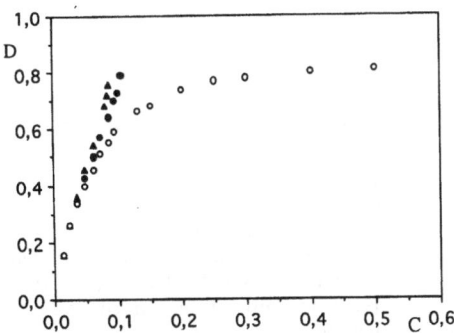

Fig. 5 Deformation vs. Capillary number of an initially spherical capsule with a neo-Hookean membrane (▲), with a strain hardening membrane (●) $K_s = 0.05$, (○) $K_s = 0.25$

The Capillary number is based on E_{s0}. The parameter K_s measures the amount of strain hardening. The stiffening effect under increasing levels of strain limits the deformation of the capsule and prevents it from bursting through the continuous elongation process (Fig. 5). This agrees with what is observed experimentally for a Nylon capsule [6, 7].

A similar numerical model has also been devised for capsules with an area incompressible membrane [14]. In that case, the area incompressibility limits the particle deformation and prevents break-up. Recently, the first 3D numerical study of a capsule in simple shear flow has been presented for $\lambda = 1$, [15, 16]. This model shows that the deformation predicted by the analytical perturbation solution is within 10% of the exact value, for $C \leq 0.05$. It is also useful to assess the role of the initial geometry of the capsule as it can allow for non spherical shapes.

Flow of a capsule in a pore

The objective is to model situations where a capsule has to flow into a pore with a diameter R smaller or of the same order as the capsule typical dimension. Equations (9)-(13) must then be completed with a no-slip boundary condition

on the pore wall, and with proper entrance and exit conditions for the channel. Such conditions depend on whether the flow is driven by a constant pressure drop (then the capsule lowers the flow rate) or by a constant flow rate (then the capsule increases the pressure drop). The cases of a hyperbolic pore [17] and of a cylindrical pore [18, 19] have been considered for $\lambda = 1$. For example, an initially spherical capsule of radius A ($A/R = 0.8$), with a Mooney-Rivlin membrane (3), flowing in a cylindrical tube of radius R, can assume parachute shapes similar to those observed for red blood cells in microcapillary vessels (Fig. 6). When the capsule membrane is an elastomer with constitutive law (3), the model predicts that the pressure drop depends mostly on the size ratio A/R and weakly on the membrane stiffness, as measured by C. It also predicts the burst of the capsule when C exceeds a critical value that depends on A/R. When the capsule membrane is area incompressible (4), pore plugging may occur but burst has not been observed since the deformation is limited by the area incompressibility constraint. Such models allow the interpretation of filtration experiments in terms of the particle physical properties.

Conclusion

There is still much work to be done to understand the respective roles that the different intrinsic parameters play in the determination of capsule deformability and resistance to stress. It is important to have a direct determination of the membrane elastic properties. Except for the micropipette, which has been around for some time now, flat film stretching, spinning rheometry or squeezing experiments are relatively new and should be developed. On the theoretical side, a general three-dimensional model of the motion of non spherical capsules undergoing large deformations should be developed. This will allow prediction of the motion of a capsule in shear flow and most importantly, may help to understand the physics of burst, in view of preventing this effect.

Fig. 6 Successive half-profiles of an initially spherical capsule ($A = 0.8$, $R = 1$) enclosed by a Mooney-Rivlin membrane (3), for $C = 0.04$. From [18]

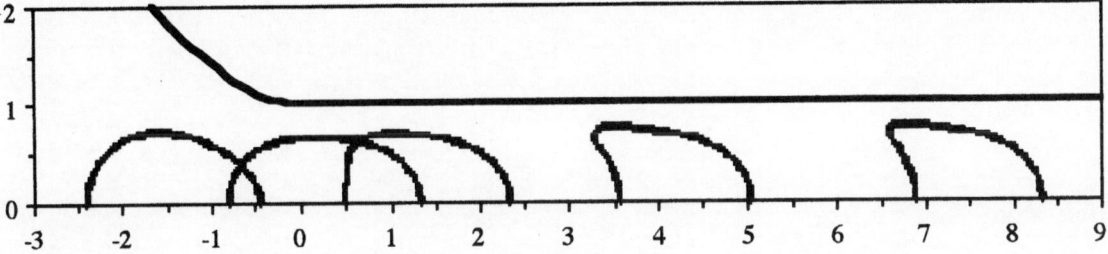

References

1. Hochmuth RM, Berk DA (1984) J Biomech Eng 106:2–9
2. Hochmuth RM, Waugh RE (1987) Ann Rev Physiol 49:209–219
3. Burger A, Rehage H (1992) Ang Makromol Chem 202–203:31–44
4. Lardner TJ, Pujara P (1980) Mech Today 5:161–176
5. Hiramoto Y (1970) Biorheology 6: 201–234
6. Chang KS, Olbricht WL (1993) J Fluid Mech 250:587–608
7. Chang KS, Olbricht WL (1993) J Fluid Mech 250:609–633
8. Pieper G, Rehage H, Barthes-Biesel D (1998) Colloid Interface Sci, to appear
9. Barthes-Biesel D (1991) Physica A 172:103–124
10. Barthes-Biesel D, Rallison JM (1981) J Fluid Mech 113:251–267
11. Barthes-Biesel D, Sgaier H (1985) J Fluid Mech 160:119–135
12. Skalak R, Tozeren A, Zarda RP, Chien S (1973) Biophys J 13:245–264
13. Li XZ, Barthes-Biesel D, Helmy A (1988) J Fluid Mech 187:179–196
14. Pozrikidis C (1990) J Fluid Mech 216: 231–254
15. Pozrikidis C (1995) J Fluid Mech 297: 123–152
16. Zhou H, Pozrikidis C (1995) J Fluid Mech 283:175–200
17. Leyrat-Maurin A, Barthes-Biesel D (1994) J Fluid Mech 279:135–163
18. Quéguiner C, Barthes-Biesel D (1995) Boundary Elements XVII:661–668. Brebbia CA, Kim S, Oswald TA, Power H (eds) Computational Mechanics publications
19. Quéguiner C, Barthes-Biesel D (1997) J Fluid Mech 348:349–376

Progr Colloid Polym Sci (1998) 111:65–73
© Steinkopff Verlag 1998

Stability of hydrophilic/hydrophobic silica particles in binary liquids: Adsorption, rheological and SAXS experiments

I. Dékány
T. Haraszti
L. Turi
Z. Király

Prof. Dr. I. Dékány (✉)
T. Haraszti · L. Turi · Z. Király
Department of Colloid Chemistry
Attila József University
Aradi Vt. 1
H-6720 Szeged
Hungary
E-mail: i.dekany@chem.u-szeged.hu

Abstract Dispersions of hydrophilic and hydrophobic SiO_2 particles in binary mixtures (ethanol–cyclohexane, ethanol–toluene, benzene–n-heptane) were studied in order to gain information on the interaction between the solid particles and the dispersion medium as well as on interparticle interactions.

Knowledge of interparticle interactions and wetting on the solid/liquid interface allows valuable information to be obtained regarding the stability of disperse systems. Adsorption excess isotherms need to be determined in binary mixtures for determination of the composition and thickness of the adsorption layer. Heat of wetting and free energy of wetting are also indicative of the strength of the solid/liquid interfacial interaction. Interparticle interactions were characterized by the rheological properties of the suspensions (yield value, energy of separation). The structure and the aggregated/disaggregated state of the suspensions were inferred from light and X-ray scattering data. It may be established that interactions between dispersed SiO_2 particles are determined by the composition of the binary mixture, the thickness of the adsorption layer and the Hamaker constants of the dispersion medium and the adsorption layer.

Key words Adsorption – colloid stability – light scattering – SAXS rheology

Introduction

In the field of the quantitative description of disperse systems in non-electrolytic media, results regarding steric and electrostatic interactions are best known [1–6]. The interaction between dispersed particles may be regulated not only by changing the amount of polymer, tenside or electrolyte added but may also be influenced merely by altering the polarity of the medium, e.g. the composition of the binary mixture [7–9]. In partially miscible liquid pairs, in the vicinity of the solubility limit a 3–7 nm thick adsorption layer is built up on the surface of the particles, the mutual attraction of which leads to their aggregation [10, 11]. Edwards et al. studied the stability of silica particles organophilized by octadecanol and dispersed in n-alkanes and found that the dispersion could be reversibly flocculated by varying the temperature. They established that interparticle interactions are determined by the difference between the Hamaker constant of the adsorption layer and that of the medium [12]. Vincent et al. studied organophilic SiO_2 particles dispersed in ethanol–cyclohexane mixtures and found that the organosol stable in cyclohexane is flocculated by the addition of ethanol [11–13]. On the other hand, when cyclohexane is added to the hydrophilic silicasol (alcosol) stable in ethanol, the hydrophilic particles are aggregated at a critical concentration of cyclohexane [13].

66

I. Dékány et al.
Stability of silica particles in binary liquids

We have previously shown that the interparticle interactions are determined not only by the Hamaker constants of the medium, the particle and the adsorption layer but also by the structure (composition and thickness) of the latter. According to Vold et al. [6, 10, 11], interaction pair potential functions are determined by the thickness of the adsorption layer. In addition to the thickness of the adsorption layer as a geometrical factor, in the case of binary mixture media the polarity of the adsorption layer developing on the solid/liquid interface is also very important, since the Hamaker constant of the adsorption layer may significantly differ from that of the mixture medium [7–9]. The primary basis for the calculation of the former are excess isotherms determined on the particle–liquid interface, since in disperse systems solid particles are in equilibrium with the mixture medium and, therefore, an adsorption layer of a given thickness t^s and composition x_1^s is formed on their surface.

The interaction on the solid/liquid interface is associated with a decrease in free enthalpy, which means that the decrease in free enthalpy accompanying adsorption ($\Delta_{21}G$) may be calculated from the adsorption excesses with the help of the Gibbs equation, given the knowledge of the activity coefficients of the bulk phase. A similarly important, directly measurable experimental value is the heat of immersion $\Delta_w H$ characteristic of the wetting of solid particles, the magnitude of which may also be indicative of stability [9, 14, 15].

Another essential task for the characterization of stability is the elucidation of interparticle interactions. In the case of disperse systems, quantitative information may be obtained from rheological properties, i.e. the so-called flow curves and the flow limit and separation energy calculated therefrom [7, 8]. Interparticle interactions will naturally determine the structure of the disperse system, the interparticle distance and the extent of aggregation. Changes in the structure of disperse systems may also be monitored by light scattering and small-angle X-ray scattering (SAXS) measurements [16, 17]. Thus, light and X-ray scattering data measured on the more or less aggregated structures greatly help the interpretation of the interaction pair potential functions calculated with knowledge of adsorption thermodynamic and Hamaker constants.

Theoretical part

Interparticle interactions

The van der Waals interaction pair potential V_A between two spherical particles of radius R, each surrounded by an adsorption layer of thickness t^s, may be calculated at a given surface separation of h (i.e. the distance of the centres of mass is $h + 2R + 2t^s$) as in [6, 10, 11]:

$$-12V_A = (\sqrt{A_m} - \sqrt{A_s})^2 H_s + (\sqrt{A_s} - \sqrt{A_p})^2 H_p + 2(\sqrt{A_m} - \sqrt{A_s})(\sqrt{A_s} - \sqrt{A_p}) H_{ps}, \quad (1)$$

where subscripts m, s and p refer to the dispersion medium, the adsorption layer and the particle, respectively (Fig. 1). The geometric factor H is defined in Refs. [6, 10, 11].

The Hamaker constants have been tabulated for a number of solid particles and pure liquids [11, 15]. For liquid mixtures, the Hamaker constant can be calculated on the basis of the characteristic frequency (ν_v) and the intrinsic dielectric constant, extrapolated from the visible light range (ε_0) [15]:

$$A = \frac{27}{64} h\nu_v \left(\frac{\varepsilon_0 - 1}{\varepsilon_0 + 2}\right)^2. \quad (2)$$

The parameters ν_v and ε_0 are determined experimentally by the minimum deviation method [7, 8], based on a linear representation of the wavelength (or frequency, ν) dependence of the refractive index n of the liquid mixtures [15]

$$\frac{n^2 + 2}{n^2 - 1} = \frac{\varepsilon_0 + 2}{\varepsilon_0 - 1}\left(1 + \frac{\nu^2}{\nu_v^2}\right). \quad (3)$$

The slope and the intercept of the linear function give access to ν_v and ε_0. If the Hamaker constant of the dispersion medium is known as a function of the composition of the medium (A_m vs. x_1), then the Hamaker constant of the interfacial layer as a function of the dispersion medium (A_s vs. x_1) can also be determined via the adsorption equilibrium diagram x_1^s vs. x_1 calculated from the adsorption excess isotherms [7–9, 19]. Finally, the van der Waals attraction potential between the particles can be computed according to Eq. (1): either at a given composition, as a function of the distance between the particles, or at a given particle separation, as a function of the composition of the dispersion medium.

Fig. 1 Particles with adsorption layer in liquid. R: particle radius, t^s: adsorption layer thickness, A_m, A_p, A_s Hamaker constants of the dispersion medium, particles and adsorption layer, respectively

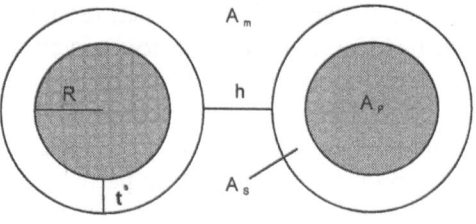

Static light scattering on colloidal particles

For characterizing colloidal particles in suspensions, the analysis introduced by Zimm is a reasonable method. From the measurements the normalized excess scattering per unit volume ($R(\Theta)$, Rayleigh-ratio) for a p-polarized incident beam is given. In the Zimm approximation a linearized form is given by

$$\frac{Kc}{R(h)} = \frac{1}{P(h)}\left(\frac{1}{M} + 2A_2 c + \cdots\right)$$
$$\approx \left(1 + \frac{R_g^2}{3}h^2\right)\left(\frac{1}{M} + 2A_2 c + \cdots\right), \tag{5}$$

where K is a constant:

$$K = \frac{2\pi^2 n^2}{\lambda_0^4 N_A}\left(\frac{dn}{dc}\right)^2, \tag{6}$$

$P(h)$ is the form factor approximated by the Guinier approximation where R_g is the radius of giration of the particle, M is the weight averaged "molecular mass", A_2 is the second virial coefficient, c is the concentration, n is the refractive index of the dispersion liquid, λ_0 is the wavelength of the incident beam in vacuo, N_A is the Avogadro-number.

The scattering vector is given by [16, 17, 20]:

$$h = \frac{4\pi n}{\lambda_0}\sin\left(\frac{\Theta}{2}\right), \tag{7}$$

where Θ is the scattering angle, the angle between the incident and the scattered beam. In our case we used $\Gamma_2 = MA_2$ instead of A_2, that got meaning from a little model calculation.

Let us consider that the silica particles in the media being hard spheres, are interacting with a square well. The potential in this case is infinite at contact distance ($r \leq 0$), and $-\varepsilon$ deep upto a distance Δ. At larger distances the potential is zero. In this case Γ_2 has the form [20]:

$$\Gamma_2 = \frac{1}{8\rho}\left(4 + 12\frac{\Delta}{R}\left(1 - \exp\left(\frac{\varepsilon}{kT}\right)\right)\right), \tag{8}$$

where ρ is the density and R is the radius of the particles, k is the Boltzmann constant, T is the temperature.

Small angle X-ray scattering in colloidal dispersion

When X-rays are scattered by colloidal particles then, due to differences in electron density caused by inhomogeneities, the intensity of scattered radiation is a function of the angle of scattering (Θ) or the scattering vector (h):

$$I(h) = w_1 w_2 (\Delta\rho_e)^2 V \int_0^{\omega} 4\pi r^2 \gamma_0(r)\frac{\sin hr}{hr}dr, \tag{9}$$

where V is the volume of the system in which X-rays are scattered by electrons. In Eq. (9) $\Delta\rho_e = \rho_{e1} - \rho_{e2}$ is the average electron density difference of the phases and w_1 and w_2 are the volume fractions of the solid and fluid phases, respectively. The $\gamma_0(r)$ is the correlation function and $h = 4\pi\sin\Theta/\lambda$.

Correlation length (l_c) can also be calculated directly from the integral of and the Q invariant scattering function [21–23]. The distance distribution function, calculated from the scattering function with inverse Fourier-transformation and the electron density function are characteristic for the particle–particle interaction and the structure of the dispersion [21–25].

Materials and methods

Hydrophilic A200 and hydrophobic R-972 (Degussa AG, Germany) SiO_2 particles were dispersed in binary mixtures (ethanol–cyclohexane, ethanol–toluene, benzene–n-heptane). The suspensions were studied at various concentrations by the methods described below. Monodisperse SiO_2 particles obtained by Stöber's procedure, via hydrolysis of TEOS were also studied [18].

1. Rheological measurements were carried out in a HAAKE Rotovisco CV100/RV 20 rotational viscosimeter at $25 \pm 0.1°C$. Flow curves [$\tau = f(D)$] and the so-called Bingham yield value (τ_B) [7, 8] were determined in the low shearing gradient range ($D = 0–100 s^{-1}$).

2. Immersion microcalorimetric measurements were done in an LKB type 2107 sorption calorimeter in various organic solvents, at $25 \pm 0.01°C$. The result of the measurements is the enthalpy of wetting ($\Delta_w H$), a value directly characterizing wetting on the solid/liquid interface [14].

3. Adsorption excess isotherm [$n_1^{\sigma(n)} = f(x_1)$] in the various binary mixtures were determined on the surface of hydrophilic and hydrophobic SiO_2 particles. After equilibrium had been established (24 h), the suspension was centrifuged, the composition of the supernatant was determined in a Zeiss interferometer and adsorption excesses were calculated as a function of equilibrium composition (x_1).

4. Hamaker constants were determined on the basis of the optical characteristics of binary mixtures by the method of minimal deviation. The method basically measures the wavelength dependence of the refractive index of the mixtures [16, 17].

68

I. Dékány et al.
Stability of silica particles in binary liquids

5. Light scattering was measured using a Sematech type 633 He–Ne laser at $25 \pm 0.1\,°C$ in binary mixtures of various compositions and in suspensions of various concentrations.

6. Small-angle X-ray scattering (SAXS) measurements were carried out using a so-called compact Kratky camera and CuK_α X-rays ($\lambda = 0.154\,nm$) obtained by a Philips PW 1830 generator. The measurements were performed in the same binary mixtures as those studied by light scattering. In order to obtain well-detectable X-ray scattering intensities, the concentration of the suspensions was raised about 10-fold as compared to concentrations acceptable for light-scattering measurements.

Results and discussion

Solid–liquid interfacial interactions were first studied in benzene-n-heptane mixtures. The adsorption excess isotherm [$n_1^{\sigma(n)} = f(x_1)$] was determined on hydrophobized aerosil R972 particles. The U-shaped excess isotherm is shown in Fig. 2A; knowing the activity of the benzene-n-heptane liquid pair [7–9], the change in free enthalpy associated with adsorption [$\Delta_{21}G$: from n-heptane(2) to benzene(1)] as a function of bulk composition was calculated from the isotherm according to the Gibbs equation, by integration [7–9, 14] (Fig. 2B).

The thickness of the adsorption layer was calculated by a computer iteration program, knowing the value of $\Delta_{21}G$ and the activities [18]. The results shown in Fig. 3 demonstrate that the thickness of the adsorption layer t^s is a function of mixture composition. This means that in n-heptane and in mixtures rich in n-heptane, the adsorption layer is thinner, practically monomolecular. In benzene and in mixtures rich in benzene, $t^s = 1.0–1.2\,nm$, meaning that benzene is better adsorbed on the surface of hydrophobic silica and multilayer adsorption occurs, a conclusion also verified by the function $\Delta_{21}G = f(x_1)$. The previous statements are also supported by enthalpies of wetting measured by microcalorimetry (batch procedure), since the value of $\Delta_w H$ is considerably higher in mixtures rich in benzene that in n-heptane (see Fig. 4).

Interparticle interactions were characterized by rheological measurements on suspensions prepared in binary mixtures. The flow curves were used for the determination of the Bingham yield value τ_B (see Fig. 5) from which the energy of separation can be calculated [27, 28]. It is revealed by Fig. 6. that in mixtures rich in heptane the particles may be extensively aggregated as the suspensions behave in a non-Newtonian manner and possess a flow limit. In mixtures rich in benzene from τ_B, the calculated energy of separation (E_{sep} on Fig. 6) is decreased since

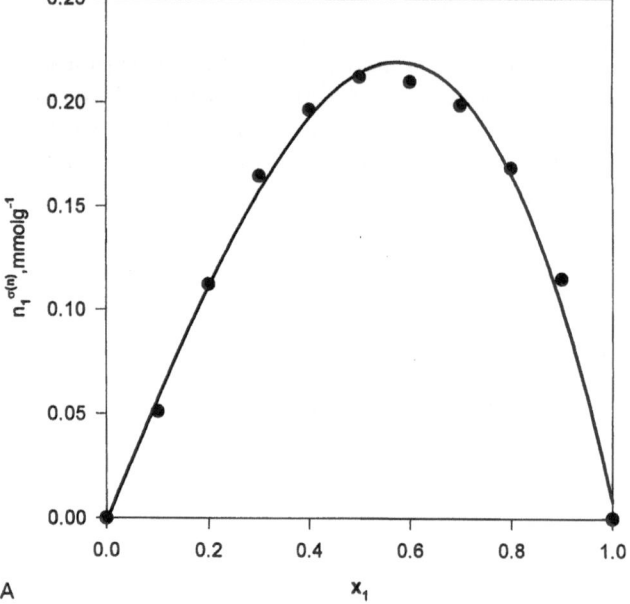

A

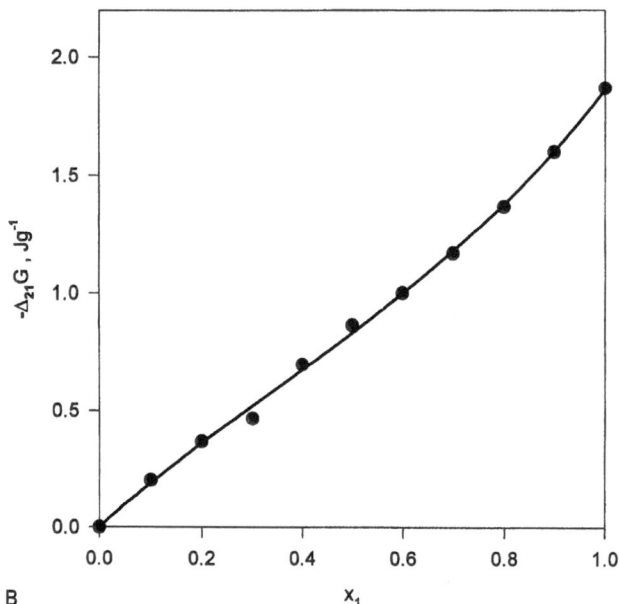

B

Fig. 2 (A) Adsorption excess isotherm in benzene(1)–n-heptane(2) mixtures on hydrophobic SiO_2 (Aerosil R972). (B) Free enthalpy of adsorption in benzene(1)–n-heptane(2) mixtures on hydrophobic SiO_2 (Aerosil R972), calculated from the Gibbs equation [7–9, 14]

wetting of the particles ($\Delta_w H$) is increased and the surface of the particles is enveloped in a thicker adsorption layer.

Interparticle attraction is also demonstrated in monodisperse systems: flow curves of Stöber silica suspensions in ethanol(1)–cyclohexane(2) mixtures are shown in Fig. 7A. Hydrophilic silica is well wetted in ethanol, interparticle interactions are therefore not significant at $x_1 = 0.7–0.9$ mixtures and the values of shear stress are

Progr Colloid Polym Sci (1998) 111:65–73
© Steinkopff Verlag 1998

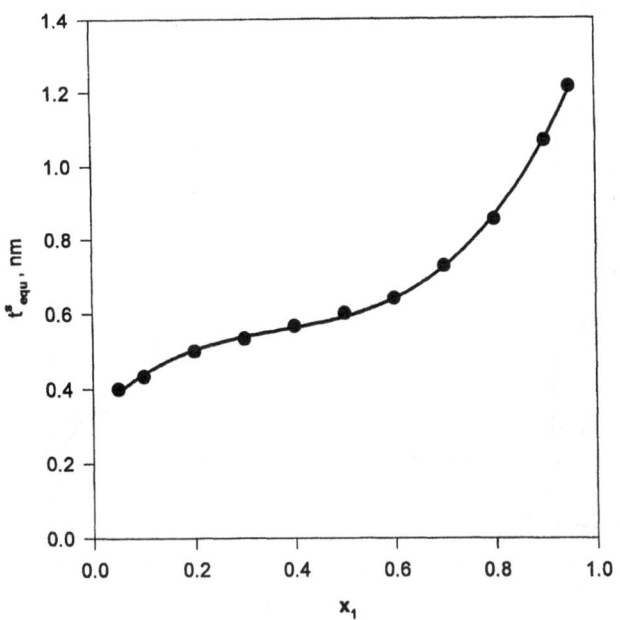

Fig. 3 Adsorption layer thickness on hydrophobic SiO_2 (Aerosil R972) in benzene(1)–n-heptane(2) mixtures

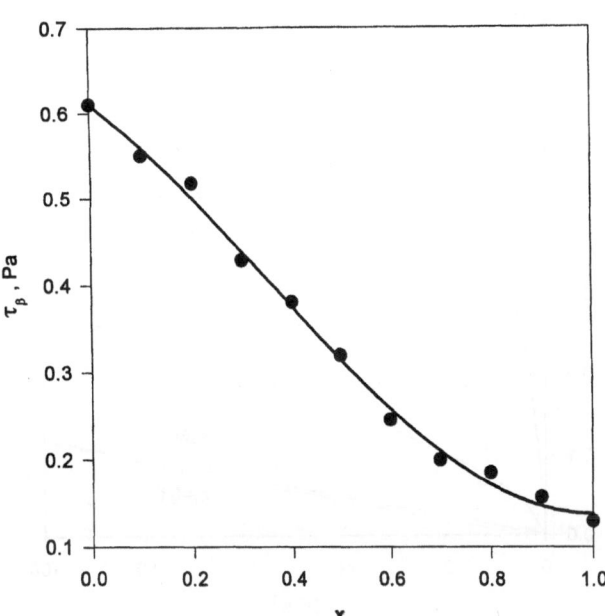

Fig. 5 Bingham yield stress as a function of the bulk composition in benzene(1)–n-heptane(2) mixtures on hydrophobic SiO_2 (Aerosil R972) calculated from the rheological flow curves of hydrophobic SiO_2 (R972) suspensions at $c = 2.48$ w/v% concentration

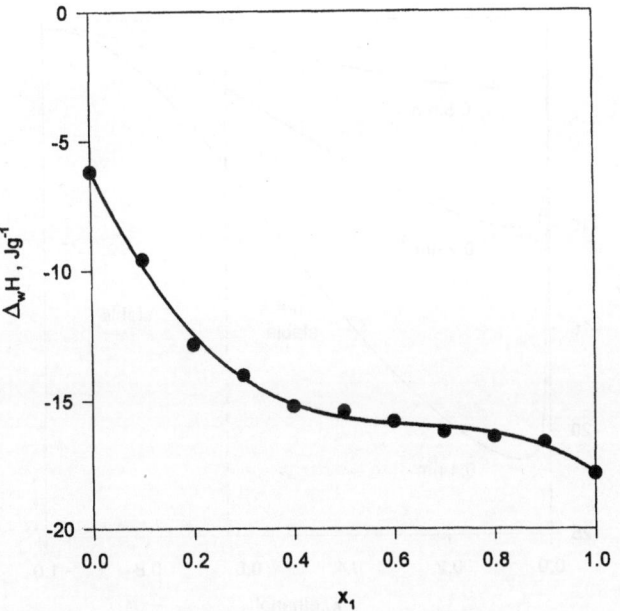

Fig. 4 Enthalpy of wetting on hydrophobic SiO_2 (Aerosil R972) in benzene(1)–n-heptane(2) mixtures

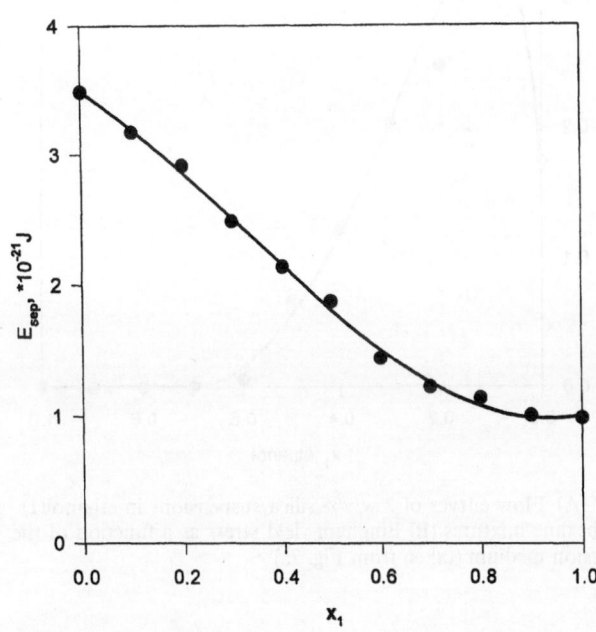

Fig. 6 The energy of separation E_{sep} as a function of the mixture composition on hydrophobic SiO_2 in benzene(1)–n-heptane(2) mixtures

low. The τ_B values presented in Fig. 7B clearly demonstrate that extensive aggregation is possible only in mixtures rich in cyclohexane. In ethanol or in mixtures rich in ethanol, stable suspensions of newtonian character are obtained. Adsorption excess isotherms determined in

ethanol–cyclohexane mixtures indicate preferential adsorption of ethanol on the surface of SiO_2 particles. This means that the adsorption layer is made up of ethanol in the entire concentration range [19]. The same is

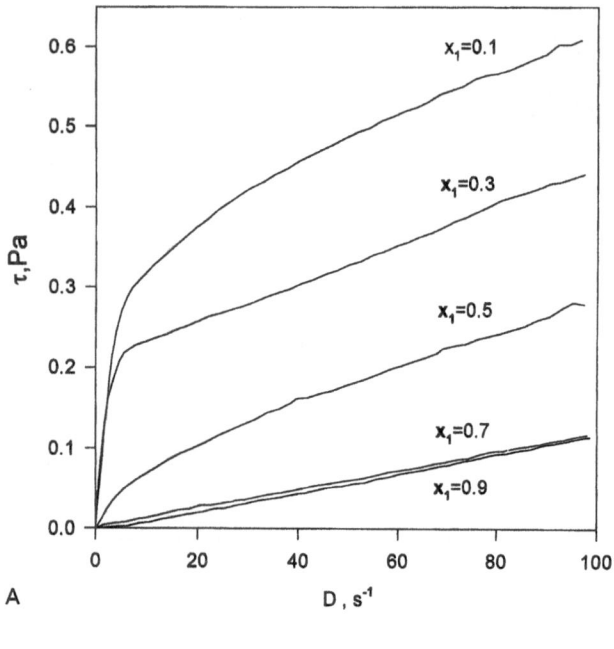

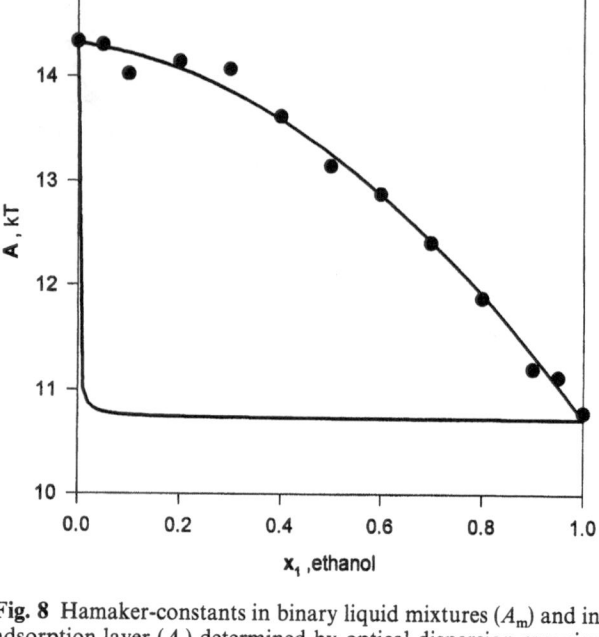

Fig. 8 Hamaker-constants in binary liquid mixtures (A_m) and in the adsorption layer (A_s) determined by optical dispersion experiments using Eqs. (1–3)

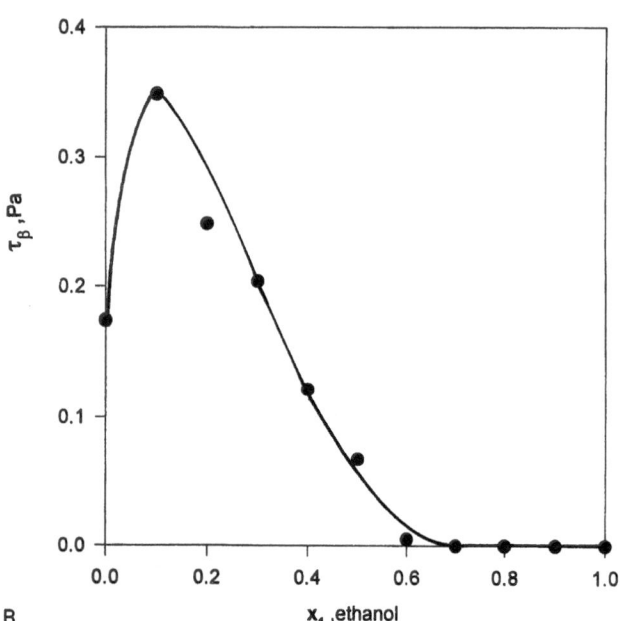

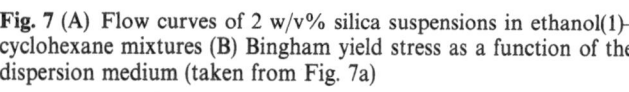

Fig. 7 (A) Flow curves of 2 w/v% silica suspensions in ethanol(1)–cyclohexane mixtures (B) Bingham yield stress as a function of the dispersion medium (taken from Fig. 7a)

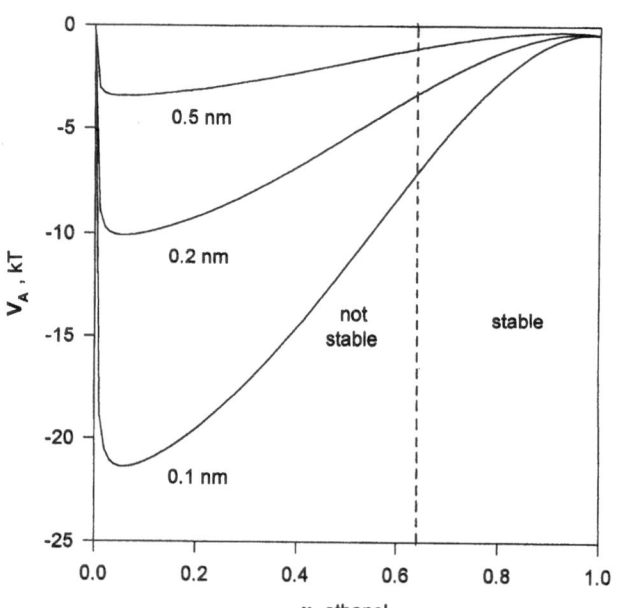

Fig. 9 Attraction potentials between hydrophilic silica particles in ethanol(1)–cyclohexane(2) mixtures

demonstrated, on the one hand, by the values of the Hamaker constants in the bulk phase (A_m) which were experimentally determined and are shown in Fig. 8 and, on the other hand, by values of A_s calculated on the basis of the composition of the adsorption layer (x_1^s). Thus, it has to be pointed out that an entirely different Hamaker constant is operative in the vicinity of the particles which considerably alters the shape of the attraction potential

functions V_A calculated according to Eq. (1). The results of the calculation are presented in Fig. 9 showing the magnitude of attraction between the silica spheres at various interparticle distances. In accordance with the results of the rheological measurements, the calculations clearly

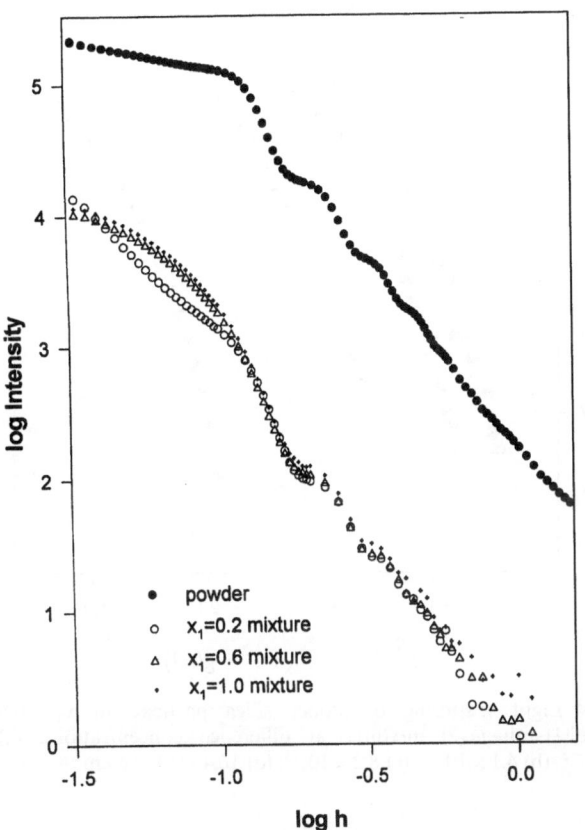

Fig. 10 SAXS curves of Stöber silica particles: dry powder, $x_1 = 0.2$, $x_1 = 0.6$, $x_1 = 1.0$ in ethanol(1)–toluene(2) mixtures

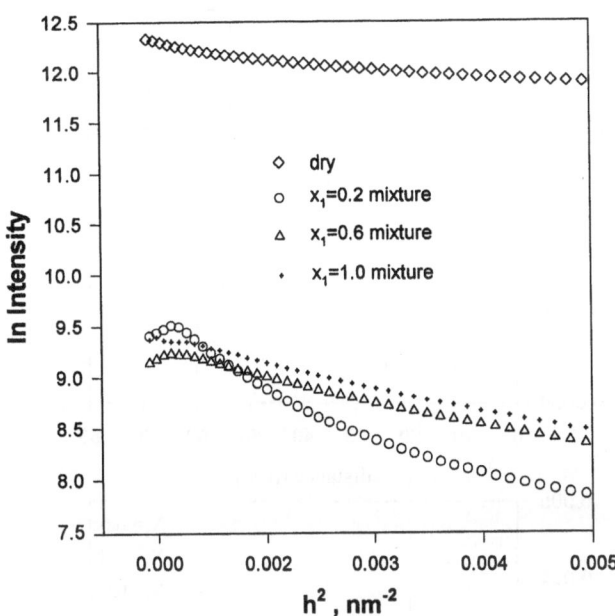

Fig. 11 The Guinier plot of SAXS curves for Stöber silica particles. Notation as in Fig. 10

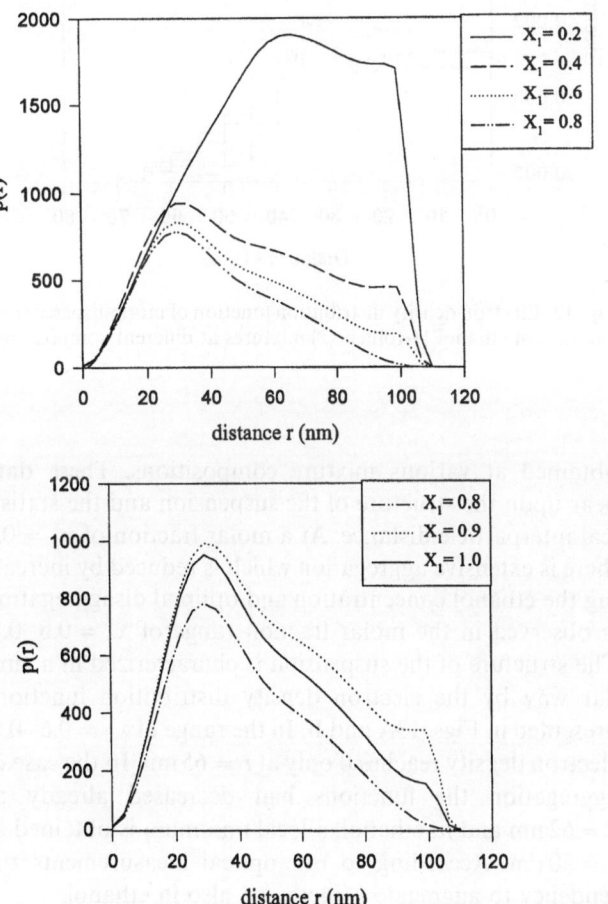

Fig. 12 Distance distribution functions calculated by inverse Fourier-transformation from SAXS curves of monodisperse silica particles in ethanol(1)–toluene(2) mixtures at different compositions

indicate that in mixtures rich in cyclohexane the system is in aggregated state.

The scattering of electromagnetic waves on colloid particles, measured in suspensions of monodisperse silica particles with a diameter of $d = 60$ nm (TEM) prepared in ethanol–toluene mixtures is next described. Due to the large difference in electron density, X-rays are efficiently scattered by the SiO_2/ethanol–toluene system. Scattering curves are shown in Fig. 10. The local maxima represent the intensities characteristic of the monodisperse system. Based on the initial section of the scattering curve, the radius of the particles may be determined from the so-called Guinier plot. It is well demonstrated by the data in Fig. 11 that at a molar fraction of ethanol of $x_1 = 0.2$ (i.e. at a composition rich in toluene) the Guinier plot is not straight, the particles are aggregated ($R_G = 65$, $R = 110$ nm). As soon as ethanol concentration is increased, straight lines are obtained, indicating more efficient wetting by the binary mixture medium. The value of the surface fractal dimension also stays within the range of $D_s = 2.02$–2.05, characteristic of planar surfaces.

If distance distribution functions are calculated from scattering curves by inverse Fourier tranformation [24–26], the curves with maxima shown in Fig. 12 are

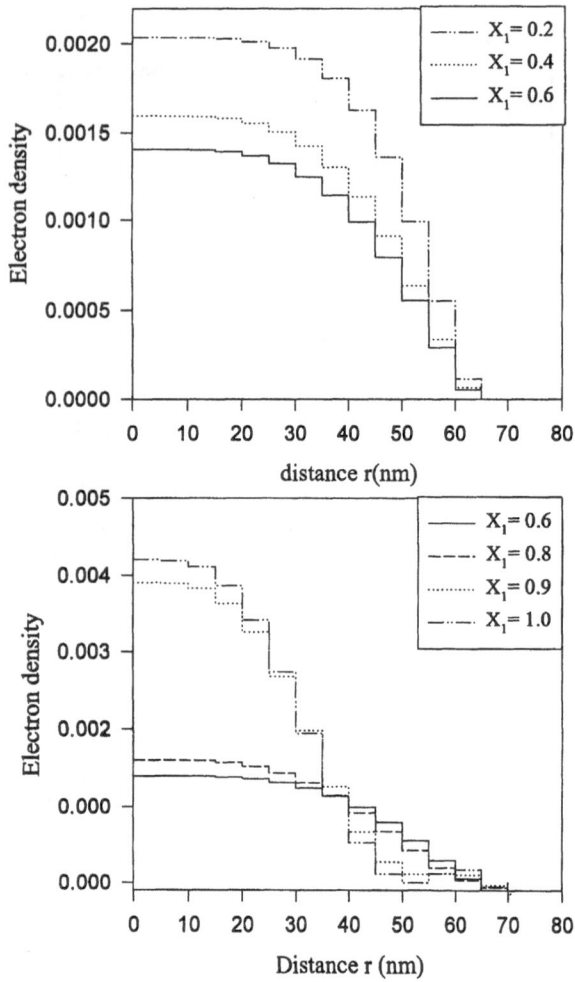

Fig. 13 Electron density distribution function of monodisperse silica particles in ethanol(1)–toluene(2) mixtures at different compositions

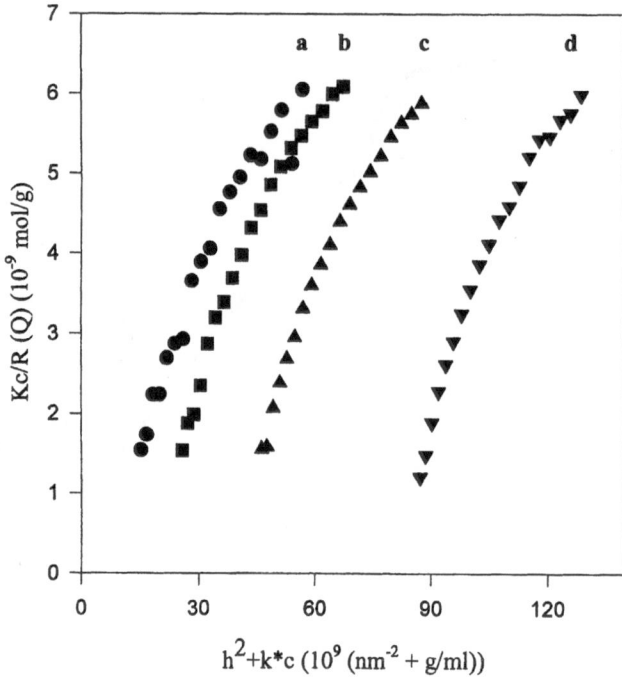

Fig. 14 Light scattering of Stöber silica particles in $x_1 = 0.8$ ethanol(1)–toluene(2) mixtures at difference concentrations: (a) 2×10^{-5}, (b) 4.1×10^{-5}, (c) 8.2×10^{-5}, (d) 16.4×10^{-5} g/cm³

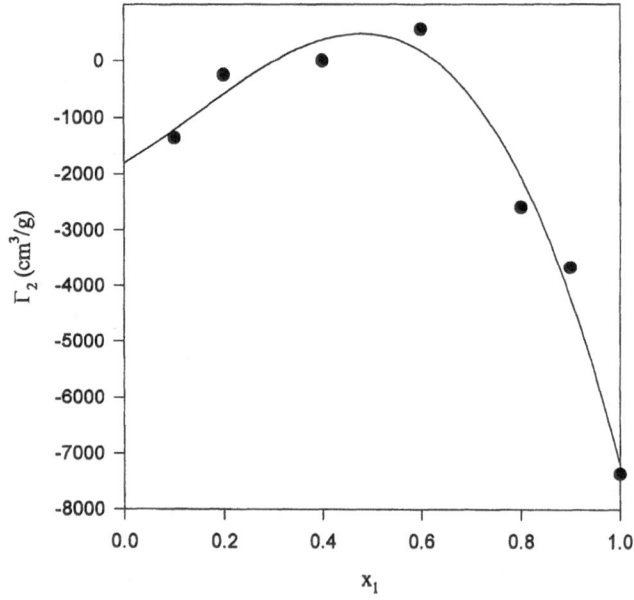

Fig. 15 Second virial coefficient in ethanol(1)–toluene(2) binary mixture suspension calculated by Eq. (8) from the light scattering experiment

obtained at various mixture compositions. These data bear upon the structure of the suspension and the statistical interparticle distance. At a molar fraction of $x_1 = 0.2$ there is extensive aggregation which is reduced by increasing the ethanol concentration and optimal disaggregation is observed in the molar fraction range of $x_1 = 0.6–0.8$. The structure of the suspension is characterized in a similar way by the electron density distribution functions presented in Figs. 13A and B. In the range of $x_1 = 0.6–0.8$, electron density reaches 0 only at $r = 65$ nm. In the case of aggregation the functions had decreased already at $r = 62$ nm and in ethanol, a local minimum is obtained at $r = 50$ nm. According to our optical measurements the tendency to aggregate is increased also in ethanol.

Monodisperse silica suspensions were also subjected to light-scattering measurements at various suspension concentrations and the Zimm-plot shown in Fig. 14 was

obtained. Based on the experiments carried out in different mixture series, the second virial coefficient characteristic of the solid/liquid interaction may be calculated according to

Progr Colloid Polym Sci (1998) 111:65–73
© Steinkopff Verlag 1998

Eq. (8); its dependence on composition is presented in Fig. 15. It may be established that the extent of solvation interaction is most favorable at a mixture composition of $x_1 = 0.6$–0.8 and in pure ethanol enhances the turbidity of the system, indicating aggregation of the silica particles in the dispersion.

References

1. Ottewill RH (1973) In: Everett DH (ed) Colloid Science, Specialist Periodical Reports, Vol 1. The Chemical Society, London, Chap 5
2. Overbeek TTh (1982) In: Goodwin JW (ed) Colloidal Dispersions. The Royal Society of Chemistry, London, Chap 1
3. Kline SR, Kaler EW (1994) Langmuir 10:412
4. Beysens D, Esteve D (1985) Phys Rev Lett 54:2123
5. Gurfein V, Beysens D, Perrot F (1989) Phys Rev A 40:2543
6. Vold MJ (1961) J Colloid Sci 16:1
7. Machula G, Dékány I (1991) Colloid Surf 61:331
8. Machula G, Dékány I, Nagy LG (1993) Colloid Surf 71:241
9. Dékány I (1993) Pure Appl Chem 65: 901
10. Osmond DWJ, Vincent B, Waite FAW (1973) J Colloid Interface Sci 42:262
11. Vincent B (1973) J Colloid Interface Sci 42:270
12. Edwards J, Everett DH, O'Sullivan T, Pangalou I, Vincent B (1984) J Chem Soc Faraday Trans I 80:2599
13. Vincent B, Király Z, Emmett S, Beaver A (1990) Colloid Surf 49:121
14. Everett DH (1973) In: Ref. [1], Chap 2
15. Gregory J (1969) Adv Colloid Interface Sci 2:396
16. van Helden AK, Jansen JW, Vrij A (1982) J Colloid Interface Sci 81:354
17. JW Jansen, CG de Kruif, Vrij A (1986) J Colloid Interface Sci 114:492
18. Stöber W, Fink A, Bohn E (1968) J Colloid Interface Sci 26:62
19. Király Z, Turi L, Dékány I, Bean K, Vincent B (1996) Colloid Polym Sci 274:779–787
20. Rouw PW, Vrij A, de Kruif CG (1988) Progr Colloid Polym Sci 76:1–15
21. Guinier A, Fournet G (1955) Small-Angle Scattering of X-rays. Wiley, New York
22. Glatter O, Kratky O (1982) Small-Angle X-ray Scattering. Academic Press, New York
23. Jánosi A (1991) Monatschafte für Chemie 124:815–822
24. Glatter O (1977) Acta Phys Austriaca 47:83
25. Glatter O (1981) J Appl Crystallogr 14:101
26. Glatter O (1984) J Appl Crystallogr 17:435
27. Firth BA, Hunter RJ (1976) J Colloid Interface Sci 57:248
28. Firth BA, Hunter RJ (1976) J Colloid Interface Sci 57:266

Progr Colloid Polym Sci (1998) 111:74–77
© Steinkopff Verlag 1998

M. Mielke
R. Zimehl

The reverse behavior of negatively and positively charged poly(*N*-isopropylacrylamide latex dispersions in alcohol-water mixtures at temperatures below their phase transition

M. Mielke · Dr. R. Zimehl (✉)
Institut of Inorganic Chemistry
University of Kiel
Olshausenstr. 40
D-24098 Kiel
Germany

Abstract Results of size control experiments on both negatively and positively charged poly(*N*-isopropylacrylamide) (PNIPAAM) latex dispersions are reported. In alcohol–water mixtures containing low weight fractions of alcohols higher than methanol, negatively and positively charged PNIPAAM particles exhibit the opposite swelling and shrinking behavior at 20 °C. Negatively charged particles shrink due to cononsolvency. Conversely, positively charged particles swell dramatically. Above 40 wt% alcohol, this opposite behavior is cancelled out. At 40 °C, the negatively charged particles collapse and show a similar behavior to those with a positive charge. The general idea that positively charged particles are more hydrophobic is also underlined by other clues.

Key words *N*-isopropylacrylamide – microgel – electrosteric – stabilization – polymer – swelling – water structure – hydrophobic hydration

Introduction

Aqueous PNIPAAM solutions are characterized by a lower critical solution temperature (LCST) at approximately 32 °C. Below the LCST the polymer is soluble but undergoes phase separation upon heating. The question arises, if the properties of PNIPAAM in the colloidal state are governed by the same scaling laws as the properties of macroscopic PNIPAAM. We suspect that important parameters are likely to be particle size, charge density and the nature of the surface charge present.

Materials and methods

Chemicals

Ion-exchanged water was used throughout all the experiments. The alcohols and the surfactants sodium dodecylsulfate and cetylpyridiniumchloride were all of AnalR quality *N*-isopropyl acryl amide NIPAAM, kindly donated by Acros-chemicals, was used as received. The radical initiators potassium peroxodisulfate, KPS (Riedel de Haen) and azo-*N*,*N*'-dimethylene isobutyramidine dihydrochloride, ADMBA (Wako) were also used as received.

Latex preparation

The PNIPAAM dispersions were prepared by emulsion polymerization in the presence of different emulsifiers and initiators (Table 1). Polymerizations were performed in a 1 l round bottomed three-necked flask according to the procedure given by Zhou and Wu [1]. Positively charged particles were obtained by varying the recipe, as shown in Table 1, and the particles were crosslinked by 1.8 wt% *N*,*N*'-methylenbis(acrylamide) (Fluka). The latex dispersions were diluted to give a weight fraction of 1 g ml^{-1} and were used without any further purification. The surface charge densities were determined by polyelectrolyte titration using a microelectrophoresis apparatus (PenKem

501). The amount of polyelctrolyte required to reach the point of zero charge was determined, from which the surface charge for a latex diameter and volume fraction was calculated.

Particle sizing

Particle sizes were measured using a BI-ZetaPlus Photon Correlation Spectrometer (Brookhaven Instruments) at a fixed angle of 90°. Before measurements, the PNIPAAM

Table 1 Preparation properties of PNIPAAM latices

Latex	PNIPAAM [−]	PNIPAAM [+]
Initiator, surface group	KPS, sulfate sulfonate	ADMBA, amidinium
Surfactant	SDS	$C_{16}PyCl$
Surfactant/monomer	0.007	0.01
React. temp.	70 °C	60 °C
Surface charge density	$-12\ \mu C\ cm^{-1}$	$6\ \mu C\ cm^{-1}$

dispersions were diluted with alcohol–water mixtures to a final weight fraction of approximately $10^{-3}\ g\ ml^{-1}$. These dispersions were then equilibrated for 3 h. The results were verified by measuring the diameters after 24 h equilibration time.

Results

At room temperature

Figures 1A–D illustrate the change in diameter of the PNIPAAM particles in various alcohol–water mixtures at 20 °C. In general, anionic particles show a minimum in particle diameter between 20 and 40 wt% alcohol. At low alcohol fractions always deswelling of the PNIPAAM particles can be observed. Alcohol weight fractions larger than approximately 80% causes the latex particles to take up liquid in excess: the polymer particles swell considerably. The general trend does not appear to alter by using alcohols of higher chain length.

Fig. 1 (A) Cationic and anionic PNIPAAM versus methanol at 20 °C; (B) cationic and anionic PNIPAAM versus ethanol at 20 °C; (C) cationic and anionic PNIPAAM versus n-propanol at 20 °C; (D) cationic and anionic PNIPAAM versus iso-propanol at 20 °C

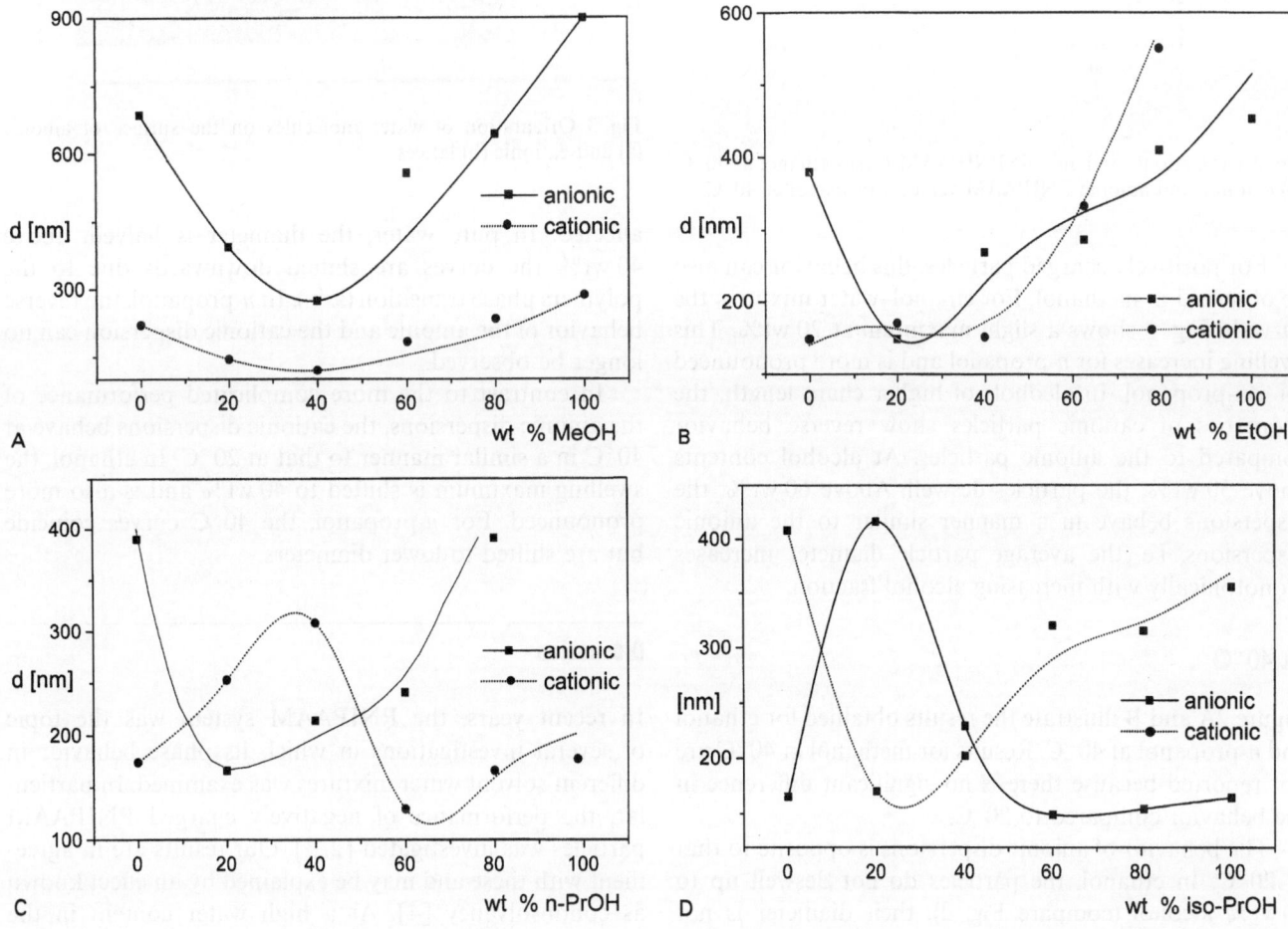

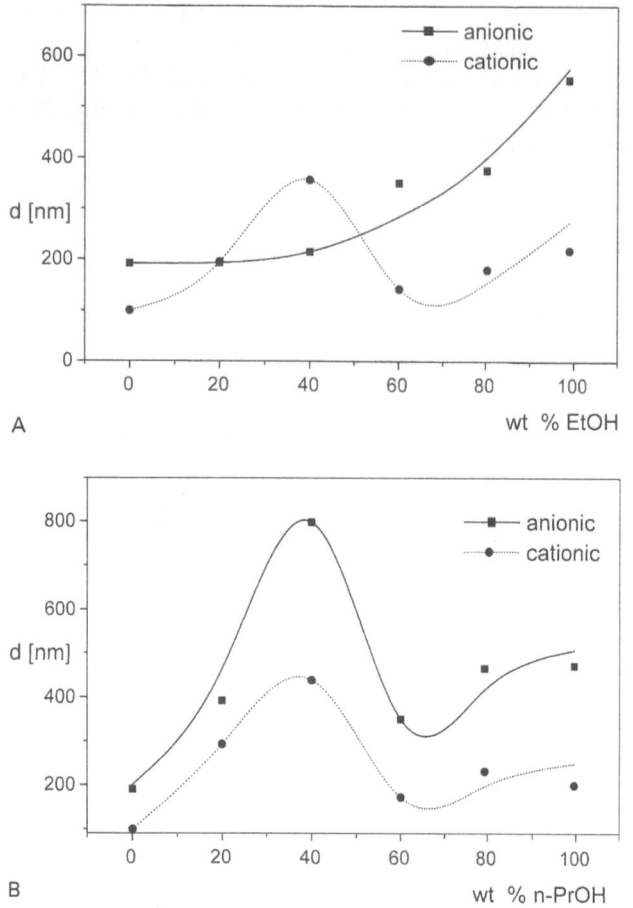

A

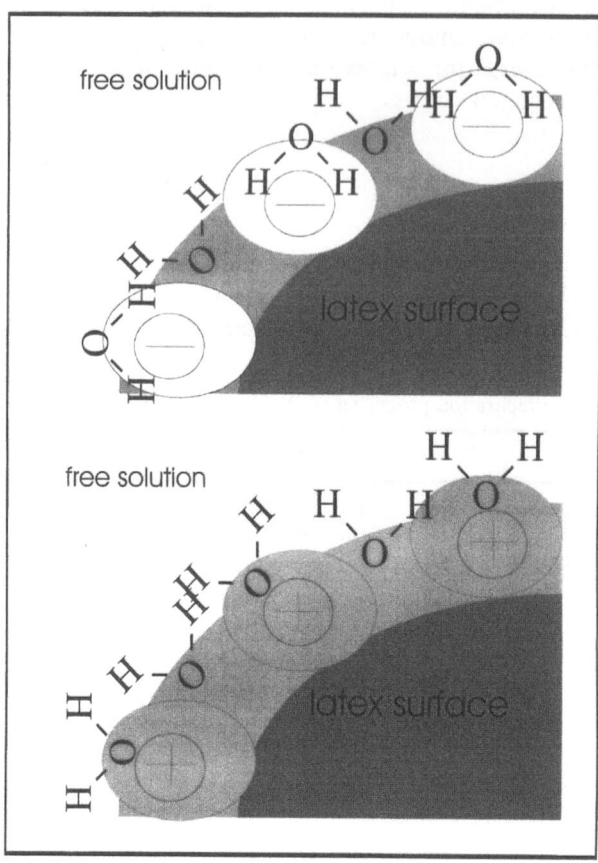

Fig. 3 Orientation of water molecules on the surface of anionic (a) and cationic (b) latices

Fig. 2 (A) Cationic and anionic PNIPAAM versus ethanol at 40 °C; (B) cationic and anionic PNIPAAM versus *n*-propanol at 40 °C

For positively charged particles, this behavior can also be observed in methanol. For ethanol–water mixtures the curve in Fig. 2 shows a slight maximum at 20 wt%. This swelling increases for *n*-propanol and is more pronounced for iso-propanol. In alcohols of higher chain length, the properties of cationic particles show reverse behavior compared to the anionic particles. At alcohol contents above 50 wt%, the particles deswell. Above 60 wt%, the dispersions behave in a manner similar to the anionic dispersions, i.e. the average particle diameter increases monotonically with increasing alcohol fraction.

At 40 °C

Figure 2A and B illustrate the results obtained for ethanol and *n*-propanol at 40 °C. Results for methanol at 40 °C are not reported because there is no significant difference in the behavior compared to 20 °C.

The behavior of anionic dispersions is opposite to that at 20 °C. In ethanol, the particles do not deswell up to 20 wt% alcohol (compare Fig. 2), their diameter is not

affected. In pure water, the diameter is halved, above 40 wt% the curves are shifted downwards due to the polymers phase transition (s. b.). In *n*-propanol, the reverse behavior of the anionic and the cationic dispersion can no longer be observed.

In contrast to the more complicated performance of the anionic dispersions, the cationic dispersions behave at 40 °C in a similar manner to that at 20 °C. In ethanol, the swelling maximum is shifted to 40 wt% and is also more pronounced. For *n*-propanol, the 40 °C curves coincide but are shifted to lower diameters.

Discussion

In recent years, the PNIPAAM system was the topic of several investigations in which its phase behavior in different solvent water mixtures was examined. In particular, the performance of negatively charged PNIPAAM particles was investigated [2, 3]. Our results are in agreement with these and may be explained by an effect known as cononsolvency [4]. At a high water content in the solvent, the swollen PNIPAAM brushes are dehydrated due to changes in the water structure at the interface and in solution. These changes are induced by alcohol molecules competing for free water, and the balance of free and clustered water is altered. In contrast, the addition of more alcohol causes the PNIPAAM to reswell due to direct hydrophobic interaction of alcohol molecules with the hydrophobic groups of PNIPAAM.

The reverse behavior of crude positively (and highly) charged PNIPAAM particles in alcohols with two or more CH_3-groups can be explained by specific alcohol adsorption which forces swelling of the PNIPAAM even at very low alcohol contents. Cononsolvency cannot take place. At slightly higher alcohol fractions (above 40 wt%), free alcohol molecules appear in solution competing for water. In this case, deswelling occurs. The behavior of positively and negatively charged particles is then similar in nature.

Alcohol adsorption, and therefore counter-ion displacement, is also believed to explain the increase in the zetapotential of negatively charged latices at low alcohol contents [5]. At higher alcohol contents, the zetapotential decreases; water molecules are packed in the interface to hydrate the OH-groups of the adsorbed alcohol layers [6]. We found the same results with PMMA dispersions. However, the decrease in zetapotential for the cationic dispersions is stronger [7].

In order to explain why the alcohol adsorption for cationic particles is much more pronounced, we assume that positively charged particles are more hydrophobic. This is underlined by transmission electron micrographs showing that positively charged particles form aggregates whereas negative particles remain isolated during drying [7]. This is supported by the observation that adsorption of tetraphenylborate ions onto a positive polystyrene latex is 1000 times stronger than the adsorption of tetraphenyl phosphonium ions onto a negatively charged PS latex [8].

The behavior of the particles is governed by hydrophobic hydration and therefore hydrophobic interactions. Hydrophobic hydration entails the formation of a water structure around the particles (several water molecules thick) in which the O–H bonds point away from the hydrophobic surface [9]. The water molecules around the cationic charges adopt a similar orientation. Negative charges force the opposite orientation of the water molecules. The entropy is therefore increased by the presence of negative charges on a hydrophobic surface whereas positive charges do not destroy the balance of free and clustered water molecules [10, 11] at the interface (Fig. 3).

The temperature dependence of the dispersions shows that above the phase transition [3] of PNIPAAM, the negatively charged particles swell at low alcohol fractions. The reverse behavior of positively and negatively charged particles is no longer observed. The collapsed PNIPAAM particles became hydrophobic above 32.5 °C. It can therefore be concluded that the positively charged particles are more hydrophobic at all temperatures.

Conclusions

The stabilization of both electrostatically (hydrophobic latices) and electrostatically (PNIPAAM microgels) stabilized particles is relatively complex. We propose that the phase behavior of hydrophilic dispersions, and also aggregation of more hydrophobic dispersions in the presence of organic matter (counterions [8], long-chain alcohols, etc.) is controlled mainly by a balance of different forces. These forces include coulombic, hydrophobic and van der Waals attractions. They are affected by the water structure at the hydrophobic domains of the particle, as well as at the hydration layer surrounding the charges on the particles. Our conception can explain the higher hydrophobicity of positively charged polymer colloids. In the case of highly charged particles (e.g. crude dispersions), the opposite behavior of positively and negatively charged latex dispersions can be understood.

Furthermore, because the hydrophobic nature of the surface changes with the sign of the surface charges, the degree of hydrophobicity of certain latices (poly-2-vinyl pyridine [12]) can be adjusted by changing the pH.

Acknowledgements We are grateful to Prof. G. Lagaly for helpful discussions.

References

1. Zhou S, Wu C (1996) Macromolecules 29:4998–5001
2. Schild HG, Muthukumar M, Tirell DA (1991) Macromolecules 24:948–952
3. Zhu PW, Napper DH (1996) J Colloid Interf Sci 177:343–352
4. Winnik FM, Ringsdorf H, Venzmer J (1990) Macromolecules 23:2415–2416
5. Vincent B (1992) Adv Colloid Interf Sci 42:279–302
6. Seebergh JE, Berg JC (1997) Colloids Surfaces A: Physiochem Eng Asp 121:89–98
7. Zimehl R, Lagaly G, Mielke M (1997) XX. Hamburger Makromolekulares Symposium 22–23. September 97, Universität Hamburg
8. Mielke M, Lagaly G, Zimehl R (1996) Proc 7th Conf on Colloid Chemistry 23–26 September 96, Eger, Hungary
9. Holz M, Sörensen M (1992) Ber Bunsenges Phys Chem 96:1441–1447
10. Luck WAP (1978) Progr Colloid Polym Sci 65:6–28
11. Ueberreiter K (1982) Colloid Polym Sci 260:37–45
12. Loxley A, Vincent B (1997) Colloid Polym Sci 275

Progr Colloid Polym Sci (1998) 111:78–81
© Steinkopff Verlag 1998

M. Wagener
B. Günther

High pressure DC-magnetron sputtering on liquids: A new process for the production of metal nanosuspensions

Dipl.-Ing. M. Wagener (✉) · B. Günther
Fraunhofer-Institut für angewandte
Materialforschung IFAM
Lesumer Heerstrasse 36
D-28717 Bremen
Germany
E-mail:wa@ifam.fhg.de

Abstract Vacuum evaporation on running liquids (VERL) is an established method for the production of metal nanoparticles in low vapor pressure carrier liquids like silicone oils or resins. Such metal suspensions may be useful, e.g. as additives for functional metal/polymer composites or for sintering additives in thick film pastes for microelectronics. In this paper we present a modified VERL-process employing high gas pressure dc magnetron sputtering for the preparation of suspensions of metal nanoparticles in various carrier liquids. By using magnetron sputtering, materials having high melting points can be evaporated without harming the liquid substrate. The method was tested for Ag and Fe-suspensions by varying the carrier liquid and the pressure of the Argon sputtering atmosphere in the range of 1–30 Pa. A narrow particle size distribution is obtained with the mean particle size $\langle d \rangle$ ranging from 2 to 20 nm. For Ag $\langle d \rangle$ increases with increasing gas pressure following roughly a power law type $p^{1/3}$. The variation of particle size with sputtering gas pressure is consistent with in a model where particle formation takes place in the gas phase exclusively before introduction into the carrier liquid. In the case of Fe agglomeration could be prevented effectively by adding two different surfactants to the carrier liquid before starting the sputtering process.

Key words Metal colloid – nano-suspension – sputtering – VERL – metal-polymer composites

Introduction

Vacuum evaporation on running liquids (VERL) is an established method for the preparation of metal nanosuspensions in a non-aqueous, low-vapor pressure liquid, like various oils and resins [1]. In this method a metal is evaporated by Joule heating onto a liquid surface that is continuously renewed by the action of a rotating drum or spinning disk.

Recently, the formation of particulate nanomaterials on solid substrates by using the sputtering method was also reported [2, 3]. Here in contrast to the above-mentioned method gas pressures $P > 20$ Pa are used in order to induce nucleation and grain growth within the gas atmosphere (Inert gas condensation technology: IGC).

In this work we describe a method that combines the two principles of IGC and VERL-technique employing high-pressure magnetron sputtering on organic liquids for the in situ preparation of Ag- and Fe-nanosuspensions.

This Sputter-VERL-method has the following advantages compared to the conventional VERL employing Joule heating:

Progr Colloid Polym Sci (1998) 111:78–81
© Steinkopff Verlag 1998

Fig. 1 Sketch of the modified VERL-process used for the preparation of metal nanosuspensions. The vacuum chamber has a diameter of 700 mm. The size of the rotating steel drum is Ø200 × 400 mm

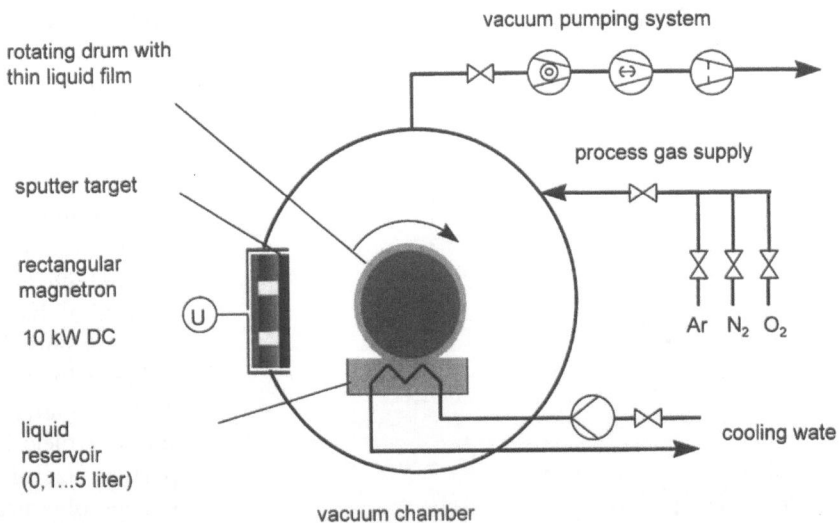

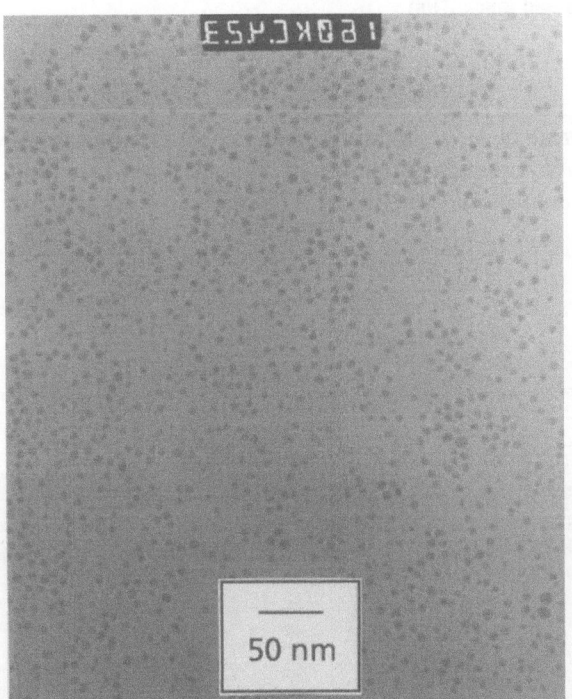

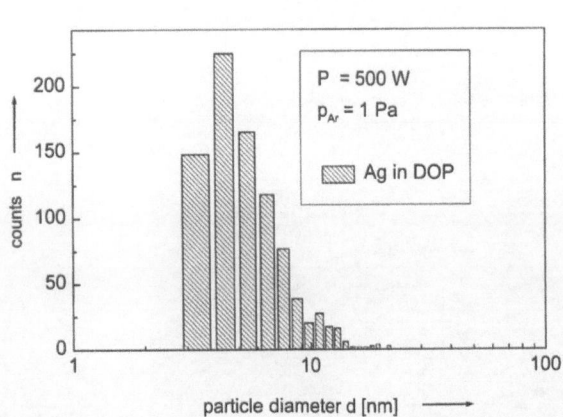

Fig. 2 TEM bright field image (left) and respective particle size histogram (right) of Ag-nanoparticles in diethylhexylphthalate (DEHP) prepared at a sputtering pressure of 1 Pa and a sputtering input power of 500 W. A mean particle size $\langle d \rangle$ = 4.9 nm with a standard deviation of $s = 1.34$ was derived by fitting the size histogram to a log-normal distribution function [4]

– High melting point materials can be evaporated.

– Evaporation conditions can be held stable over longer periods of time.

– Much less radiation heating of the liquid substrate.

Experimental

In the experimental setup (Fig. 1) a planar 400 × 130 mm magnetron sputtering cathode was used with a maximum

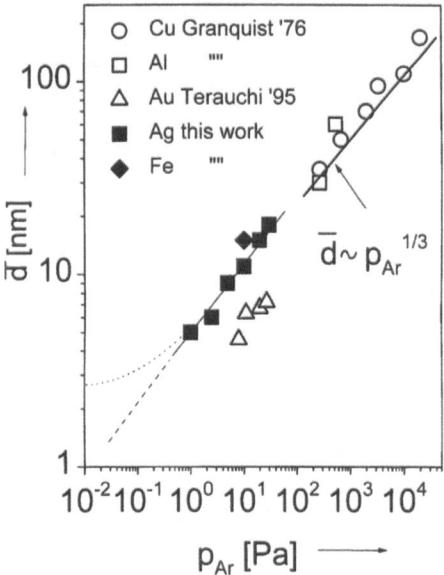

Fig. 3 Mean size $\langle d \rangle$ of Ag particles in the prepared suspensions as a function of Ar gas pressure. The mean particle size of Cu and Al particles obtained by Granquist (1976) using an IGC process and solid substrates show similar behavior at a higher-pressure regime. Au nanoparticles prepared on solid substrates were obtained by Terauchi (1995) using a diode sputtering device in a similar pressure regime than in this work

input power of 10 kW. Experiments were performed at a fixed input power of 500 and 5000 W for Ag and Fe, respectively. The pressure of the sputtering gas (Ar) was varied in the range of 1–30 Pa. The substrate consisted of a thin liquid film on a rotating steel drum. The thin liquid film was permanently renewed due to the rotation of the drum dipping into a reservoir of the respective liquid.

As liquid substrates we used silicone oil based on polyphenylmethyldisiloxan (DC702 DowCorning), mineral oil based on naphthalene (L9 Edwards), and diethylhexylphthalate (DEHP) and polytetrahydrofurane (PTHF). In order to prevent agglomeration of Fe particles we used a fatty acid condensation polymer (LP4, ICI) and sarcosyl oleic acid (Korantin, BASF) as additives. The particle size distributions were derived from bright-field images obtained in a transmission electron microscope (Philips CM30).

Results and discussion

Ag suspensions

Sputtering of Ag was performed at a constant input power of 500 W and at gas pressures p_{Ar} ranging from 1 to 30 Pa on different liquid substrates. The mean particle size was

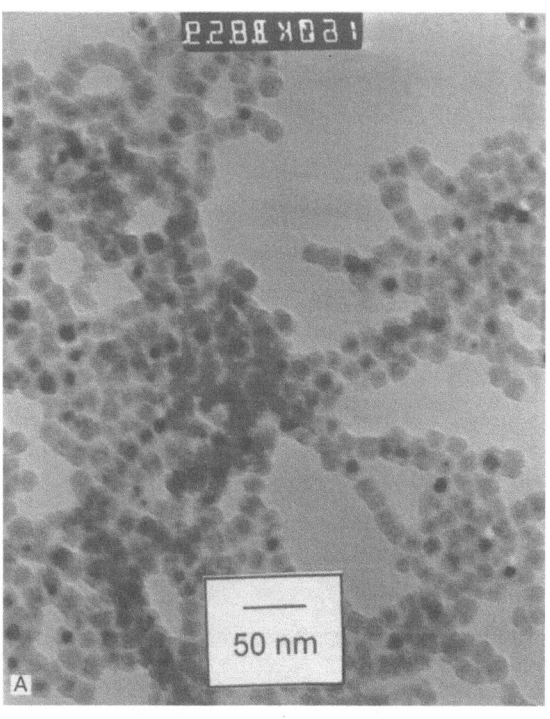

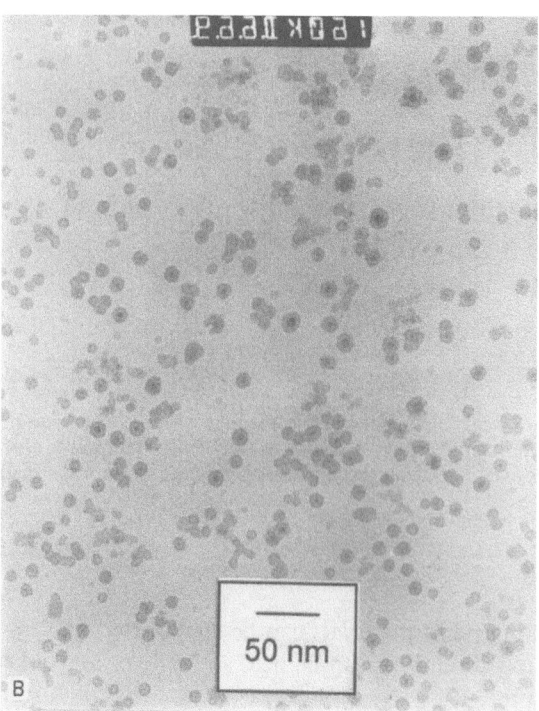

Fig. 4 (A) TEM bright field image of Fe nanoparticles in silicone oil prepared at an Ar sputtering pressure of 10 Pa and a sputtering input power of 5 kW. (B) Deagglomeration effect obtained by adding a surfactant (1 wt% LP4) to the liquid carrier before sputtering

Progr Colloid Polym Sci (1998) 111:78–81
© Steinkopff Verlag 1998

determined from fits of the respective size histograms to log-normal size distributions [4]. Figure 2 gives an idea of quality of the homogenity of distribution and size of the Ag nanoparticles in DEHP.

Similar particle size distributions were obtained when using other substrate liquids instead of DEHP (DC702, L9, PTHF). The effect of increasing gas pressure on $\langle d \rangle$ is shown in Fig. 3. Obviously in the pressure range under investigation $\langle d \rangle$ of the Ag particles increases with increasing gas pressure with a power-law type behavior that is similar to the results obtained in the IGC process employing thermal evaporation and solid substrates [4].

Further the results of Terauchi [3] who also used a sputtering device but solid substrates show in the case of Au particles a similar dependence of $\langle d \rangle$ vs. p_{Ar} at somewhat lower values of $\langle d \rangle$. Summarizing this suggests that with our experimental conditions particle formation and growth takes place in the gas phase, i.e. before deposition on the carrier liquid.

Fe suspensions

Sputtering of Fe was performed at a constant input power of 5 kW and a constant Ar gas pressure of 10 Pa. Silicone oil (DC702) was used as a substrate liquid. Agglomerated particles having a mean particle size of $\langle d \rangle = 12$ nm were obtained (Fig. 4A)). Figure 4B demonstrates the effect of adding a surfactant (LP4). Adding 1 wt% of LP4 prevents agglomeration of the Fe particles effectively.

Conclusion

It has been shown that it is possible to produce suspensions of metal nanoparticles in organic liquids via the sputtering method. By changing the sputtering gas pressure it was possible to shift the mean size of Ag-particles in the range of 5–20 nm. A comparison with respective data that have been obtained without employing liquid substrates suggests that particle formation takes place in the gas phase exclusively.

Acknowledgments Financial support by the German Federal Ministry for Education and Research (BMBF) under contract no. 03N 2004D is gratefully acknowledged.

References

1. Yatsua S, Tsukasaki Y, Mihama K, Uyeda R (1978) J Crystal Growth 45:490–494
2. Hahn H, Averback RS (1990) J Appl Phys 67:1113–1115
3. Terauchi S, Koshizaki N, Umehara H (1995) Nanostruct Mater 5:71–78
4. Granquist CG, Buhrman RA (1976) J Appl Phys 47:2200–2219
5. Yatsuya S, Hayashi T (1978) Japan J Appl Phys 17:355–359
6. Wagener M, Murty BS, Günther B (1996) Mat Res Soc Symp Proc 457:149–154

Progr Colloid Polym Sci (1998) 111:82–84
© Steinkopff Verlag 1998

SUSPENSIONS AND MICROCAPSULES

F. Bagusat
O. Seidel
H.-J. Mögel

Time periodic changes of viscosity in kaolin suspensions as an example for modelling concentrated suspensions

F. Bagusat (✉) · O. Seidel
H.-J. Mögel
Freiberg University of Mining
and Technology
Leipziger Str. 29
D-09596 Freiberg
Germany

Abstract We observed time periodic viscosity fluctuations in concentrated kaolin suspensions. This behavior can be described assuming a complex system of shear-induced agglomeration and deglomeration processes. For modelling this kinetics we used the Lotka–Volterra scheme including autocatalytic cluster formation steps.

The rate constants of the Lotka–Volterra model are fitted applying Fourier transformation to the experimental data set.

Key words Concentrated kaolin suspensions – viscosity fluctuations – Lotka–Volterra model

Introduction

Concentrated suspensions occur in various industrial processes such as fabrication of concrete, ceramics, insulating and high-temperature-resistant materials. Transport, mixing, seperation, storage and dosage of suspensions are subprocesses, which require the knowledge of flow properties. The flow behavior of suspensions is very complex and highly non-Newtonian [1]. In particular, the apparent viscosity depends on shear conditions and on time.

In this paper we present a special kind of this complexity, observed in concentrated kaolin suspensions and investigated with a rotational viscometer. At constant shear rate, apparent viscosity does periodically fluctuate. Obviously, this behavior is caused by structural changes within the suspension during shear flow. We suppose these changes are due to agglomeration and deglomeration processes, which we have modelled using the Lotka–Volterra scheme [2].

Experimental

Suspensions with 30% mass content of kaolin in 0.5 M NaSCN solution at pH 11 have been used in viscometric measurements with the rotational viscometer Haake RS100. The concentric cylinder system is specified by DINZ20. Experiments have been performed at constant shear rates and at a temperature of 20 °C. It was painfully taken care that the suspensions were not modified by solvent evaporation. Sedimentation could not be observed for weeks.

Results

A typical viscosity–time curve is represented in Fig 1. This curve shows fluctuations of viscosity that extend over a range of several thousand seconds. Large fluctuations occur with a period length of about 2500 s. In addition, fluctuations with smaller period lengths are present. Similiar behavior was found for any experiment performed at constant shear rates in the range from 9 to 60 s^{-1}. For all experiments the magnitude of the fluctuations was found to be about 2% of the average viscosity. We approved by comparative measurements with calibration liquids that these fluctuations are not caused by the experimental setup.

Modelling

We interpret the viscosity fluctuations as a result of agglomeration and deglomeration of solid particles in

Progr Colloid Polym Sci (1998) 111:82–84
© Steinkopff Verlag 1998

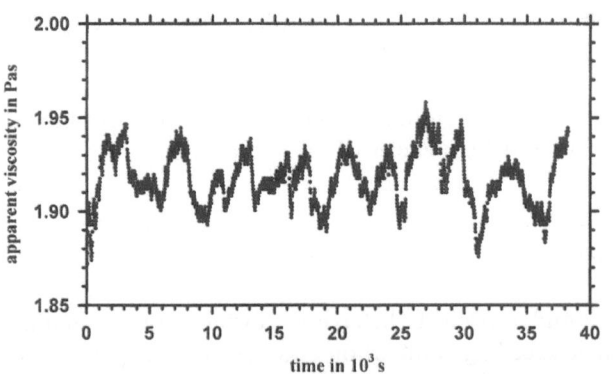

Fig. 1 Apparent viscosity of 30% kaolin suspension at constant shear rate 23 s^{-1} as function of time

concentrated suspensions. These suspensions are considered as homogeneous liquids with a constant viscosity η_0. Agglomerates formed by shear flow increase the viscosity by $\eta_1(t)$ proportional to their concentration. The suspension viscosity is given by

$$\eta(t) = \eta_0 + \eta_1(t) . \tag{1}$$

The agglomerate concentration as a function of time is needed to model the time dependency of viscosity. The agglomerate concentration is derived from the agglomeration and deglomeration kinetics. We apply the Lotka–Volterra model as a simple kinetic scheme showing oscillation behavior [3]

$$A + A^* \xrightarrow{a} 2A^* , \tag{2a}$$

$$A^* + C \xrightarrow{b} 2C , \tag{2b}$$

$$C \xrightarrow{c} \text{products} . \tag{2c}$$

The two steps (2a) and (2b) are autocatalytic processes, representing the kinetics of agglomerate formation. Particles of species A are permanently available from a reservoir (homogeneous suspension). Therefore, concentration of particles A is constant, whereas the concentration of A* and C changes as a function of time. Step (2a) is considered to be an activation process only, such that viscosity is not effected by the concentration of A*. Agglomerates C are uniquely responsible for viscosity alternation in time according to

$$\eta_1(t) = \text{const.} \ Y(t) , \tag{3}$$

where $Y(t)$ is the concentration of C. Agglomerate formation is induced by the shear field energy. Step (2c) describes the decay of agglomerates C due to energy and impulse

transfer. Decay products can finally be transformed into particles A. From (2a)–(2c) follows the system of differential equations for the concentrations X of species A* and Y of species C

$$\frac{dX}{dt} = aX - bXY , \tag{4a}$$

$$\frac{dY}{dt} = bXY - cY , \tag{4b}$$

where a, b and c are adjustable parameters, which correspond to the rate constants in (2a)–(c). Solutions $X(t)$ and $Y(t)$ of Eqs. (4a) and (b) are periodic functions of time. Thus, concentration of C and $\eta_1(t)$ change periodically. Initial conditions are chosen according to the experimental value of the viscosity. In order to fit the rate constants in Eq. (4) we applied Fourier transformation to the experimental viscosity data. A significant part of the Fourier spectrum of the data presented in Fig. 1 is shown in Fig. 2. Therein, intensity is decreasing towards higher frequencies. Any amplitudes below a threshold value are treated as noise. Using the frequency and amplitude of the highest peak the parameters of the Lotka–Volterra model are adapted. Curve 1 in Fig. 3 was calculated with these

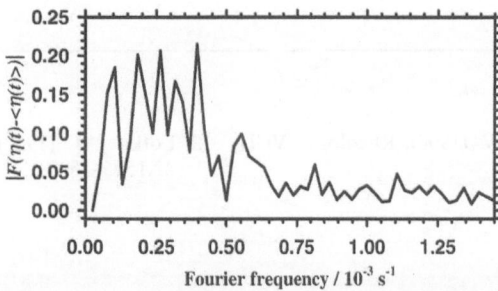

Fig. 2 Part of the Fourier spectrum of experimental data shown in Fig. 1

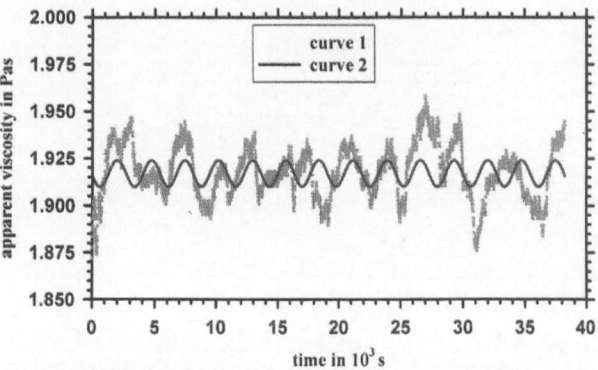

Fig. 3 Experimental viscosity curve (1) and calculated viscosity curve (2) based on a single Lotka–Volterra process

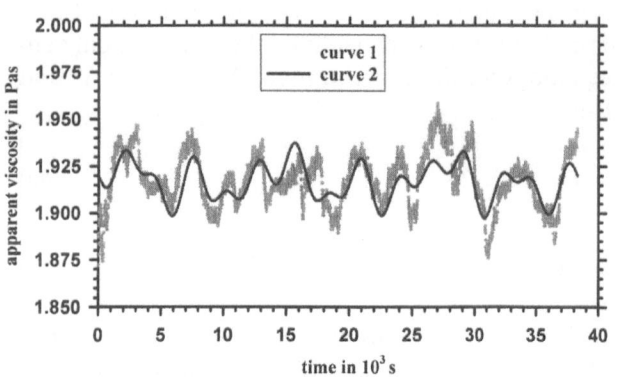

Fig. 4 Experimental viscosity curve (1) and calculated viscosity curve (2) based on four Lotka–Volterra processes

parameters. Comparison of curve 1 with curve 2 (experimental data) shows an excellent correspondence of experimental and fitted periods. The amplitudes, however, are not well fitted by this model. It is necessary to account for more frequencies of the spectrum to improve the amplitude fitting in our model. Henceforth, a single Lotka–Volterra process was related to each frequency. In our model, different Lotka–Volterra processes are distinguished by variation of species C into C_i ($i = 1, \ldots, n$) in Eq. (2b). The assumption of the existence of different species C_i is in accordance with the polydispersity of the suspension. Lotka–Volterra processes with different C_i are described by mutually independent differential equation systems. Figure 4 shows the curve derived from our model by using four frequencies of the spectrum.

Conclusions

The complex viscosity behavior is explained by assuming shear-induced agglomeration and deglomeration processes. Applying the Lotka–Volterra model, agglomeration kinetics can be described such that aggregate concentrations change periodically in time. Because of the polydispersity of suspensions, superimposition of several Lotka–Volterra processes is required to optimize the fit of experimental viscosity–time curves. In principle, the Lotka–Volterra model can be replaced by other reaction schemes including autocatalytic steps. The advantage of the Lotka–Volterra model is the small number of parameters.

Acknowledgement This work was supported by funding from the Deutsche Forschungsgemeinschaft through the Grant SFB 285 "Particle Technology". We further acknowledge financial support from the Fonds der Chemischen Industrie.

References

1. Macosco CW (1994) Rheology. VCH, Weinheim

2. Lotka AJ (1920) J Am Chem Soc 42:1595–1599

3. Haken H (1983) Synergetics, An Introduction. Springer, Berlin

Progr Colloid Polym Sci (1998) 111:85–91
© Steinkopff Verlag 1998

L. Fei
M. Szymula
S.E. Friberg
P.A. Aikens

Vapor pressure of phenethyl alocohol and phenethyl acetate in the system with water and nonionic surfactant – Polyoxyethylene 4 lauryl ether (Brij®30)

L. Fei · M. Szymula · Prof. S.E. Friberg (✉)
Department of Chemistry
Clasrkson University
Box 5810
Potsdam, New York 13699-5810
USA

P.A. Aikens
ICI Surfactants
Concord Plaza, Bedford Bldg.
3411 Silverside Road
Wilmington, DE 19850-5391
USA

Abstract The limit of water solubilization in solutions composed of a commercial surfactant, Laureth 4 (Brij®30), and a fragrance combination of phenethyl acetate (PEAc) and phenethyl alcohol (PEA) was determined. The vapor pressures of both the fragrances along these borders and along the fragrance-surfactant axis were measured. The variation of the fragrance vapor pressures for the surfactant/fragrance solution series with or without water was related to surfactant self-association, to formation of inverse micelles, and to the competition between PEAc and PEA molecules for the surfactant polar groups. The results show that PEA and PEAc almost ideally dissolved in each other, but in the presence of the surfactant, the addition of PEA significantly increases the chemical potential of PEAc.

Key words Surfactants – fragrances – vapor pressure – liquid crystals – emulsions

Introduction

Vapor pressure and vapor composition in equilibrium with a colloidal solution provides important theoretical and practical information. This equilibrium, as a reflection of the chemical potential of volatile compounds, provides a means of studying their thermodynamic properties and for developing and testing model colloid systems. A number of investigations have been performed on this important topic [1–9].

The practical uses of vapor pressure measurements in surfactant systems include: evaluation of liquid–vapor equilibrium partitioning of "hydrophobic" organic contaminants for soils and aquifers remedy [10]; investigation for the development of aqueous fire-resistant diesel fuel microemulsions [4]; and estimation of vapor pressure of fragrance components in model systems for personal care products [11–13].

The present publication expands the evaluation to fragrance mixtures in a water–nonionic surfactant system.

Since nearly all the personal care products are actually colloid systems and the fragrance perception is decisive for consumer acceptance, the relationship between the colloidal phenomena in the formulated product and the fragrance vapor pressure is essential.

However, the literature concerning fragrance compund vapor pressure versus composition is surprisingly limited. Some studies are concerned with the conditions to prepare fragrance solutions without volatile organic solvents [14–16], while others focus on the vapor pressure variation in an entire water–nonionic surfactant-single fragrance system [11, 12]. Abe et al. have studied the solubilization of several systems of water, surfactant, and synthetic perfumes [17], clarifying the distribution coefficient between micelles and the bulk phase [18]. They have recently related the partition between dissolved and solubilized perfume components to their volatility [19].

In this article, we investigate the influence on the vapor pressure of one fragrance compound by the addition of another fragrance compound to a system of water and nonionic surfactant as a first introduction to the clarification

86

L. Fei et al.
Vapor pressure of fragrance combinations

of the behavior of the complex combinations of compounds which form a fragrance.

Experimental

Materials

Phenethyl alcohol (PEA) and phenethyl acetate (PEAc), 99%, Aldrich Chemical Co., Milwaukee, WI; Laureth 4 ($C_{12}EO_4$) (Brij®30), ICI Surfactants, Wilmington, DE; water, doubly distilled, deionized.

Isotropic liquid solution regions

The isotropic liquid solution border was determined by visual observation of samples during gradual addition of one liquid component.

Vapor pressure

The equilibrium vapor pressure of phenethyl alcohol and phenethyl acetate in each sample was measured by Headspace Gas Chromatography, which has been described in detail in previous articles [11, 12]. The retention time of

phenethyl alcohol and phenethyl acetate was 7.6 and 10.4 min, respectively.

Data analysis

Experimental data were fit with polynomial functions and for some selected series cofidence band curves at 90% confidence are also shown in Fig. 2A, B and 5B calculated by function data analysis 4.1, Microcal Origin 4.1, on a Quantex 486 computer.

Results

The water solubilized, surfactant/fragrance(s) isotropic liquid solution regions of five systems of water, surfactant-Laureth 4, and phenethyl alcohol; water, Laureth 4, and phenethyl acetate; as well as water, Laureth 4, and phenethyl alcohol/phenethyl acetate mixtures with different alcohol/acetate weight ratio, 3/1, 1/1, or 1/3 are shown in Fig. 1. The solubility of water in the surfactant is about 12%. Water is soluble in phenethyl alcohol to 7.5%, but only has insignificant solubility in phenethyl acetate. When 25% of the alcohol was replaced by acetate, the water solubility decreased to about 5%; when half of the alcohol was replaced, fragrances mixture dissolved up to

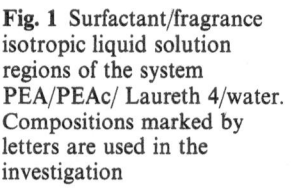

Fig. 1 Surfactant/fragrance isotropic liquid solution regions of the system PEA/PEAc/ Laureth 4/water. Compositions marked by letters are used in the investigation

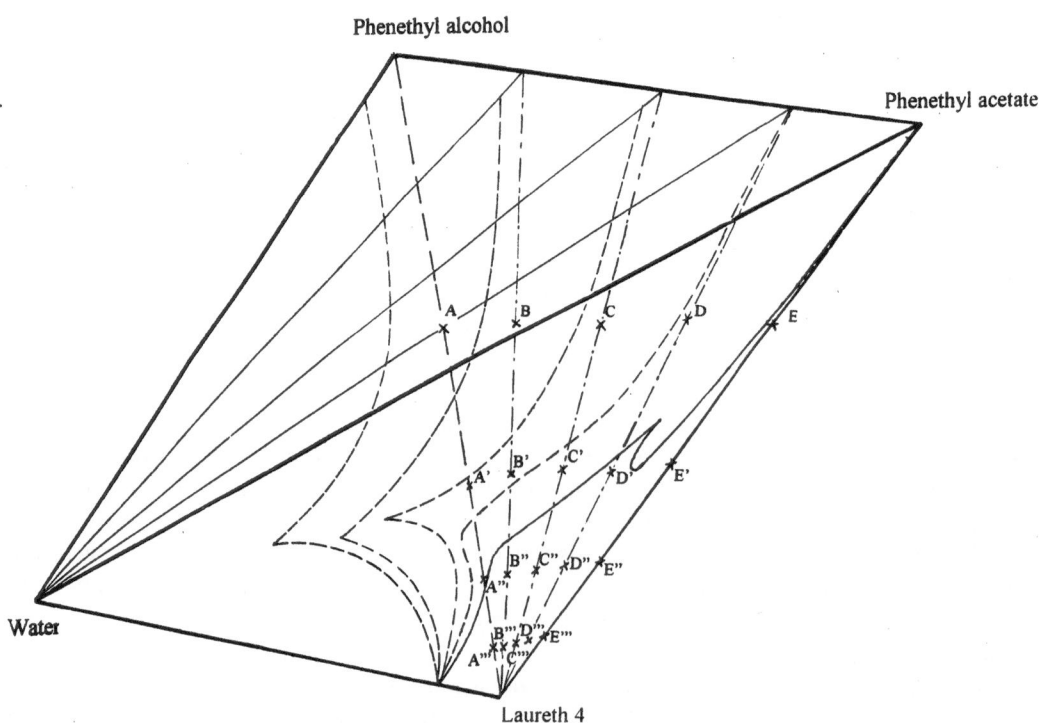

Progr Colloid Polym Sci (1998) 111:85–91
© Steinkopff Verlag 1998

3% water by weight; when 75% of alcohol was replaced, only about 1% of water would saturate the fragrances mixture. The water was initially dissolved in the fragrance or fragrances mixture without surfactant; with increased amount of surfactant it became solubilized into inverse micelles at higher surfactant amounts [11]. The maximum value, appearing at fragrance/surfactant ratio of 3/7, was 46% of water by weight in the system with only phenethyl alcohol as the fragrance. When alcohol/acetate weight ratio becomes 3, this value was reduced to 35%. When fragrances mixture has the same amount of phenethyl alcohol and phenethyl acetate, this value became 33%, and the maximum shifted to a fragrance mixture/surfactant weight ratio of 3.5/6.5. The maximum solubility of water, which appears at fragrances/surfactant ratio of 3.25/6.75, was 25% when three quarters of phenethyl alcohol was replaced by phenethyl acetate; and 21% for pure phenethyl acetate standing as the fragrance at a PEAc/surfactant ratio in the range of 3/7–2.5/7.5.

The vapor pressure of phenethyl alcohol or phenethyl acetate in their mixtures is found in Fig. 2A and B. The statistical treatment shows the relationship between the vapor pressure and the mole fraction in each case to be slightly greater than that of an ideal solution.

In Fig. 3A, four curves present the vapor pressure behavior of phenethyl alcohol, when the surfactant was added to the fragrance mixtures. The pure phenethyl alcohol was used as the standard state pressure p_0. All the curves except for the one with PEA/PEAc weight ratio of 1/3 are characterized by an initial slightly negative deviation from the ideal solution, while in the surfactant-rich part vapor pressures slightly higher than those in an ideal solution were found. The combination with a weight ratio of 1/3 showed only negative deviation from ideal solution. The pressures for ideal solutions are shown as a dashed line.

The results of PEAc vapor pressure measurements in the same system as above are described in Fig. 3B. Each curve shows an initial negative deviation at fragrance(s) rich part, followed by a positive deviation when the surfactant becomes the main component.

The vapor pressures of PEA were measured along the borders of maximum water content in the fragrance(s)/surfactant one phase regions, and the results are given in Fig. 4A–D. In these cases, the surfctant-free, water-saturated samples were set as the standard solutions, and the mole fraction of PEA is calculated on the organic compounds only. For the PEA only, the vapor pressure values, shown in Fig. 4A, follow those for an ideal solution except for the range of high water content in which they are significantly higher. When 25% of PEA (by weight) was replaced by PEAc, Fig. 4B, the vapor pressure was close to ideal at fragrance mole fractions close to one and

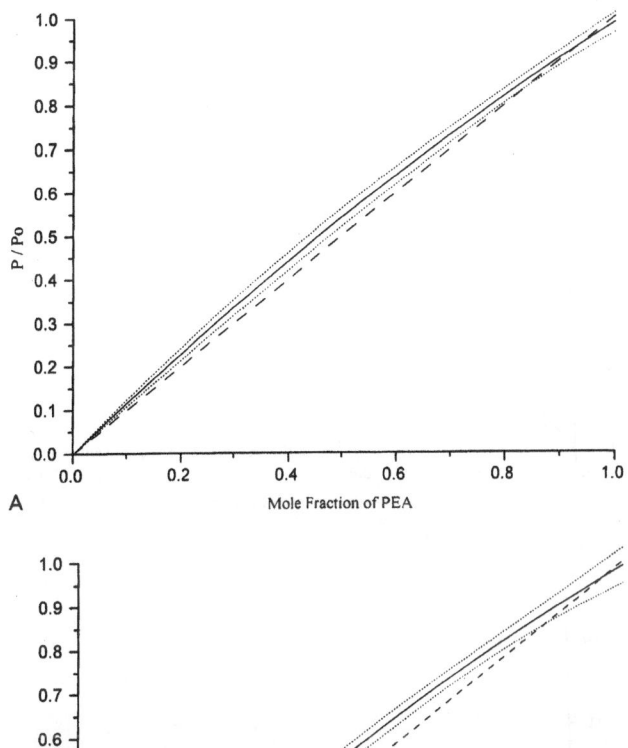

Fig. 2 (A) The vapor pressures of PEA in PEA/PEAc solutions. p_0 is the vapor pressure of pure PEA. The dashed line represents the vapor pressure behavior of an ideal solution. The point lines show the range for 90% confidence interval. (B) The vapor pressures of PEAc in PEA/PEAc solutions. p_0 is the vapor pressure of pure PEAc. The dashed line represents the vapor pressure behavior of an ideal solution. The point lines show the range for 90% confidence interval

close to zero with a negative deviation in between. The vapor pressures of PEA in the mixtures with the same amount of PEA and PEAc are reported in Fig. 4C, and those for 25% PEA, 75% PEAc mixtures are reported in Fig. 4D. The variation is similar to that in Fig. 4B except the negative deviation becomes more pronounced.

The vapor pressure of PEAc is reported in Fig. 5A–D. For PEAc only, the vapor pressures were close to those in an ideal solution. Figure 5B–D describes the vapor pressure variations of PEAc in the systems of water–Laureth 4–PEA/PEAc mixtures with decreasing PEAc/PEA ratio of 3/1, 1/1, 1/3. When the surfactant is added to each of the

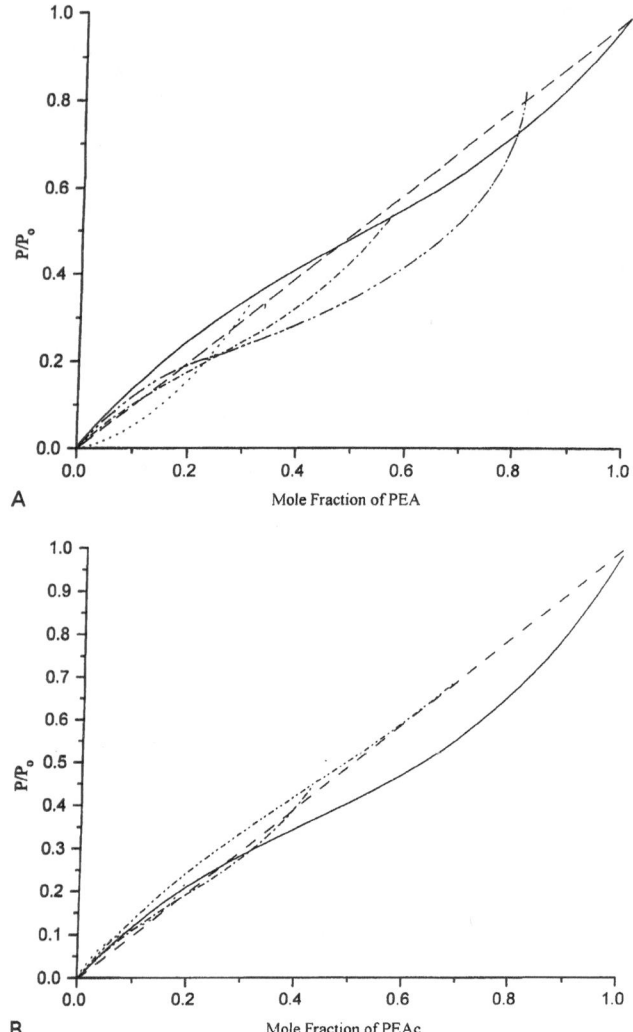

A

B

Fig. 3 (A) The vapor pressures of PEA in fragrance combination/Laureth 4 solutions. p_0 is the vapor pressure of pure PEA. The dashed line represents the vapor pressure behavior of an ideal solution. Fragrance combination: —— PEA; – ·· – 75% PEA, 25% PEAc (by weight); – · – · – 50% PEA, 50% PEAc (by weight); ···· 25% PEA, 75% PEAc (by weight). (B) The vapor pressures of PEAc in fragrance/Laureth 4 solutions. p_0 is the vapor pressure of pure PEAc. The dashed line represents the vapor pressure behavior of an ideal solution. Fragrance combination: —— PEAc; – ·· – 75% PEAc, 25% PEA; – · – · – 50% PEAc, 50% PEA; ···· 25% PEAc, 75% PEA

PEA/PEAc mixtures, the pressure variation is characterized by an initial tiny part of ideal behavior, followed by a highly positive deviation. The ideal behavior region became smaller with increased content of PEA.

Discussion

The present results provide information about the importance of intermolecular forces of specific groups on the

activity of one fragrance compound in a system of a mixture of fragrance compounds, water and an amphiphile, and also illustrate the influence of the self-association taking place in amphiphilic systems and their combination with water.

The vapor pressure of phenethyl acetate/phenethyl alcohol combinations showed an almost ideal behavior, Fig. 2A and B, due to the structural similarity between these two molecules.

As far as the interaction between the fragrance and the surfactant, Fig. 3A and B, is concerned, the hydrogen bond from phenethyl alcohol to the polyoxyethylene group in the surfactant molecule has some influence. A comparison between the vapor pressures in Fig. 3A and B reveal a much stronger negative deviation from ideal vapor pressure for the PEAc combination with surfactant than the PEA-surfactant pair, when only one fragrance compound is present. The reason for this difference is that the PEA–PEA molecular interactions is stronger than the PEAc–PEAc one and, hence, the reduction in chemical potential when dissolved with surfactant is smaller for the PEA.

On the other hand, the variation of vapor pressure with the surfactant content for a single fragrance compound in compositions without water, Fig. 3A and B, is understood using Christenson's results [20]. The initial PEA and PEAc vapor pressure reduction, Fig. 3A and B, by addition of the surfactant to the fragrance single compound is related to the effect of the surfactant in its monomeric form. After initiation of the self-association of the surfactant [20], the vapor pressure curves become more horizontal, and at high surfactant content the vapor pressures are in excess of those of an ideal solution. Those higher vapor pressures are referred to the fact that the polar groups of the surfactant are interassociated [20], leaving the interaction with the phenethyl compound mainly to the aliphatic chains of the surfactant. As a corollary it should be noted that solutions of phenethyl acetate in decane show an activity factor at the level of 8 [12] for very dilute solutions. Hence, the present vapor pressure variation of the surfactant fragrance compound solutions are satisfactorily explained by taking the weak surfactant–surfactant association into consideration.

The mutual influence of the two fragrance compounds is also illustrated by the results in Fig. 3A and B. The addition of PEA to the surfactant–PEAc combination, Fig. 3B, leads to PEAc vapor pressures more close to those for an ideal solution; that is the reduction of vapor pressure became less. This is an expected result: the PEA interaction with the surfactant polar group is stronger due to its OH group and the PEAc molecules are left to interact with the non-polar part of the surfactant and with the PEAc molecules.

Progr Colloid Polym Sci (1998) 111:85–91
© Steinkopff Verlag 1998

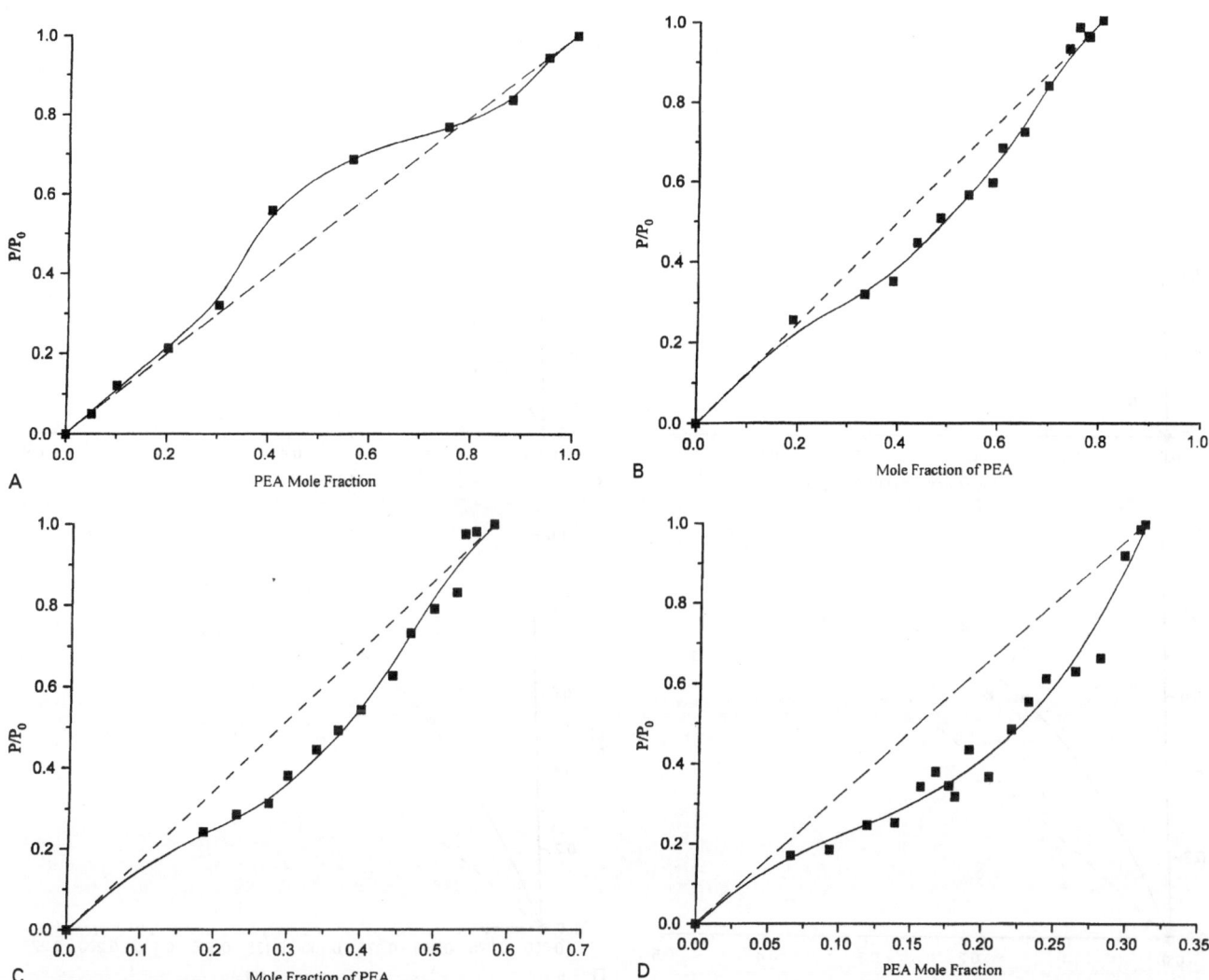

Fig. 4 (A) The vapor pressure of PEA in PEA/Laureth 4 solutions saturated with water. The mole fraction is calculated on the non-aqueous compounds only. p_0 is the vapor pressure of the fragrance saturated with water. The dashed line represents the vapor pressure behavior of an ideal solution. (B) The vapor pressure of PEA in the fragrances combination (75% PEA, 25% PEAc)/Laureth 4 solutions saturated with water. The mole fraction is calculated on the non-aqueous compounds only. p_0 is the vapor pressure of PEA in the fragrances solution saturated with water. The dashed line represents the vapor pressure behavior of an ideal solution. (C) The vapor pressure of PEA in the fragrance combination (50% PEA, 50% PEAc)/Laureth 4 solutions saturated with water. The mole fraction is calculated on the non-aqueous compounds only. p_0 is the vapor pressure of PEA in the fragrances solution saturated with water. The dashed line represents the vapor pressure behavior of an ideal solution. (D) The vapor pressure of PEA in the fragrances combination (25% PEA, 75% PEAc)/Laureth 4 solutions saturated with water. The mole fraction is calculated on the non-aqueous compounds only. p_0 is the vapor pressure of PEA in the fragrances solution saturated with water. The dashed line represents the vapor pressure behavior of an ideal solution

The behavior of the PEA vapor pressure, Fig. 3A, is opposite; addition of PEAc caused a more pronounced negative deviation from ideal solution behavior. Obviously, the presence of the less polar and hydrogen bonding PEAc allowed the PEA to partition more towards the polar part of the surfactant molecule resulting in a stronger interaction and consequently a lowered vapor pressure. For the higher PEAc ratios this trend remained also to the combinations with highest surfactant content.

The vapor pressure of PEA for solutions with maximum water content, Fig. 4A–D, shows an identical trend; enhanced PEAc/PEA ratio leads to increased reduction of the PEA vapor pressure in the range of maximum water content. The vapor pressure of PEAc, Fig. 5A–D, shows the opposite trend; a significant positive deviation from the ideal solution behavior enhanced at higher PEA/PEAc ratios. The similarity to the behavior in decane solutions [12] lends creditability to an explanation of the PEAc

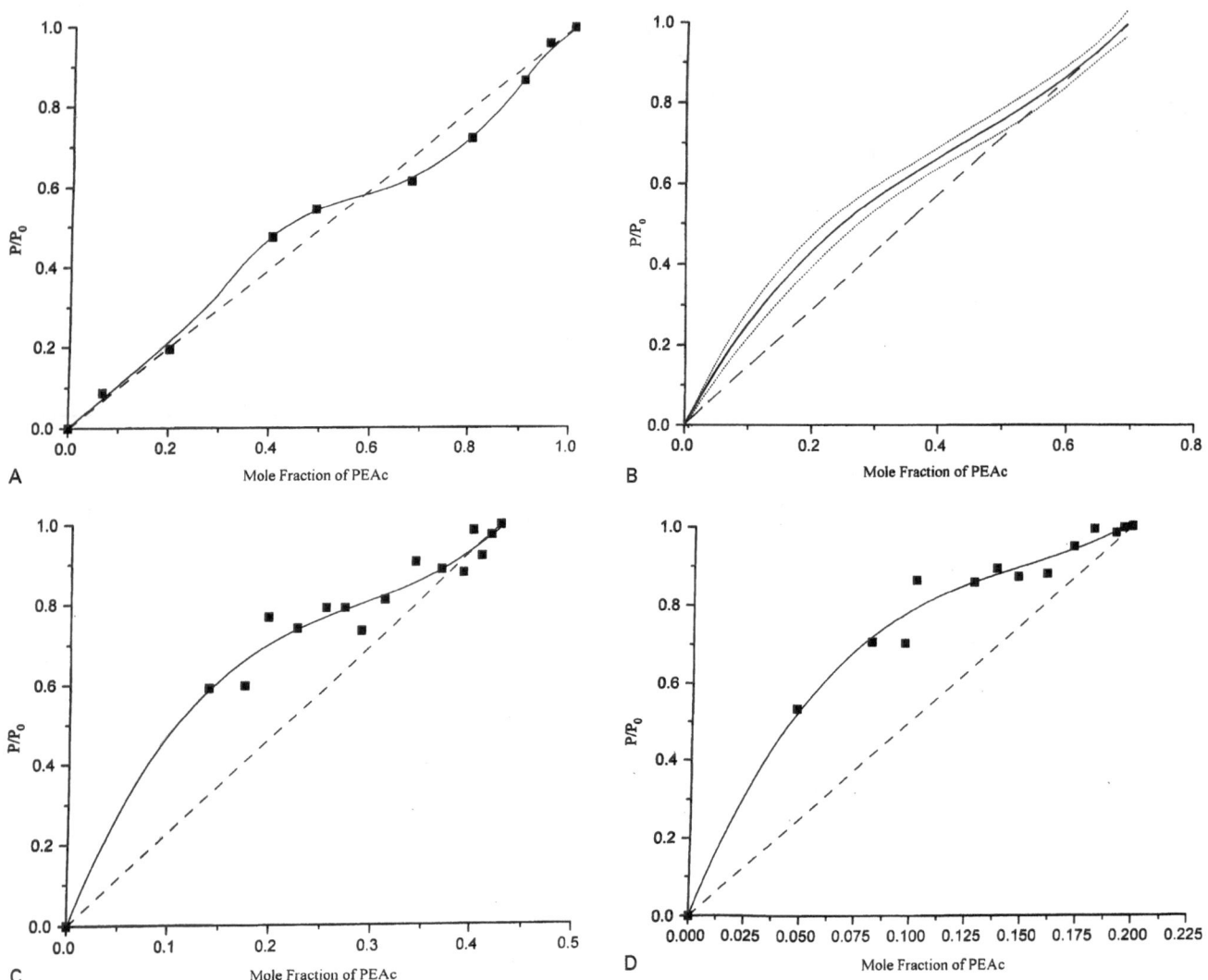

Fig. 5 (A) The vapor pressure of PEAc in PEAc/Laureth 4 solutions saturated with water. The mole fraction is calculated on the non-aqueous compounds only. p_0 is the vapor pressure of the fragrance saturated with water. The dashed line represents the vapor pressure behavior of an ideal solution. (B) The vapor pressure of PEAc in the fragrances combination (75% PEAc, 25% PEA)/Laureth 4 solutions saturated with water. The mole fraction is calculated on the non-aqueous compounds only. p_0 is the vapor pressure of PEAc in the fragrances solution saturated with water. The dashed line represents the vapor pressure behavior of an ideal solution. (C) The vapor pressure of PEAc in the fragrances combination (50% PEAc, 50% PEA)/Laureth 4 solutions saturated with water. The mole fraction is calculated on the non-aqueous compounds only. p_0 is the vapor pressure of PEAc in the fragrances solutions saturated with water. The dashed line represents the vapor pressure behavior of an ideal solution. (D) The vapor pressure of PEAc in the fragrances combination (25% PEAc, 75% PEA)/Laureth 4 solutions saturated with water. The mole fraction is calculated on the non-aqueous compounds only. p_0 is the vapor pressure of PEAc in the fragrances solution saturated with water. The dashed line represents the vapor pressure behavior of an ideal solution

molecular interaction being limited mainly to the non-polar part of the surfactant molecule.

The trends are well illustrated in Fig. 6 showing the fractional deviation from ideality at the surfactant content for maximum water solubilization, Fig. 1. Increasing the PEA/PEAc ratio leads to an increase of the vapor pressure both for PEA and PEAc. The increase in the vapor pressure of PEAc is expected; the added PEA interacts strongly with the surfactant polar groups leaving the interaction

with the more aliphatic parts for PEAc and a consequence of increase in its vapor pressure. What is note worthy is the rapid reduction of the PEA vapor pressure by addition of even comparatively small amounts of PEAc. Without PEAc one finds an increase of vapor pressure by 0.22 units; for a PEAc mole fraction of 0.2 counted only on PEA + PEAc the change is −0.20 units.

One should observe that this change is not a consequence of the reduced amount of PEA. The results are

Progr Colloid Polym Sci (1998) 111:85–91
© Steinkopff Verlag 1998

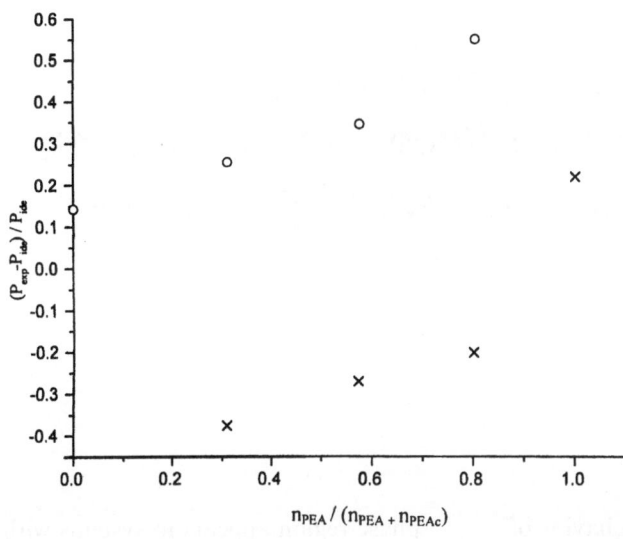

Fig. 6 The vapor pressure deviation of PEA (×) and PEAc (○) from the ideal solution value in the fragrance combination/Laureth 4 solutions with maximum solubilized water

presented as the ratio with the vapor pressure in an ideal solution and the reduction due to changed mole fraction is included. Instead, the reduction must be considered from the point of view of influence by the added PEAc.

The proof of the explanation will be presented elsewhere [21] but the conclusions are rational; the added PEAc molecules associate with the PEA ones and the molecular interaction from association structure is more attuned to the aliphatic parts of the colloidal structure relaxing the vapor pressure increase due to the water–polar group interface in the inverse micelles.

References

1. Biais J, Bortherel P, Clin B, Lalanne P (1981) J Colloid Interface Sci 80: 136–145
2. Biais J, Ödberg L, Stenius P (1982) J Colloid Interface Sci 86:350–357
3. Damaszewski L, Mackay RA (1984) J Colloid Interface Sci 97:166–175
4. Weatherford Jr WD, Naegeli DW (1984) J Disp Sci Technol 5:159–177
5. Weatherford Jr WD (1985) J Disp Sci Technol 6:467–488
6. Sjöblom E, Jönsson B, Jönsson A, Stenius P, Saris P, Ödberg L (1986) J Phys Chem 90:119–124

7. Linse P (1986) J Phys Chem 90: 6821–6828
8. Chew CH, Wong MK (1991) J Disp Sci Technol 12:495–501
9. Li P, Han B, Yan H, Liu R (1996) J Chem Eng Data 41:285–286
10. Anderson MA (1992) Environ Sci Technol 26:2186–2194
11. Friberg SE, Huang T, Fei L, Vona Jr SA, Aikens PA (1996) Progr Colloid Polym Sci 101:18–22
12. Friberg SE, Fei L, Aikens PA (1997) J Molecular Liquids 72:31–53
13. Friberg SE, Szymula M, Fei L, Borber J, Al-Bowaab A, Aikens PA (1997) Int J Cosmet Sci 19:259–270

14. US Patent 5,283,056, 1994
15. Gareiss J, Fussbroich P, Ghyczy M (1994) SOFW J 120:93–97
16. Japan Patent 07,238,086, 1993
17. Tokuoka Y, Uchiyama H, Abe M (1993) Colloid Polym Sci 272:317–323
18. Tokuoka Y, Uchiyama H, Abe M (1995) Langmuir 11:725–729
19. Saito Y, Miura K, Tokuoka T, Kondo Y, Abe M, Sato T (1996) J Disp Sci Technol 17:567–576
20. Christenson H, Friberg SE (1980) J Colloid Interface Sci 75:276
21. SE Friberg, Marin V (in preparation)

Progr Colloid Polym Sci (1998) 111:92–99
© Steinkopff Verlag 1998

C. Stubenrauch
S.K. Mehta
B. Paeplow
G.H. Findenegg

Microemulsion systems based on a $C_{8/10}$ alkyl polyglucoside: A reentrant phase inversion induced by alcohols?

C. Stubenrauch · S.K. Mehta* · B. Paeplow
G.H. Findenegg (✉)
Iwan-N.-Stranski-Institut für Physikalische
und Theoretische Chemie
TU Berlin
Straße des 17. Juni 112
D-10623 Berlin
Germany
E-mail: findenegg@chem.tu-berlin.de

* Permanent address
Department of Chemistry
Punjab University
Chandigarh 160014
India

Abstract The phase behavior of quaternary microemulsion systems with a technical-grade nonionic surfactant ($C_{8/10}$ alkyl polyglucoside) and different medium-chain alcohols (including the monoterpene geraniol) as cosurfactants is studied at 25 °C. Contour diagrams of the three-phase body at constant oil-to-water ratio have been mapped in terms of the mass fraction variables γ (surfactant + alcohol) and δ (relative amount of alcohol in the surfactant + alcohol mixture). Significant differences in the phase behavior of systems based on the pure alkyl glucoside β-C_8G_1 and those with the technical-grade $C_{8/10}$APG are observed. Specifically, at high δ an anomalous shape of the three-phase region, or a second narrow three-

phase region appears in systems with $C_{8/10}$APG, which is absent in systems based on β-C_8G_1. Whereas at $\delta = 0.5$ all systems exhibit a phase inversion $\underline{2}$–3–$\bar{2}$ on increasing γ, a retrograde transition $\underline{2}$–3–$\underline{2}$ is observed at high δ. The anomalous shape of the three-phase body can lead to a phase sequence $\underline{2}$–3–$\bar{2}$–3–$\underline{2}^*$ with a reentrant $\underline{2}^*$ state, on increasing δ at constant γ. It is argued that the reentrant $\underline{2}^*$ state is due to a preferential extraction of the more lipophilic oligomers of the technical-grade surfactant into the oil phase, causing the interfacial film to become more hydrophilic again, as δ is increased.

Key words Microemulsions – alkyl polyglucosides – retrograde phase inversion

Introduction

In three-component water–oil–surfactant systems with nonionic surfactants of the alkyl polyglycol ether (C_nE_m) family, a phase inversion $\underline{2}$–3–$\bar{2}$ can be promoted by raising the temperature as shown in Fig. 1a. At low temperatures ($T < T_l$) a water-rich microemulsion is in equilibrium with an excess oil phase ($\underline{2}$) whereas at high temperatures ($T > T_u$) an oil-rich microemulsion coexists with an excess water phase ($\bar{2}$). At surfactant concentrations (mass fractions) $\gamma < \bar{\gamma}$ this phase inversion is connected with the appearance of a three-phase region (3) in

which a middle-phase microemulsion coexists with excess oil and water phases. The surfactant concentration $\bar{\gamma}$ (see point X in Fig. 1a) represents the lowest concentration needed to solubilize equal amounts of water and oil in one single phase and thus $\bar{\gamma}$ is a measure of the surfactant's efficiency. The pronounced temperature dependence of the phase behavior in this class of systems can be attributed to the temperature-dependent hydration of the head-group of C_nE_m surfactants. An increase in temperature causes a partial dehydration of the oxyethylene chain of C_nE_m molecules, which has two closely related effects: (i) a decrease of the effective size of the polar head group of the surfactant, and thus a decrease of the spontaneous curvature of the

Progr Colloid Polym Sci (1998) 111:92–99
© Steinkopff Verlag 1998

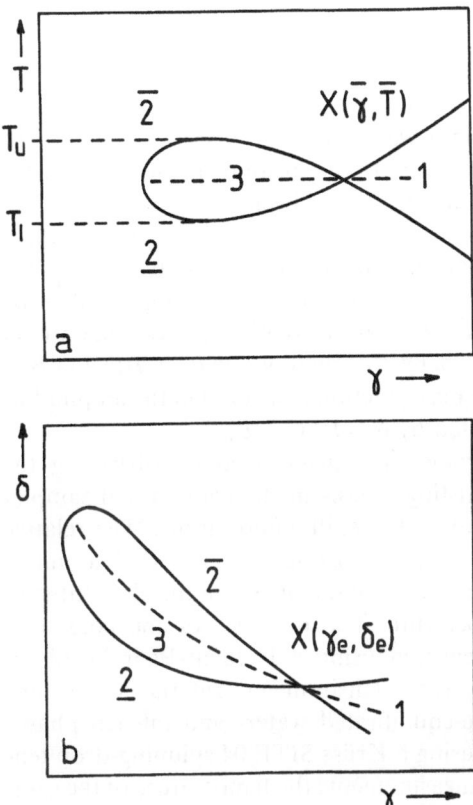

Fig. 1 Contour of the three-phase region at a 1:1 oil-to-water ratio, i.e. at constant α, for microemulsion systems: (a) γ vs. temperature T for a ternary system; (b) γ vs. δ for a quaternary system ($T =$ const.). Point X marks the lowest surfactant concentration ($\bar{\gamma}$ or γ_e) needed to solubilize equal amounts of water and oil in a microemulsion phase. The broken lines mark the locus of mid-points of the three-phase bodies and correspond to the HLB line of the systems for the given α

surfactant film at the oil/water interface; (ii) an increase of the effective hydrophobicity of the surfactant, which promotes its oil solubility. However, the resulting temperature sensitivity of microemulsion systems containing C_nE_m surfactants can be a disadvantage in certain fields of application.

Microemulsions exhibiting only a weak temperature sensitivity are obtained by using sugar-based surfactants such as alkyl mono- or polyglucosides (C_nG_m with $m \geq 1$), or sucrose fatty acid esters [1, 2]. In microemulsion systems containing C_nG_m-surfactants, a phase inversion $\underline{2}$–3–$\bar{2}$ can occur when cosurfactants such as alcohols [1, 3, 4] or alkyl glycerol ethers [5] are added. In the resulting four-component systems, temperature has only a weak influence on the width of the three-phase region [1] as well as on other properties of the system [5, 6]. Presumably, the temperature insensitivity of C_nG_m systems results from the strength of the hydrogen bonds between the

hydroxy groups of the glucose moiety and water molecules, which prevents any significant dehydration of the head group in the relevant temperature range. In this case, the hydrophile–lipophile balance of the surfactant film can be tuned by the addition of an appropriate cosurfactant. The phase diagram resulting by addition of increasing amounts of a lipophilic cosurfactant is shown schematically in Fig. 1b; here the variable γ represents the sum of the mass fractions of surfactant and alcohol in the quaternary system, and δ is the mass fraction of alcohol in the surfactant plus alcohol mixture. Lipophilic alcohols have been shown to act both as a cosurfactant and as a cosolvent in such systems [3, 4, 7–9]. Here, a *cosurfactant* is defined as an additive that is incorporated into the surfactant film and thus affects its spontaneous curvature, while a *cosolvent* remains in the bulk phase and improves its ability to solubilize further components. Qualitatively, the addition of an alcohol to a water–oil–C_nG_m system has the same effects as an increase of temperature in a ternary water–oil–C_nE_m system. A striking difference between the two schematic phase diagrams shown in Fig. 1 is the "distortion" of the three-phase body in the γ, δ-plane of the quaternary system (Fig. 1b), which is a signature of such quaternary microemulsion systems [1, 3, 4, 7, 8], and is due to the competition between the incorporation of the alcohol into the interfacial film and its solubility in the bulk phases. Lipophilic alcohols have a tendency to remain dissolved in the oil and thus they are not fully available as cosurfactant when the amphiphile mixture (surfactant + alcohol) is diluted with equal amounts of water and oil. In other words, as γ is decreased an increasing fraction of the alcohol is extracted from the surfactant film into the oil phase. This selective extraction of the alcohol from the amphiphile mixture leads to a decreasing lipophilicity of the interfacial film. In order to compensate for this effect, more alcohol is needed in the amphiphile mixture (i.e. larger δ) to obtain a *balanced* interfacial film, which is believed to exist in the three-phase region.

Recently, we have presented a quantitative analysis of balanced microemulsions in the quaternary system water–cyclohexane–octyl monoglucoside (β-C_8G_1)–geraniol (trans-3,7-dimethyl-2,6-octadien-1-ol) [4]. In that study, the contour of the three-phase body at a 1:1 oil-to-water volume ratio has been mapped at 25 °C in terms of the variables γ and δ. The phase behavior found for this system is exactly of the type sketched in Fig. 1b. It was possible to determine the composition of the balanced surfactant film in the middle of the three-phase body by taking into account the different solubilities of β-C_8G_1 and geraniol in the oil phase. For the balanced interfacial film a ratio of geraniol to β-C_8G_1 molecules of about 2:5 was found [4].

Microemulsions supported by $C_n G_m$-surfactants are of interest for several applications due to the favourable properties of these surfactants [10]. Meaningful quantitative investigations of the physicochemical properties of $C_n G_m$ systems require chemically pure surfactants. Stereoselective methods [11] have been used on a laboratory scale to synthesize a variety of model substances for this purpose. On the other hand, technical alkyl polyglucosides such as $C_{8/10}APG$ represent complex mixtures of different chain length of the alkyl group and different degree of glucosidation. Such technical mixtures commonly exhibit favourable properties for formulations in common products [10]. Kahlweit et al. found that the efficiency of the expensive pure β-$C_{10}G_1$ is about the same as that of the inexpensive technical product $C_{10/12}G_{1.3}$ [1]. Such a correlation between the properties of certain pairs of pure and technical surfactants is of considerable importance for an optimization of commercial products.

In this paper we present a study of the phase behavior of microemulsions containing the technical alkyl polyglucoside $C_{8/10}APG$ as the surfactant and different alcohols as cosurfactants. We shall discuss the effect of substituting pure β-C_8G_1 in a selected four-component system (water–cyclohexane–β-C_8G_1–geraniol) [4] by the technical $C_{8/10}APG$. Furthermore, the influence of different oils and alcohols on the efficiency of the $C_{8/10}APG$/alcohol mixture will be investigated. Finally, the shape of the three-phase body in different water–oil–$C_{8/10}APG$–alcohol systems will be presented, which is found to be strikingly different in microemulsions with pure $C_n G_m$-surfactants and with the technical product $C_{8/10}APG$.

Experimental

Materials

The alkyl polyglucoside $C_{8/10}APG$ (Glucopon 225 [10]), synthesized by Henkel KGaA, Düsseldorf, was used as received. This technical product consists of 70% surfactant and 30% water, with the surfactant being a mixture of alkyl mono- and oligoglucosides with an average composition of $C_{8/10}G_{1.7}$. Decane ($>99\%$) and n-pentanol ($>99\%$) were purchased from Fluka, n-butanol ($>99.5\%$), n-hexanol ($>98\%$) and n-heptanol ($>99.5\%$) from Merck-Schuchardt, geraniol (trans-3,7-dimethyl-2,6-octadien-1-ol; $M = 153.25$; 98%) from Aldrich-Chemie and cyclohexane ($>99\%$) from Riedel-de Haën. These materials were used without further purification. The water used in this study was distilled and passed through a Milli-Q pure-water system.

Methods

Samples of the four-component systems were prepared by weight in Teflon-sealed glass tubes and allowed to equilibrate at 25 °C in a water bath for at least one week. To characterize the overall composition of the samples the notations introduced by Kahlweit and coworkers were used [13]. Thus, the mass fraction of oil in the binary system water (A) plus oil (B) is denoted by the symbol $\alpha = B/(A + B)$, the mass fraction of the amphiphile mixture, i.e. surfactant (C) plus alcohol (D), in the quaternary system is denoted by the symbol $\gamma = (C + D)/(A + B + C + D)$, and the mass fraction of alcohol in the amphiphile mixture is denoted by $\delta = D/(C + D)$.

Phase diagrams were determined by observing the number of coexisting phases in the equilibrated samples for a large number of overall compositions. The relative phase volumes of the coexisting phases were determined by measuring the height of the phases in the glass tubes at the given temperature. The densities of the coexisting phases were measured using a Heraeus-Paar DMA 40 digital vibrating-tube densitometer. Interfacial tensions between the pre-equilibrated water- and oil-rich phases were measured using a Krüss SITE 04 spinning-drop tensiometer. In all measurements the temperature of the samples was controlled to within ± 0.2 K.

Results and discussion

The system water–cyclohexane–$C_{8/10}APG$–geraniol

The phase diagram of the system water–cyclohexane–$C_{8/10}APG$–geraniol was investigated and will be compared with that of the system water–cyclohexane–β-C_8G_1–geraniol [4] in order to assess the effect of replacing a pure surfactant (β-C_8G_1) by a technical product ($C_{8/10}APG$). Figure 2a shows a section through the phase prism at an 1:1 oil-to-water volume ratio (corresponding to a mass ratio $\alpha = 0.44$) in terms of the composition variables γ and δ at 25 °C. At the chosen value of α, three liquid phases coexist for γ values in the range $0.03 \leq \gamma \leq 0.15$; the corresponding range of δ varies significantly with γ, viz. from $0.87 \leq \delta \leq 0.42$. As outlined in the Introduction the distorted shape of the three-phase body in Fig. 2a is a direct consequence of the solubility of the alcohol in the bulk oil phase.

In order to clarify the nature of the coexisting phases, the relative phase volumes and densities of the coexisting phases, as well as the interfacial tension between the water- and the oil-rich phases, were measured as a function of the tuning variable δ at the chosen oil-to-water ratio, $\alpha = 0.44$, and at a fixed overall concentration of amphiphile, viz.

Progr Colloid Polym Sci (1998) 111:92–99
© Steinkopff Verlag 1998

Fig. 2 Phase diagrams of four quaternary systems water + oil + $C_{8/10}$APG + alcohol in terms of the variables γ and δ, at a 1:1 oil-to-water volume ratio at 25 °C. The diagrams refer to the following combinations of oil and alcohol components: (a) + (b) cyclohexane, (c) + (d) decane; (a) + (c) geraniol, (b) + (d) hexanol. A 1:1 oil-to-water volume ratio corresponds to $\alpha = 0.44$ for cyclohexane and $\alpha = 0.42$ for decane

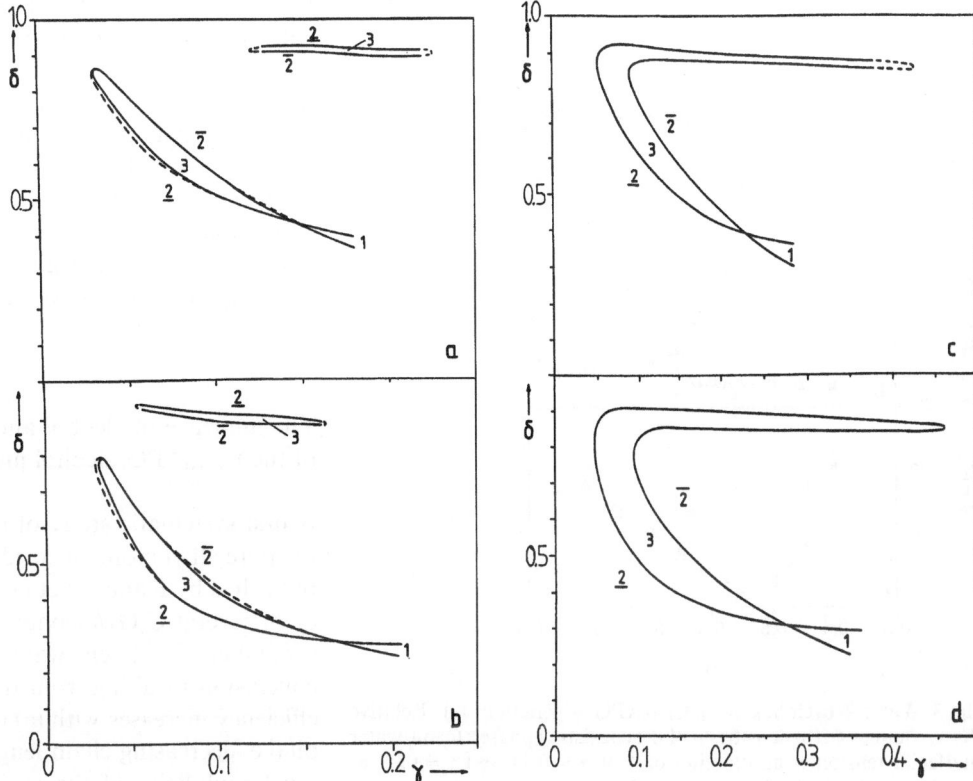

$\gamma = 0.09$. Figure 3a shows the resulting relative phase volumes, Φ, as a function of δ: at the chosen value of γ three phases coexist in a range $0.55 < \delta < 0.63$ and two phases outside this range. At $\delta < 0.55$ the volume of the lower phase, and at $\delta > 0.63$ the volume of the upper phase amounts to *ca.* 60%, indicating a transition from a water-rich to an oil-rich microemulsion ($\underline{2}$–3–$\overline{2}$) as expected from earlier studies [3, 4]. Figure 3b shows the densities of the coexisting liquid phases as a function of δ ($\alpha = 0.44$, $\gamma = 0.09$). These results are consistent with the presumed phase inversion: At low relative content of alcohol in the amphiphile mixture ($\delta < 0.55$), the density of the oil-rich phase nearly coincides with the density of pure cyclohexane ($\rho_B = 0.773 \times 10^3$ kg m^{-3} at 25 °C) while the density of the water-rich phase varies with δ and becomes significantly less than the density of pure water ($\rho_A = 0.997 \times 10^3$ kg m^{-3} at 25 °C) as the three-phase region is approached from below. Such a behavior is expected for an oil-in-water microemulsion in contact with an excess oil phase (two-phase region $\underline{2}$). At high relative content of alcohol ($\delta > 0.63$), the density of the water-rich phase is almost equal to that of pure water while the density of the oil-rich phase is significantly greater than that of pure oil, and increases as the three-phase region is approached from above. This behavior is expected for a water-in-oil microemulsion in contact with an excess

water phase (two-phase region $\overline{2}$). In the three-phase region the density of the middle phase exhibits a pronounced dependence on δ as to be expected for a gradual transformation of a water-rich into an oil-rich microemulsion. Thus, the dependence of the densities of the coexisting phases on the tuning variable δ shown in Fig. 3b is a signature of the phase inversion $\underline{2}$–3–$\overline{2}$. Figure 3c shows the interfacial tension, σ, between the coexisting water-rich and oil-rich phases plotted as a function of δ along the same linear path ($\alpha = 0.44$, $\gamma = 0.09$). One notes that the tension exhibits a minimum within the region of three-phase coexistence, with $\sigma_{min} \approx 2$–5×10^{-6} N m^{-1} at $\delta = 0.57 \pm 0.02$. The appearance of such a minimum in σ is analogous to the minimum in the temperature dependence of the oil/water tension in water–oil–C_nE_m microemulsions, where the temperature represents the tuning variable [14]. A minimum of σ implies a change of the mean curvature H of the amphiphile film from positive to negative and thus a phase inversion [15]. From the present results we infer that H is positive (oil-in-water droplets) at low δ but negative (water-in-oil droplets) at high δ; a mean curvature $H = 0$ corresponds to a bicontinuous microstructure of the microemulsion at and near the minimum of σ.

It is of interest to compare these results with the findings for the corresponding system with β-C_8G_1 [4] instead

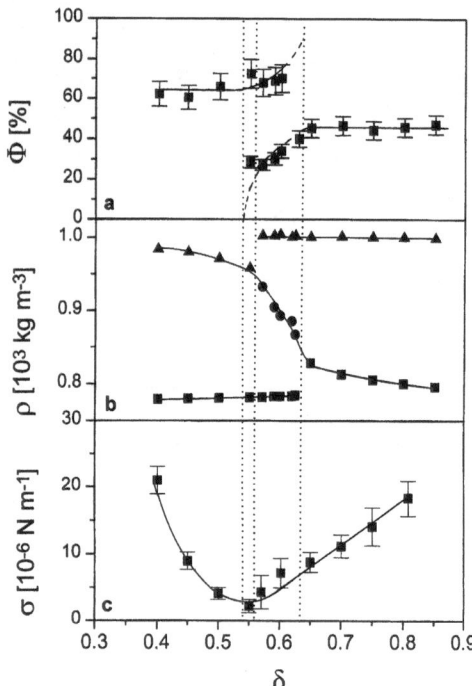

Fig. 3 Water + cyclohexane + $C_{8/10}$APG + geraniol: (a) Relative phase volumes Φ; (b) densities ρ of the coexisting phases; (c) oil/water interfacial tension σ as function of δ at $\alpha = 0.44$ and $\gamma = 0.09$ at $T = 25\,°C$. The lines connecting the data points are drawn to guide the eye, the vertical lines mark the limits of the three-phase region for the chosen α and γ

of $C_{8/10}$APG. Whereas at low δ the shape of the three-phase body and the results for the dependence of Φ, ρ and σ on the variable δ is in qualitative agreement for the two systems, a second three-phase region appears at high δ-values in the $C_{8/10}$APG system, which has not been observed with pure surfactants. We presume that the existence of this second three-phase region is caused by the fact that $C_{8/10}$APG represents a mixture of different surfactants; we return to this point later. In both systems, the single-phase microemulsion is reached at a concentration $\gamma_e \approx 0.15$. However, as $C_{8/10}$APG contains only 70% surface active material this result indicates that the active components of $C_{8/10}$APG are actually more efficient than β-C_8G_1, although the hydrophilicity of the two materials is similar. A higher efficiency of the surfactant/cosurfactant mixture is expected to cause a shrinkage of the three-phase region in the γ–δ plane (i.e. a decrease in its extension in both the directions γ and δ), and a lowering of the oil/water interfacial tension. These trends are indeed observed when the present system is compared with the corresponding system with pure β-C_8G_1 [4]. Specifically, the minimum of the interfacial tension of the present system ($\sigma_{min} \approx 2\text{--}5 \times 10^{-6}\,N\,m^{-1}$) is nearly one order of magnitude lower than for the system with β-C_8G_1

($\sigma_{min} \approx 20 \times 10^{-6}\,N\,m^{-1}$) [4]. This result complies with the well-known fact that surfactant mixtures are more efficient than pure surfactants with regard to their interfacial performance and other physicochemical properties, and this behavior is commonly attributed to synergistic effects among the different components of the mixture [16]. In spite of these differences, the present comparison shows that quantitative studies of microemulsion systems containing pure alkyl glucosides can help to predict the properties of systems based on technical-grade APG surfactants and vice versa.

The influence of alcohol and/or oil on the efficiency of the $C_{8/10}$APG/alcohol mixture

A first systematic study of microemulsions supported by the pure alkyl monoglucosides β-$C_{10}G_1$ and β-$C_{12}G_1$ was published by Kahlweit et al. [1]. In that work the efficiency of several C_nG_1/alcohol mixtures was investigated for a number of different water + oil systems. An important conclusion resulting from these investigations is that the efficiency increases with (a) decreasing lipophilicity of the oil (i.e. decreasing chain length for n-alkanes); (b) increasing lipophilicity of the surfactant and cosurfactant (i.e. increasing chain length of the surfactant and/or n-alkanol). In order to check if microemulsions supported by $C_{8/10}$APG conform to this pattern we have investigated the influence of two different oils and two different alcohols on the phase diagram. Accordingly, the phase diagrams of four different systems were studied as a function of γ and δ at a constant 1:1 oil-to-water volume ratio at $T = 25\,°C$. The results are shown in Fig. 2, where the phase diagrams for a given oil ((a)/(b) for cyclohexane and (c)/(d) for decane) but different alcohols are plotted together. Starting with the systems with cyclohexane as the oil, the graphs (a) and (b) show the effect of replacing the alcohol geraniol by hexanol. Because of the shorter alkyl chain length of hexanol, this substitution was expected to cause a decrease of the efficiency of the surfactant/ alcohol mixture, leading to a higher value γ_e. However, the changes in the phase diagram resulting from this replacement are rather insignificant as seen in Fig. 2a and b. In fact, if the mass fractions γ_e and δ_e are converted to mole fractions, the co-ordinates of point X in the phase diagram are very similar for these two systems. In other words, the hydrophilic–lipophilic balance (HLB) of the $C_{8/10}$APG/alcohol mixture is similar for these two alcohols. It is known that the HLB-value of a surfactant does depend not only on the chain length but also on other parameters. For example, surfactants with C–C double bonds in the hydrocarbon tail are more hydrophilic than the corresponding saturated compounds, and *cis*-isomers are more hydrophilic than the corresponding *trans*-isomers [17]. It appears that similar trends apply also to unsaturated alcohols like geraniol, i.e. a higher chain length is needed to compensate for the effect of two double-bonds, so that geraniol is about as hydrophilic as hexanol.

Figures 2c and d show the effects of replacing cyclohexane by a more hydrophobic oil like decane. As to be expected, this replacement causes the three-phase body to grow in width (extension in δ) and the entire three-phase body is shifted towards higher values of γ (note the different scale of γ for the two oils). The decreasing efficiency of a given surfactant or a surfactant/alcohol mixture with increasing lipophilicity of the oil is a well-known feature of microemulsions with nonionic surfactants [1, 18]. Thus, in this respect quaternary systems of the present type can be considered as "normal" nonionic microemulsions. However, a novel and striking feature of the present systems with technical $C_{8/10}APG$ is the appearance of a three-phase region at high δ-values. This three-phase region is observed for both oils. In the case of cyclohexane it appears as a separate narrow three-phase region, but for decane these two three-phase regions appear to be connected. Some aspects of this novel phase behavior are discussed below.

The "anomalous" shape of the three-phase body

The contour of the three-phase body has been mapped for a series of systems water + decane + $C_{8/10}APG$ + cosurfactant, with different *n*-alkanols (butanol, pentanol, hexanol and heptanol) as the cosurfactant. Figure 4 shows the contour map in the γ–δ plane at $\alpha = 0.5$ (1:1 decane-to-water mass ratio) at 25 °C. Only the region of $\delta > 0.5$, i.e.

Fig. 4 Contour of the three-phase region in quaternary systems water–decane–$C_{8/10}APG$–alcohol for $\alpha = 0.50$ and $\delta \geq 0.50$ at $T = 25$ °C: butanol (---), pentanol (–·–·–·), hexanol (——) and heptanol (––··––). The horizontal lines ($\delta = 0.50$ and 0.88) in the phase diagram of the hexanol system mark the γ scans of the results shown in Fig. 5

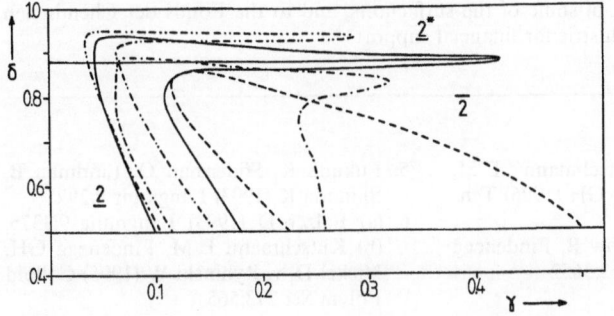

high alcohol content of the surfactant + alcohol mixture, is shown. For δ up to ca. 0.75 the contour diagrams exhibit the expected trends, viz. for each system the three-phase region shifts to higher δ as γ decreases, as to be expected for lipophilic alcohols. Furthermore, as the chain-length of the alcohol increases in the series of quaternary systems, the location of the entire three-phase region shifts to lower γ and its width $\Delta\gamma$ decreases. For the system with butanol the contour of the entire three-phase region conforms with the general diagram sketched in Fig. 1b. For pentanol and the higher alcohols, however, the contour graph develops a bulge toward higher γ at $\delta > 0.75$, which swells in the γ-direction and shrinks in its width in the δ-direction as the chain length of the alcohol increases. In order to clarify this anomalous behavior, the phase volumes, densities and refractive indices of the coexistent phases were measured along several linear paths across the three-phase region. Results of such scans for the hexanol system are shown in Fig. 5 where the respective properties are plotted as a funtion of γ for fixed values of δ. For $\delta = 0.50$ the resulting scans exhibit the characteristic features of a phase inversion $\underline{2}$–3–$\overline{2}$ with increasing γ, as discussed in the Introduction. The scans for $\delta = 0.88$, however, signify a qualitatively different behavior, reminiscent of a retrograde transition $\underline{2}$–3–$\underline{2}$: Starting in the two-phase region $\underline{2}$ at low γ, a slender layer of a middle phase appears at $\delta \approx 0.05$. As γ increases, the density of this middle phase at first decreases – as in the scan at $\delta = 0.50$ – but soon reaches a minimum and then increases steadily, approaching the density of the initial aqueous phase. The relative amount of this middle phase is small (phase volume $\Phi < 10\%$) and decreases steadily for higher γ. Visual observation indicates that this phase disappears at ca. $\gamma = 0.45$ for the $\delta = 0.88$ scan of the hexanol system, but it is difficult to locate this point due to the slow decrease of Φ as γ increases. The compositions of both the water-rich and oil-rich phase are also changing along the γ-scan across the three-phase region, as indicated by the gradual increase of the density of the oil phase and the refractive index of the water phase.

The retrograde transition documented for the $\delta = 0.88$ scan in Fig. 5 corresponds to the alcohol-to-surfactant ratio at which the bulge of the three-phase contour reaches its maximum extension for the hexanol system (see Fig. 4). It must be emphasized, however, that such a retrograde behavior is not a signature of this "anomalous" bulge in the three-phase contour. Indeed, a retrograde transition was also observed for the butanol system (in a γ-scan at $\delta = 0.75$) which exhibits a regular three-phase contour (see Fig. 4). Such retrograde transitions have been reported previously by Salager et al. [19] for related four-component microemulsions based on commercial nonionic ethoxylated surfactants. For this type of quaternary systems

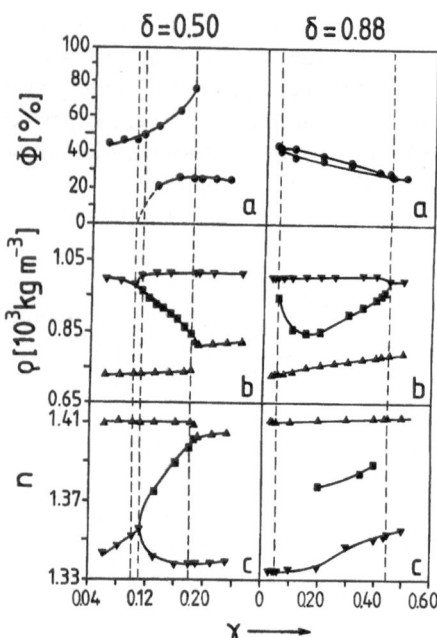

Fig. 5 Water–decane–$C_{8/10}$APG–hexanol: (a) Relative phase volumes Φ; (b) densities ρ, and (c) refractive indices n of the coexisting phases for the γ scans indicated in Fig. 4: $\delta = 0.50$ (left), $\delta = 0.88$ (right), at $\alpha = 0.50$, 25 °C. The lines connecting the data points are drawn to guide the eye, the vertical lines mark the limits of the three-phase region for the chosen α and δ

retrograde transitions are expected to occur whenever the chosen scan across the three-phase region does not cross the HLB plane of the system. In other words, the criterion for a retrograde transition of the type $\underline{2}$–3–$\underline{2}$ or $\overline{2}$–3–$\overline{2}$ is that the chosen path enters and leaves the three-phase body on the same side of the HLB plane. Accordingly, at constant α – as in the present work – the criterion for a retrograde transition is that the chosen path enters and leaves the three-phase body on the same side of the HLB line; this line is schematically drawn in Fig. 1.

The factors causing the anomalous bulge in the three-phase contour of the systems with more hydrophobic alcohols such as pentanol, hexanol and heptanol (see Fig. 4) are not fully understood. Obviously, this behavior is due to the nature of the present technical-grade surfactant, since such anomalous phase behavior has not been observed for the related systems with pure β-C_8G_1 [4] or $C_{10}G_1$ [3]. The observation that the scan across the anomalous bulge of the three-phase body corresponds to

a retrograde transition implies formally that the two-phase region at the tip of the bulge is again of type $\underline{2}$, i.e. an oil-in-water microemulsion with an excess oil phase. Accordingly, the two-phase states near the tip and above the bulge are denoted by the symbol $\underline{2}^*$. Systems exhibiting this anomalous behavior at high δ (but regular three-phase behavior at lower δ), may undergo a phase sequence $\underline{2}$–3–$\overline{2}$–3–$\underline{2}^*$ in a scan of δ at constant $\gamma > \gamma_{\min}$, where $\gamma_{\min}$ denotes the minimum value of γ at which two phase behavior $\overline{2}$ occurs below the bulge of the three-phase contour 3. A possible explanation for such a *reentrant* $\underline{2}$ behavior is as follows. Addition of a lipophilic alcohol to a three-component water + oil + nonionic surfactant system causes a phase inversion $\underline{2}$–3–$\overline{2}$, as the alcohol partitions between the interfacial film and the oil phase. The incorporation of the alcohol into the interfacial film causes the film to become more lipophilic as well as a decrease of the mean curvature H. These two effects dominate at comparatively low alcohol concentrations and induce the regular phase inversion as explained in the Introduction. At high δ, large amounts of alcohol are taken up by the oil phase which becomes increasingly polar and thus the affinity of the surfactant for the oil is enhanced. Specifically, for technical-grade surfactants consisting of components of different polarity, the lipophilic surfactant species will be preferentially extracted from the interfacial film into the oil phase, causing the interfacial film to become more hydrophilic again. It can be imagined that this selective extraction of hydrophobic surfactant components into the oil phase can cause the system to return to a state of type $\underline{2}$, as conjectured for the present systems with more hydrophobic alcohols. From the above arguments it follows that such a mechanism may cause a reentrant $\underline{2}$ phase inversion behavior only in the case of surfactant mixtures consisting of components of different polarity, but not for pure surfactants. Thus, the observed anomalous phase behavior appears to be characteristic of technical-grade surfactants consisting of components of widely different HLB. However, the microscopic nature of the coexisting phases in the state $\underline{2}^*$ needs to be clarified.

Acknowledgement Part of this work was made possible by a DAAD fellowship granted to one of the authors (S.K.M.). The authors are also indebted to the Henkel company, Düsseldorf, for the gift of some of the surfactants, and to the Fonds der Chemischen Industrie for financial support.

References

1. Kahlweit M, Busse G, Faulhaber B (1995) Langmuir 11:3382
2. Kunieda H, Ushio N, Nakano A, Miura M (1993) J Colloid Interface Sci 159:37
3. Stubenrauch C, Kutschmann E-M, Paeplow B, Findenegg GH (1996) Tenside Surf Det 33:237
4. Stubenrauch C, Paeplow B, Findenegg GH (1997) Langmuir 13:3652
5. Fukuda K, Söderman O, Lindman B, Shinoda K (1993) Langmuir 9:2921
6. (a) Balzer D (1993) Langmuir 9:3375; (b) Kutschmann E-M, Findenegg GH, Nickel D, v. Rybinski W (1995) Colloid Polym Sci 273:565

Progr Colloid Polym Sci (1998) 111:92–99
© Steinkopff Verlag 1998

7. Penders MHGM, Strey R (1995) J Phys Chem 99:10313
8. (a) Kunieda H, Nakano A, Pes MA (1995) Langmuir 11:3302; (b) Kunieda H, Nakano A, Akimaru M (1995) J Colloid Interface Sci 170:78
9. (a) Graciaa A, Lachaise J, Cucuphat C, Bourrel M, Salager JL (1993) Langmuir 9:3371; (b) Kahlweit M, Strey R, Busse GJ (1191) J Phys Chem 95:5344
10. Henkel KGaA (1997) Alkyl Polyglycosides. Hill K, v. Rybinski W, Stoll G (eds) VCH, Weinheim

11. (a) Koeltzow DE, Urfer AD (1984) J Am Oil Chem Soc 61:1651; (b) Rosevear P, Van Aken T, Baxter J, Ferguson-Miller S (1980) Biochemistry 19:4108
12. Fischer E (1895) Ber Chem Ges 28:1145
13. Kahlweit M, Strey R (1987) J Phys Chem 91:1553
14. (a) Bonkhoff K, Hirtz A, Findenegg GH (1991) Physica A 172:174; (b) Sottmann T, Strey R (1996) Ber Bunsenges Phys Chem 100:237
15. Strey R (1994) Colloid Polym Sci 272:1005
16. Nikas YJ, Purvada S, Blankschtein D (1992) Langmuir 8:2680

17. Marszall L (1987) In: Schick MJ (ed) Nonionic Surfactants, Surfactant Sci Series, Vol 23. Marcel Dekker, New York, p 493
18. Kahlweit M, Strey R, Haase D, Firman P (1988) Langmuir 4:785
19. (a) Salager JL, Márquez N, Antón RE, Graciaa A, Lachaise J (1995) Langmuir 11:37; (b) Ysambertt F, Anton R, Salager JL (1997) Colloids Surfaces A: Physicochem Eng Aspects 125:131
20. Kahlweit M, Strey R (1985) Angew Chem 97:655

Progr Colloid Polym Sci (1998) 111:100–106
© Steinkopff Verlag 1998

T. Beitz
J. Kötz
S.E. Friberg

Polymer-modified ionic microemulsions formed in the system SDS/water/xylene/pentanol

T. Beitz · J. Kötz (✉)
Institut für Physikalische
und Theoretische Chemie
Universität Potsdam
Am Neuen Palais 10
D-14469 Potsdam
Germany

S.E. Friberg
Chemistry Department
Clarkson University
Potsdam, New York 13699-5814
USA

Abstract The influence of various polymers on the phase behavior of the quaternary system SDS/xylene/pentanol/water was investigated by measurements of conductivity and viscosity. The properties of the different polymers were changed by the gradual introduction of positive charges into the macromolecule. Uncharged polymers poly(N-vinyl-2-pyrrolidone) (PVP), poly(N-methyl-N-vinylacetamide) (PNMVA) and a copolymer show a clear rise of the water solubilization capacity of the microemulsion at an addition of only 2% aqueous polymer solution. Thus a change of the spontaneous curvature of the surfactant film is indicated. High concentrations up to 20% of the polycation PDADMAC could be incorporated into the water-in-oil phase of the microemulsion. The addition of the polycation creates a clear change in the extent of the water-in-oil phase area in the Gibbs phase diagram. There is a distinct shift of water solubilization maximum of the microemulsion with increasing polymer concentration to higher SDS content and lower water content. The change of phase behavior after the addition of PDADMAC is a consequence of the formation of polycation–surfactant complexes.

Key words Microemulsion – surfactant – polymer – SDS – polyelectrolyte

Introduction

Mixtures of a surfactant, oil and water (or cosurfactant) show a complex phase behavior. One can observe the formation of different phases like the o/w (L1) and inverse w/o microemulsions (L2), a bicontinuous microemulsion, a sponge phase, a vesicle phase and several liquid crystalline phases. The formation of these phases, their structure as well as the influence of various parameters such as temperature, ionic strength, cosurfactant concentration and the nature of the oil, surfactant and cosurfactant have been examined thoroughly for more than 20 years. The spontaneous curvature and bending elasticity of the surfactant film are important parameters that determine the kind and structure of the phases formed.

In recent years interest has also turned to an examination of the influence of different uncharged polymers on the phase behavior and the structure of microemulsions. Originally only a few percent of polymer were added; today one finds polymer additions of up to 15% [1]. In this context frequently examined and well characterized systems with a nonionic or ionic surfactant contain the components C_nE_m/oil/water or AOT/oil/water. The ternary systems simplify the interpretation of the results in absence of a cosurfactant. The influence of different adsorbing and nonadsorbing polymers was studied especially on the phase relation of a balanced Winsor III phase of the nonionic system [2–4]. However, the ionic system permits observation of the temperature induced transition from the L2 phase over the percolation of inverse microemulsion droplets to the bicontinuous phase [5–9]. The influence on the phase behavior of the former transition or the phase structure of several noncharged polymers was examined by a combination of several methods [10–16].

Untill now the influence of polymers on the quaternary ionic system SDS/water/oil/pentanol was determined only for the water soluble polymer PEG at a polymer concentration <6% and with cyclohexane as the oil component [17, 18].

In the present work, an attempt was made to characterize the system SDS/water/xylene/pentanol with regard to a gradual change of the charge density of the added polymer up to significant higher polymer concentrations. The positive charge density was increased gradually starting from the water soluble PNMVA to a copolymer, and then to the polycation PDADMAC. Thus the starting point of our approach is the effect of the variation of some polymer qualities like the hydrophilicity of the polymers and the charge density of polycations on the phase behavior and the qualities of the microemulsion. The determination of the phase behavior was accompanied by conductivity and viscosity measurements.

Experimental section

Pentanol (99+%), xylene (99+%) and sodium dodecylsulfate (99.9%) were obtained from Fluka and Merck. Water was purified through the water purification system MODULAB™ PureOne of the Fa. Continental. Poly(N-methyl-N-vinylacetamide) (PNMVA) as well as a copolymer (CoP; consisting of three monomer units of NMVA to one unit of diallyldimethylammonium chloride) were synthesized according to [20]. Molecular weights (M_w) of 200000 and 90000 have been determined for those polymers, respectively. Poly(diallyldimethylammonium chloride) (PDADMAC) ($M_w = 380000$) and poly(N-vinyl-2-pyrrolidone) (PVP) ($M_w = 40000$ resp. $M_w = 360000$) are commercial products of Aldrich and Fluka. All polymers used were purified by ultrafiltration. The molecular weights of the polymers were determined by gel permeation chromatography.

The phase diagrams were determined optically by titration (25 °C) of the ternary component mixture with the corresponding polymer solutions. The rheological measurements were carried out with the Rheometer Physica LS100 of Fa. Paar Physica. The conductivity measurements were carried out with the titrator AT 310 (Kyoto Electronics).

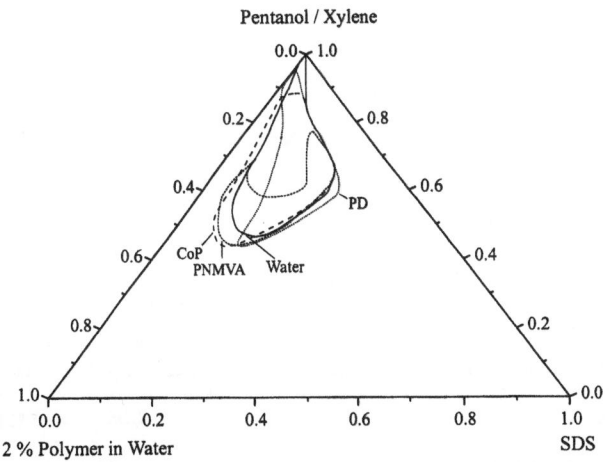

Fig. 1 L2 phase of the system SDS/water/xylene/pentanol in dependence of the different 2% aqueous polymer solutions

Results

Investigation of phase behavior

The quaternary microemulsion SDS/water/pentanol/xylene was treated as a pseudo-ternary system with a constant 50/50 ratio of xylene/pentanol. The phase diagram shows two separate areas of the microemulsion, situated in the water corner (L1 phase of normal swelled micelles) and in the oil corner (L2 phase of inverse swelled micelles). Both areas are isolated in contrast to the ternary system SDS/water/pentanol where a connection by a phase channel was observed. The influence of the added polymer on the phase behavior of the system will be discussed as follows.

In a first step the changes of the phase behavior of the microemulsion were examined by the replacement of water with 2% aqueous polymer solutions. The observed changes of the extension of the L2 phase after addition of polymers leads to a classification into two groups according to the effects observed: a first group with the uncharged polymers PVP, PNMVA and less-charged CoP and a second group with the polycation PDADMAC of high charge density (Fig. 1).

The noncharged PVP, PNMVA as well as the slightly charged CoP show a significant extension of the L2 phase in direction of the water corner. The maximum of water solubilization of the oil continuous microemulsion is increased clearly from 42% to 50% at constant surfactant concentration. This trend would be further enhanced if PVP was used as the polymer component [19]. Already at a concentration of 5% the formation of an isotropic channel starting from the L2 phase in direction of the L1

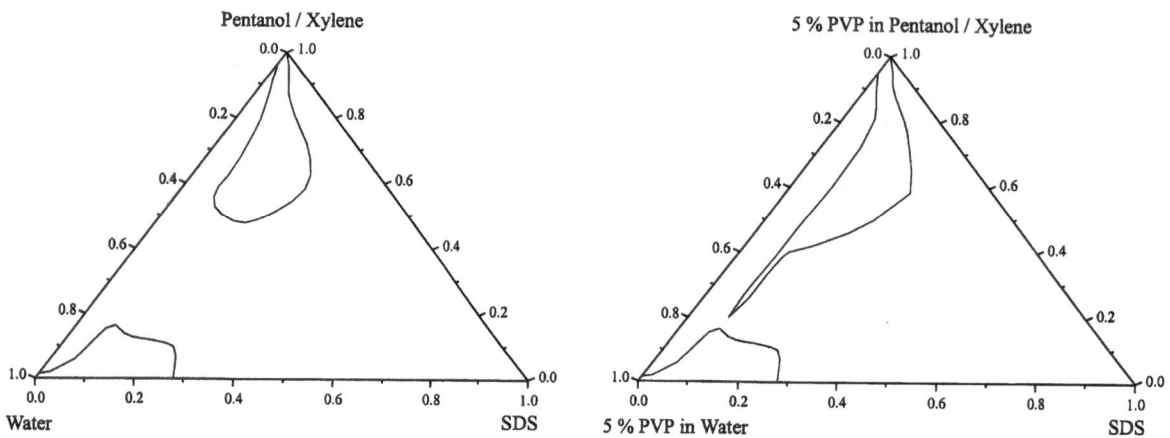

Fig. 2 PVP induced effect on the microemulsion of the system SDS/water/xylene/pentanol

phase is observed (Fig. 2). This phase channel is related to a bicontinuous structure of the microemulsion. An extension of the inverse microemulsion to lower surfactant concentration was examined at low constant water/oil relation. A phase separation was observed in the interval of low surfactant and water concentration for PNMVA and CoP, unlike the microemulsion without polymer or with PVP and PDADMAC. It has to be noted here that PNMVA shows a discontinuity inside the L2 phase area at low SDS (8–18%) and water (10–32%) content.

That means one can observe an area with two phases in equilibrium (Winsor II phase), a concentrated aqueous polymer solution and an inverse microemulsion of significant lower water content. Similar effects were observed with PNMVA in the ternary mixture SDS/water/pentanol at a water content of 15–40% [19].

The dosage of PDADMAC leads to some other effects which will be discussed as follows. In addition to the investigations described above, the polycation concentration was increased to 5, 10 and 20%. In Fig. 3 the resulting phase regions are given. It can be seen that the L2 phase is not extended in the direction of the water corner, unlike the uncharged polymers. The maximum of water solubilization of oil continuous microemulsion is even reduced with increasing polymer concentration. However, an extension of the L2 phase in direction of the SDS corner occurs which becomes more marked with increasing polymer concentration. Hence, a shift of the maximum of water solubilization of the microemulsion occurs towards higher surfactant concentration. In addition a significant shift of the L2 phase area can be observed to higher water content in the interval of low SDS concentration. Generally the L2 phase only exists in a small water content interval.

The identity of the polymer added has a quite different effect on the L1 phase. The noncharged polymers

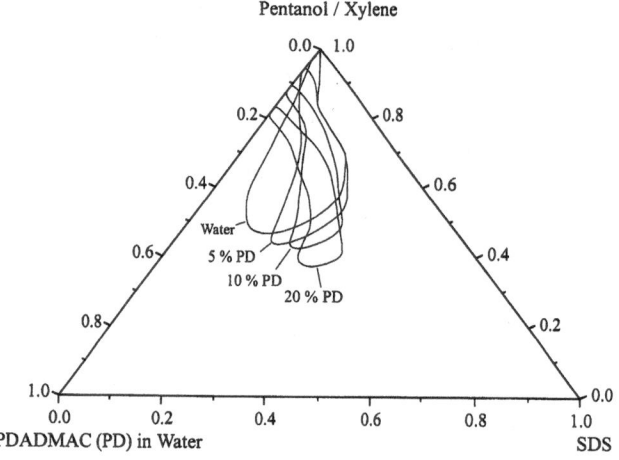

Fig. 3 Phase behavior of the inverse microemulsion after addition of high amounts of PDADMAC

PNMVA and PVP do not influence the L1 phase significantly. However the formation of the L1 phase will be completely inhibited by addition of the charged polymers PDADMAC and CoP due to the formation of nonsoluble polyelectrolyte-surfactant complexes.

Measurements of conductivity and viscosity

The L2 phase as an oil-continuous microemulsion with dispersed spherical water droplets can be characterizised by conductivity measurements because of the low conductivity of the oil and the high sensitivity with regard to structural changes, i.e. the dynamics and coalescence of droplets. The conductivity of the system remains constant up to a water content of 33%. A further increase of the

Progr Colloid Polym Sci (1998) 111:100–106
© Steinkopff Verlag 1998

Fig. 4 Conductivity measurements in several polymer modified inverse microemulsions by titration of 1% aqueous polymer solutions

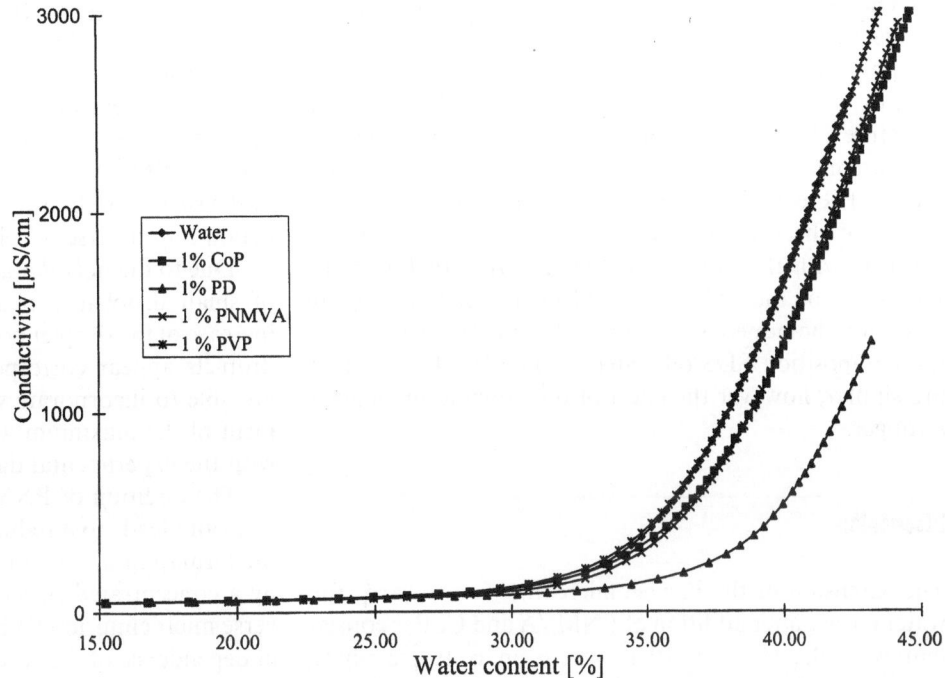

Fig. 5 Viscosity measurements in the polymer modified inverse microemulsion in dependence of the polymer concentration at the composition SDS/water/oil = 20/20/60

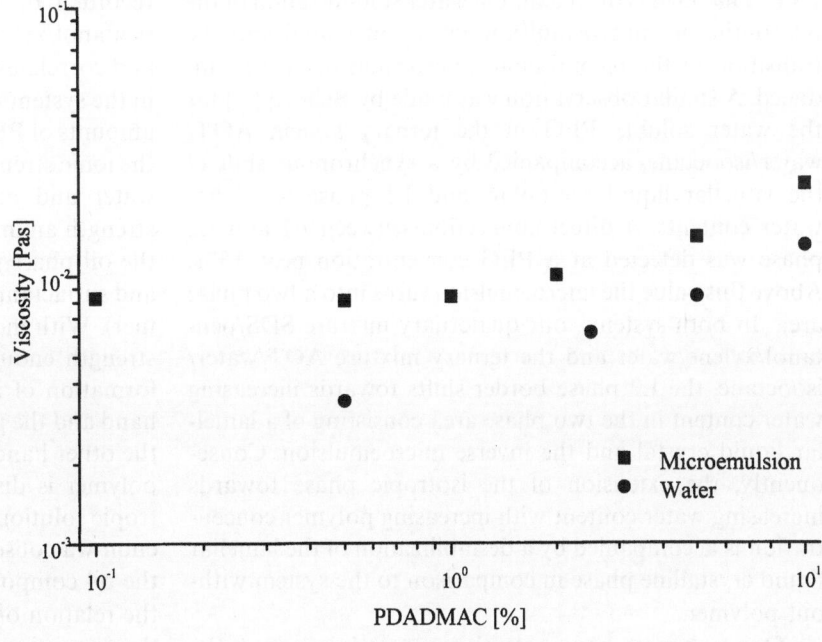

water content was accompanied by a steep increase of the conductivity. This jump in the conductivity is a characteristic feature of a w/o microemulsion and indicates the percolation of droplets (dynamic process of temporary cluster formation) or the transition to the bicontinous phase. Furthermore, measurements with other methods are needed to distinguish between these possibilities.

The shift of percolation boundaries was measured for the three polymers and PVP. Thus the solutions with a fixed ratio of oil/SDS were titrated with 1% aqueous polymer solutions and the values of conductivity were recorded (Fig. 4). The following order was found for the polymers: microemulsion without polymer ~ 1% PVP < 1% PNMVA ~ 1% CoP ≪ 1% PDADMAC.

104
T. Beitz et al.
Polymer-modified ionic microemulsions

The results show a shift of percolation boundaries to higher water contents from the noncharged polymers to the polycation at high charge density PDADMAC.

For a more comprehensive characterization of the structure of the inverse microemulsion the conductivity measurements were accompanied by rheological measurements. These could provide information about the interaction of the surfactant layer of inverse microemulsion droplets with the polycation PDADMAC. In Fig. 5 the viscosities of the polymer modified microemulsions are given for the aqueous polymer solutions (L2 phase with the composition SDS/oil/water = 20/60/20). Both curves are similar, however the effect of the polymer in water is stronger.

Discussion

The extension of the inverse microemulsion towards the water corner after addition of PNMVA and CoP is consistent with the effect of PVP, where an isotropic phase channel has been observed from the L2 phase to the L1 phase after an increase of polymer concentration to 5% PVP. That means the maximum water solubilization of the oil continuous microemulsion will be increased and the transition to the bicontinuous microemulsion will be induced. A similar observation was made by Bellocq [1] for the water soluble PEG in the ternary system AOT/water/isooctane, accompanied by a synchronous shift of the lamellar liquid crystalline and L1 phase to higher water contents. A direct connection between L1 and L2 phase was detected at a PEG concentration near 15%. Above this value the microemulsion turns into a two phase area. In both systems, our quaternary mixture SDS/pentanol/xylene/water and the ternary mixture AOT/water/isooctane, the L2 phase border shifts towards increasing water content in the two phase area consisting of a lamellar liquid crystal and the inverse microemulsion. Consequently, the extension of the isotropic phase towards increasing water content with increasing polymer concentration is accompanied by a destabilization of the lamellar liquid crystalline phase in comparison to the system without polymer.

Our polymers have a tendency to adsorb onto the surface between oil and water onto the head groups of the surfactant molecules [19]. This leads to a weakening of the hydrophilicity of the head groups of SDS and as a result to a decrease of the spontaneous curvature of the surfactant film of the inverse droplets. The described extension of the microemulsion towards higher water content will be caused by the decrease of spontaneous curvature to zero and may explain the formation of the bicontinuous phase. Analogously, a change of the sign of the spontaneous curvature was observed by Bellocq [1] i.e. the transition from the L1 to L2 phase was shown with increasing PEG concentration. This change of the spontaneous curvature could also occur in our system, where an extension of L2 phase to lower SDS content was observed after the addition of PNMVA and CoP. Thus a reduction of the surface of the surfactant film was correlated with a water solubilization into inverse micelles of larger volume.

Due to the polymer adsorption an aggregation process of small droplets can also be induced resulting in an increase of the droplet polydispersity [1]. Next to the large droplets appear correspondingly small empty ones, that are able to incorporate water, which leads to an enlargement of the maximum water solubilization, in agreement with the experimental data given.

The addition of PNMVA differs from the other polymers and leads to a reduction of the L2 phase area and to the formation of a two phase area, which is characterized by a concentrated aqueous polymer solution and the inverse microemulsion (Winsor II). This has to be discussed in dependence of the water content of the microemulsion. The crossing of several phases starting from a L2' phase over the described two phase area to a L2'' phase has been recorded by titration of a mixture of 15-25% SDS in pentanol/xylene with the aqeous 2% PNMVA solution and correlates with the observations given by Bellocq [1] in the system AOT/water/isooctane after addition of small amounts of PEG. An explanation is based on the change of the ionic strength inside the water droplets with increasing water and at constant SDS content. At higher ionic strength an important part of polymer will be passed into the oil phase (L2'), based on the competition of polymer and surfactant for water (partition equilibrium of the polymer). With increasing water content and decreasing ionic strength enough water is available for the simultaneous formation of inverse microemulsion droplets on the one hand and the polymer in a concentrated aqueous phase on the other hand. With further increasing water content the polymer is dissolved in the inverse micelles and an isotropic solution (L2'' phase) will be observed. This phenomenon was observed in the presence and in the absence of the oil component [19]. Another explanation is given by the relation of the radius of gyration of the polymer and the dimensions of the inverse microemulsion droplets [1]. These phenomena could not be observed with PVP, CoP and PDADMAC because of their higher solubility in the oil or in the water phase.

The outstanding feature of phase behavior of the inverse microemulsion in presence of PDADMAC is the shift of the maximum water solubilization to a higher SDS and a some lower water content. This is a new quality and can be only understood on the basis of the formation of corresponding polymer–surfactant complexes between the polycation and the oppositely charged SDS. After addition of the polycation, whether phase separation or an isotropic solution is observed depends on the ratio surfactant/polymer. The phase separation started earlier in the direction of the water corner with decreasing SDS content and higher polymer concentration. An extension of the isotropic area in the direction of the SDS corner was observed with increasing polymer concentration at a defined relation of SDS/PDADMAC. Both observations might be explained by the removal of the available SDS from the component mixture by the formation of polymer–surfactant complexes i.e. the composition of components must be corrected by the term

$$m \text{ (free SDS)} = m \text{ (total amount of SDS)}$$
$$- m \text{ (complex bound SDS)} \,.$$

The complex formation can lead either to membrane stabilization of the surfactant film of microemulsion droplets (as demonstrated by our conductivity measurements) or to the formation of soluble polymer–surfactant complexes in the inverse microemulsion droplets (as demonstrated by our viscosity measurements). The formed complexes can be dissolved as virtually uncharged polymers by the additional surfactant and water in the inverse microemulsion.

In an additional experiment the stability of complexes at high values of ionic strengh in the water droplets was checked. At the observed maximum ionic strengh the complexes are quite stable in the aqueous phase. Furthermore, the independence of phase behavior from the molecular weight of polycation ($M_w = 5000$ and $380\,000$) is an indication of the dominance of electrostatic interactions. These results are in good agreement with the experimental diagnosis of Plucinski et al. [21], who investigated the mass transfer in a Winsor II system (AOT/isooctane/water). The quantitative transfer of the quaternized poly(vinylpyridine) from the water phase into the inverse microemulsion and its solution in the microemulsion was found at a ratio of AOT/monomer of polycation = 1.4:1.

Three groups of conductivity curves with the corresponding percolation points were determined. The percolation points of uncharged polymers are very similar, but are significantly smaller than the one for PDADMAC. The percolation point, which means the temporary fusion of droplets to a bigger cluster, has a direct correlation with the bending elasticity of the surface film κ, because at the fusion of two droplets the negative curvature of the surface is transformed temporary into a positive one. This effect can be used to estimate qualitatively the interaction of polymers with surfactant molecules (their head groups) at the oil–water interface. This phenomen is analogous to the temperature-induced transition from the L2 to the bicontinuous phase in the system AOT/oil/water. According to Meier [12] there exists a proportionality between dT and κ that can be transferred to our system as a proportionality between d (water content) and κ. An increase of the bending elasticity is accompanied by a stronger toughness of the surfactant film of droplets at the formed bilayer at the interaction of the two droplets. The results allow the conclusion that the interaction of polymers with the surfactant head groups increases from the uncharged polymers to the extraordinarily intensive one of PDADMAC, which supports the assumption of the formation of polymer–surfactant complexes at the phase boundary. The strong electrostatic interaction between the quaternary N of the polycation and the oppositely charged sulfate head groups of the surfactant is responsible for the especially intensive change of the bending elasticity and the resulting shift of the percolation boundary. Furthermore, a shift of the percolation limiting value to a higher water content was observed with increasing solubility of polymers in water which leads to less attractive interaction between the oil droplets according to Lang [10].

Rheological experiments give the possibility of further insight into structural changes of colloids and their interaction. Figure 4 show a similar increase of the viscosity of the L2 phase and the aqueous polymer solution. This excludes the possibility that the complex is dissolved in the oil phase with the consequence of the formation of a polymer network with connected droplets. But the slope of the viscosity of the inverse microemulsion should also be larger than in the aqueous polymer solution if the polymer cluster and droplets can be treated as a hard sphere without any interaction in the system. The slope will result from the formation of the droplet cluster, connected inside the water basin by dissolved polymers, compared with cluster formation by outside dissolved polymers, because it should not be possible to dissolve a polymer–surfactant complex with a molecular weight of more than $500\,000$ in an inverse microemulsion droplet at this composition of the microemulsion. Therefore, the rheological experiments are consistent with our conductivity measurements.

References

1. Bellocq AM (1997) Progr Colloid Polym Sci 105:290
2. Kabalnov A, Lindman B, Olsson U, Piculell L, Thuresson K, Wennerström (1996) Colloid Polym Sci 274:297
3. Kabalnov A, Olsson U, Thuresson K, Wennerström H (1994) Langmuir 10:4509

106

T. Beitz et al.
Polymer-modified ionic microemulsions

4. Kabalnov A, Olsson U, Wennerström H (1994) Langmuir 10:2159
5. Meier W, Eicke H-F (1996) Current Opinion in Colloid Interface Sci 1:279
6. Vollmer D, Vollmer J, Eicke H-F (1994) Europhys Lett 26:389
7. Eicke H-F, Meier W, Hammerich H (1994) Langmuir 10:2223
8. Lang J, Zana R, Lalem N (1990) In: Bloor DM, Wyn-Jones E (eds) The Structure, Dynamics and Equilibrium Properties of Colloidal Systems. Kluwer Academic Publishers, Dordrecht, p 253
9. Garcia-Río L, Leis JR, Mejuto JC, Peña ME (1994) Langmuir 10:1676
10. Suarez M-J, Levy H, Lang J (1993) J Phys Chem 97:9808
11. Suarez M-J, Lang J (1995) J Phys Chem 99:4626
12. Meier W (1996) Langmuir 12:1188
13. González-Blanco C, Rodríguez LJ, Velázquez MM (1997) Langmuir 13:1938
14. Vollmer D, Vollmer J, Stühn B, Wehrli E, Eicke H-F (1995) Phys Rev E 52:5146
15. Zölzer U, Eicke H-F (1992) J Phys II France 2:2207
16. Holmberg A, Piculell L, Wesslén B (1996) J Phys Chem 100:462
17. Lianos P, Modes S, Staikos G, Brown W (1992) Langmuir 8:1054
18. Papoutsi D, Lianos P, Brown W (1994) Langmuir 10:3402
19. Unpublished results
20. Ruppelt D, Kötz J, Jaeger W, Friberg SE, Mackay R (1997) Langmuir 13:3316
21. Plucinski P, Reitmeir J (1997) Colloids Surfaces A 122:75

Progr Colloid Polym Sci (1998) 111:107–112
© Steinkopff Verlag 1998

H. von Berlepsch
R. Wagner

Preparing microemulsions with silicone surfactants

H. von Berlepsch (✉) · R. Wagner
Max-Planck-Institut für Kolloid- und
Grenzflächenforschung
Rudower Chaussee 5
D-12489 Berlin
Germany

Abstract In this paper, we present the results of investigations on the microemulsification of short-chain siloxanes, in particular, hexamethyl-disiloxane (HMDS), by silicone surfactants of the polyoxyethylene and amine type. The ethoxylated surfactants exhibit a temperature-dependent phase behavior similar to that of the well-known alkyl poly(glycol ethers). The amines phase behavior is less sensitive to temperature and microemulsions emerge after addition of short-chain alkanols, thereby their carbon number being a function of the size of the surfactants hydro-phobic part. The amount of amphiphile (and alkanol) required for homogeneous mixtures of water and HMDS falls within the range of 15–25 wt%. The phase behavior is rather insensitive to added salt. Serious problems for the stability of microemulsions arise from the sensitivity of the surfactants to hydrolytic decomposition. Only the amine derivative with a $(CH_3)_3Si$-hydrophobe exhibits a necessary long-time stability.

Key words Microemulsions – silicone surfactants – phase behavior

Introduction

Silicones, based on the polydimethylsiloxane backbone, are strongly hydrophobic substances which are liquid at higher molecular weights than corresponding hydrocarbons. They have low surface tensions and a high spreading ability improving the rub out of creams, lotions, etc. All this makes them important substances for a wide range of technological applications [1, 2]. To improve the compatibility with cosmetic oils and to obtain water soluble or at least water dispersible polydimethylsiloxanes, additional lipophilic long chain alkyl groups or hydrophilic groups have been attached to the silicone chain resulting in very efficient surfactants. Despite their wide use in formulations and a series of recent studies [3–8], the knowledge of the aggregation behavior of this class of surfactants in aqueous solution is still fragmentary, particularly, when other solvents are used and for ternary

systems consisting of water, silicone-based oils and surfactants [9, 10].

Depending on their backbone chain length, one has to distinguish between two basic structures of silicone surfactants. The first is derived from low molecular weight siloxanes, preferentially trisiloxanes, where the hydrophilic group is linked to the middle of the hydrophobic trisiloxane chain. The second group comprises comblike structures with a polysiloxane backbone and functional side groups. These hydrophilic moities may include nonionic, anionic, cationic, or zwitterionic structures. In aqueous solution all the different types of aggregates known for hydrocarbon surfactants comprising micelles, lyotropic liquid crystalline phases, vesicles, and "sponge" (L_3) phases have been found. It could be demonstrated [5] that several of the surfactants follow the usual sequence of liquid crystalline phases, progressing from lamellar to the inverse hexagonal phase with the decreasing size of the hydrophilic head group. Yet the formation of highly ordered liquid crystalline phases by the bulky siloxane surfactants, especially the irregularly shaped, branched and polydispersed polymeric compounds, is not well understood. The similarity of the binary phase behavior of hydrocarbon and silicone surfactants has motivated us to explore the preparation of microemulsions with silicone surfactants.

Microemulsions are thermodynamically stable colloidal mixtures of water and oil stabilized by an amphiphile. Attempts to use siloxane surfactants as amphiphiles for solubilizing hydrocarbons are scarce in literature. It has been shown that water soluble polydimethylsiloxanes with oxyethylene side chains cannot solubilize large amounts of hydrocarbons [8]. So also the dilute L_2-phase formed by the non-ionic trisiloxane surfactant Silwet L77 [11] in cyclohexane does not form water-in-oil microemulsions [10]. Motivated by these experiences with hydrocarbons we have chosen silicone-based oils to find out whether their additional chemical affinity to the hydrophobic part of the surfactant molecule promotes solubilization. Aside from interesting scientific aspects the question of how to solubilize polymeric silicone oils, preferentially in microemulsions, is of high-technological importance.

The recipe for making microemulsions with standard amphiphiles is well known [12–14]. For a ternary water–oil–amphiphile mixture it is to find the state of three coexisting liquid phases, namely, an amphiphile-rich phase in equilibrium with a water-rich, and an oil-rich excess phase, by suitable choice of the thermodynamic parameters. Following this general line and referring to recent studies on non-conventional non-ionic systems [15–19] we were looking for the three-phase state in different water – silicone oil–(Si)-amphiphile systems. Only a part of the phase investigations were successful in making balanced microemulsions.

Experimental

Phase diagrams

To simplify the estimation of the three-phase body, defined sections through the phase body were taken and the phase boundaries were represented on pseudo-binary phase diagrams. First, we prepared ternary mixtures of amphiphile, water and oil. Thereafter, we varied the temperature, T, or titrated, depending on the selected surfactant, the mixture at fixed T and composition with the cosolvent, observing the sequence of phases by visual inspection. The oil/water fraction is denoted by

$$\alpha \equiv \text{oil}/(\text{water} + \text{oil}) \, ,$$

the fraction of amphiphile in the ternary mixture by

$$\gamma \equiv \text{amphiphile}/(\text{water} + \text{oil} + \text{amphiphile}) \, ,$$

the fraction of the cosolvent in the entire mixture by

$$\delta \equiv \text{cosolvent}/(\text{water} + \text{oil} + \text{amphiphile} + \text{cosolvent})$$

and the brine (NaCl) concentration by

$$\varepsilon \equiv \text{NaCl}/(\text{water} + \text{NaCl})$$

all in weight percent (wt%). Phase diagrams were studied exclusively. The microstructure of the phases was not investigated.

Materials

Commercial and non-commercial silicone surfactants of differing structure were utilized. The ethoxylated product $(CH_3)_3SiC_6E_4$ (**I**) was provided by the Th. Goldschmidt AG (Essen, Germany), the methoxy-terminated trisiloxane derivative $M(D'C_3E_{7.5}OCH_3)M$ (Silwet L 77, **II**) by Union Carbide Europe S.A. (Versoix, Switzerland). Here M stands for $(CH_3)_3SiO–$, D for $–(CH_3)_2Si–$, D' for $–(CH_3)Si–$ and E_n for the polyethylene group $–(OCH_2CH_2)_nOH$, where n is the number of oxyethylene segments. Both commercial products were used as received. They bear a polydisperse polyethylene oxide chain. Three strictly defined amine derivatives bearing the amine structure $–(CH_2)_3OCH_2CH(OH)NH(CH_2)_2NH_2$ have been synthesized in our laboratory [20–23]: the silaneamine $(CH_3)_3Si$-amine (**III**), the trisiloxaneamine M(D'amine) M (**IV**) and the corresponding carbosilaneamine $[(CH_3)_3SiCH_2]_2$–$Si(CH_3)$-amine (**V**). The purity is better than 98%. As standard silicone oil hexamethyldisiloxane (HMDS, Aldrich, purity 98%) was used. The alkanols and sodium chloride (p.a.) were purchased from Fluka. The mixtures were prepared with Milli-Q water.

Results and discussion

All investigated silicone surfactants are non-ionic. The classical non-ionic surfactants are the alkyl poly(glycol ethers) C_iE_j. Here the distribution of the surfactant between the aqueous and the oil-rich phase changes strongly with temperature, T. On increasing T, a ternary mixture changes from the two-phase state $\underline{2}$ (amphiphile-rich aqueous phase, Winsor I) through the three-phase state 3 (amphiphile-rich middle phase, Winsor III) into the two-phase state $\overline{2}$ (amphiphile-rich oil-phase, Winsor II). Other non-ionic surfactants as for example lecithins and alkyl monoglucosides are rather insensitive to temperature and the recipe for enforcing a phase inversion is to add cosolvents which increase the solubility of the amphiphile in either the aqueous or the oil-rich phase [15–19]. To explain the complex phase behavior of ternary systems it would be necessary, in principle, to measure the phase diagrams of all three binary mixtures. Because this approach is mostly difficult and time-consuming we started with simple experiments on the solubility of the silicone amphiphiles in water and HMDS, respectively. In the second step, phase diagrams of particular systems were systematically measured to obtain quantitative data about position and extension of the three-phase body, salt influence and long-time stability of the microemulsions, all important parameters for possible applications.

In the low concentration range (below 20 wt%) the silaneamine (**III**) and the two ethoxylated products **I** and **II** are soluble in water at room temperature. As expected [7] **I** and **II** show a liquid–liquid insolubility boundary (cloud point) at elevated temperatures. The amine derivatives **IV** and **V** with a much larger hydrophobic part as compared with **III** are water insoluble over the entire experimental window (above about $\gamma = 1$ wt%). However, the miscibility gap reduces drastically when short-chain alkanols (up to propanol) are added. The solubility in HMDS shows the opposite behavior, i.e. high solubility of **IV** and **V** and miscibility gaps for **I**, **II** and **III**. In general, the solubility of the amine derivatives is temperature insensitive. Similarities between products **I** and **II** and C_iE_j as well as between **III**, **IV** and **V** and corresponding lecithins and alkyl glucosides are obvious and have been confirmed in systematic studies.

Figure 1 shows as an example for ethoxylated silicone surfactants, a section through the H₂O–HMDS-**I** phase prism at $\alpha = 50$ wt%. The result is a classical three-phase "fish" lying in a non-horizontal position, obviously due to the polydispersity of the commercial product. The amount of amphiphile required for homogenizing water and

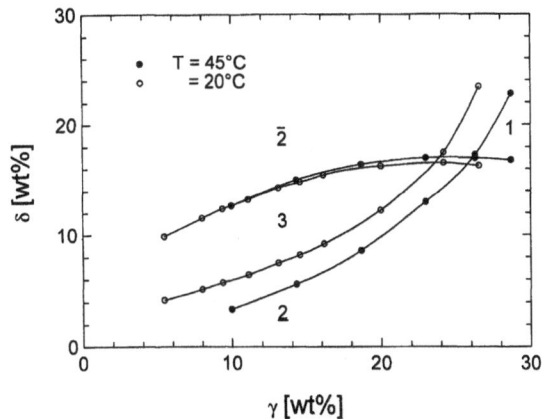

Fig. 2 Sections through the phase tetrahedra of H₂O–HMDS-**III**-C_4E_0 mixtures at $\alpha = 50$ wt% and two temperatures

HMDS (one-phase state behind the "tail" of the fish, 1) is about 15 wt% at 30 °C, and thus relatively high.

The Si-amine surfactants have identical headgroups and differ in their hydrophobic part. The water soluble product **III** shows $\underline{2}$ in the ternary system H₂O–HMDS-**III**. To test which alkanol leads to the phase sequence $\underline{2} \rightarrow 3 \rightarrow \overline{2}$, we first investigated the distribution of alcohols (C_nE_0) between HMDS and water with the result, that up to propanol ($n = 3$) the alcohol is more soluble in water than in HMDS, whereas for alcohols with $n \geq 4$ the opposite holds. This means that HMDS compared with alkanes behaves like an alkane with an effective carbon number of at least $n = 8$ (octane) [16]. Indeed, upon adding C_4E_0, we traversed the three-phase body with the sequence $\underline{2} \rightarrow 3 \rightarrow \overline{2}$, as shown for the H₂O–HMDS-**III**-C_4E_0 system in Fig. 2. The behavior is similar to that found for short-chain lecithins [16] or alkyl monoglucosides with carbon numbers between 8 and 14 [17] and long-chain alkanes. The figure shows two sections through the phase tetrahedron at 20 °C and 45 °C and $\alpha = 50$ wt%. The effect of temperature is weak because of the weak temperature dependence of the distribution of the alcohol. The amount of surfactant **III** γ to reach 1 is about 23 wt% and increases slightly with rising T, whereas the required amount of alcohol, $\delta \approx 17$ wt%, remains nearly unchanged.

The trisiloxaneamine **IV** is insoluble in water and shows $\overline{2}$ in the ternary mixture H₂O–HMDS-**IV**. Here the addition of short-chain alkanols leads to the phase sequence $\overline{2} \rightarrow 3 \rightarrow \underline{2}$ as the sections through the phase tetrahedron, Figs. 3 and 4, illustrate. Figure 3 shows the "fishes" at $T = 20$ °C and $\alpha = 50$ wt% obtained for different alkanols. Ethanol (C_2E_0) is the most efficient one in reducing the amount required for homogenizing water and HMDS. Compared with methanol (C_1E_0) it makes the water less polar, hence lowering the three-phase body. For propanol

Fig. 1 Section through the H₂O–HMDS-**I** phase prism at $\alpha = 50$ wt%

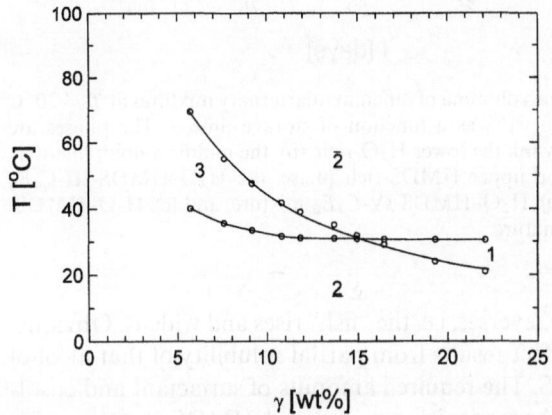

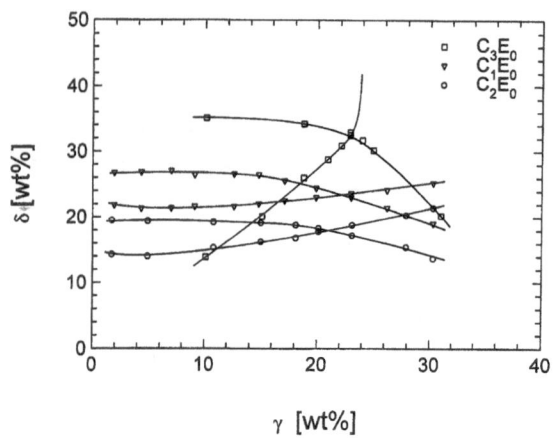

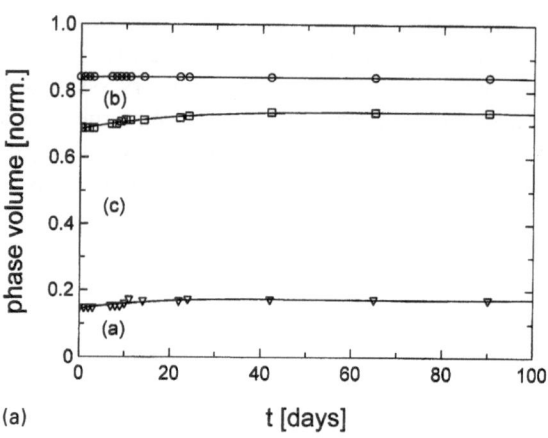

Fig. 3 Sections through the phase tetrahedra of H_2O–HMDS-**IV**-C_nE_0 mixtures at $T = 20\,°C$ and $\alpha = 50$ wt% showing the effect of the varying alcohol chain lengths

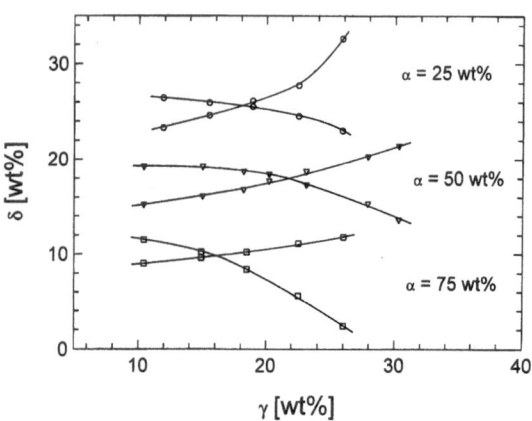

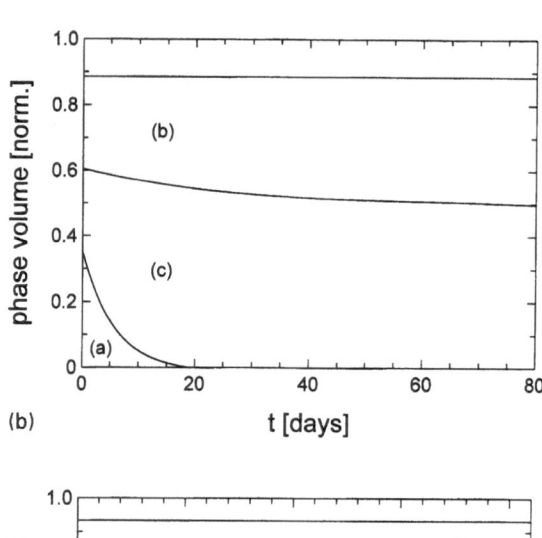

Fig. 4 Sections through the phase tetrahydron of H_2O–HMDS-**IV**-C_2E_0 at $T = 20\,°C$ and various α

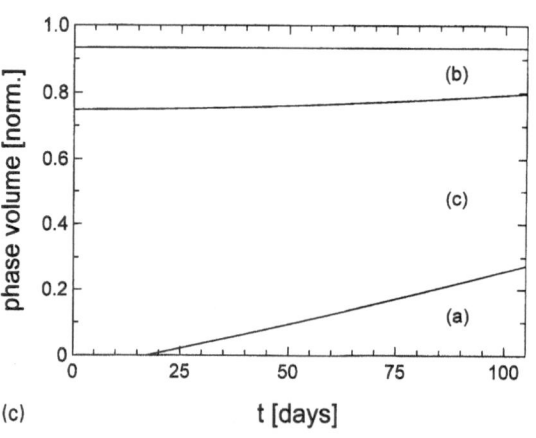

Fig. 6 Phase volumina of different quarternary mixtures at $T = 20\,°C$ and $\alpha = 50$ wt% as a function of storage time, t. The phases are denoted by (a): the lower H_2O-rich, (b): the middle amphiphile-rich, and (c): the upper HMDS-rich phase. (a): H_2O–HMDS-**III**-C_4E_0 mixture, (b): H_2O–HMDS-**IV**-C_2E_0 mixture, and (c): H_2O–HMDS-**V**-C_2E_0 mixture

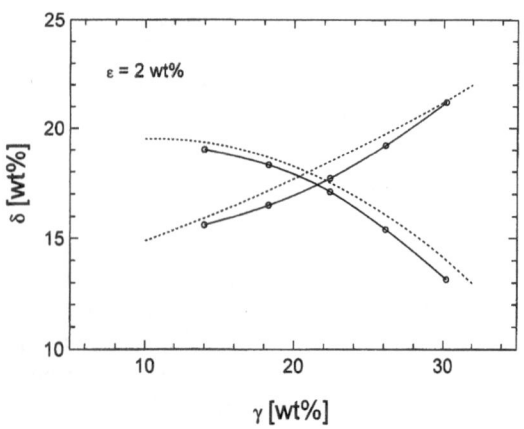

Fig. 5 Dependence of the three phase body of the H_2O–HMDS-**IV**-C_2E_0 mixture on brine concentration ε at $T = 20\,°C$ and $\alpha = 50$ wt%. The dotted line indicates the position of the salt-free system

the trend reverses, i.e. the "fish" rises and widens. Obviously, this effect results from partial solubility of that alcohol in HMDS. The required amounts of surfactant and cosolvent for homogenizing water and HMDS is at best for

Progr Colloid Polym Sci (1998) 111:107–112
© Steinkopff Verlag 1998

the H_2O–HMDS-IV-C_2E_0 system $\gamma \approx 21$ wt% and $\delta \approx$ 18 wt%, respectively. Figure 4 depicts sections through the phase tetrahedron at $T = 20\,°C$ and $\alpha = 25$, 50 and 75 wt%. At fixed temperature, increasing δ shifts the amphiphile-rich middle phase from the HMDS-rich to the water-rich side. In other words, the amount of cosolvent required for the formation of a one-phase microemulsion reduces with increasing fraction of HMDS in the mixture. The three-phase body is symmetrical with respect to $\alpha = 50$ wt%. The salt effect on the phase behavior of the quarternary H_2O–HMDS-IV-C_2E_0 system is weak, as can be seen from Fig. 5. The addition of $\varepsilon = 2$ wt% (≈ 0.35 M) NaCl leads to a minor reduction of the required amount δ of ethanol, and has also a small effect on γ.

According to its definition a microemulsion is thermodynamically stable, i.e. in particular, does not demix after preparation. But because of the hydrolytic instability of the SiOSi-bond [1], problems with long-time stability have to be considered. In order to look for such possible effects we followed the time dependence of the phase volumes of selected mixtures. The results are shown in Fig. 6. Minor volume changes are evident already after three days for all amine derivatives III–V. The effect is distinct for the trisiloxane surfactant IV (Fig. 6b) and surprisingly also for the carbosilane derivative V (Fig. 6c). In contrast, after an initial period of about 20 days with minor changes, the phase volumina of the three-phase state of the H_2O–HMDS-III-C_4E_0 system (Fig. 6a) remain constant over at least 100 days.

Finally, we present a few remarks on other silicone-based systems. A broad variety of carbohydrate-modified siloxane surfactants have been synthesized in our laboratory [20–23]. Tests with those products to prepare homogeneous mixtures with HMDS were not successful. Larger hydrophilic headgroups lead to much stronger kinetic effects, so that the time necessary for phase separation

becomes comparable with that of beginning degradation. Additionally, the hampering influence of liquid-crystalline phases grows. Emulsifying longer siloxanes than HMDS, especially polysiloxanes with chain lengths $D > 10$ would be of outstanding technological importance. Tests with surfactant IV and a siloxane of the structure $MD_{5.2}M$ gave microemulsions after ethanol addition, but failed for longer polysiloxanes due to the siloxane-insolubility of the amphiphile. Spot checks with different polyether-modified polysiloxane surfactants did not yield microemulsions.

Conclusions

We find that Si-surfactants of the polyoxyethylene and amine type alone or in combination with short-chain alkanols, respectively, are efficient emulsifiers for short-chain siloxanes. The ethoxylated surfactants I and II exhibit a temperature-dependent phase behavior similar to that of the well-known alkyl poly(glycol ethers). Microemulsions formed by the amines III to V are less temperature sensitive. Obviously a chemical affinity of the hydrophobic part of the surfactant to the silicone oil promotes a solubilization process. The amount of amphiphile (and alkanol) required for homogeneous mixtures of water and HMDS falls within the range of 15–25 wt%. But in view of the fact that in applications one is mainly interested in water-in-oil or oil-in-water dispersions the efficiency appears still sufficient. The phase behavior is rather insensitive to added salt. Serious problems for the stability of microemulsions arise from the sensitivity of the surfactants to hydrolytic decomposition. Only the amine derivative III with a $(CH_3)_3$Si-hydrophobe exhibits a necessary long-time stability.

Acknowledgment The authors wish to thank Prof. H. Möhwald for his support and Y. Wu for the synthesis of some of the surfactants.

References

1. Grüning B, Koerner G (1989) Tenside Surf Det 26:312–317
2. Schmidt G (1990) Tenside Surf Det 27: 324–328
3. Gradzielski M, Hoffmann H, Robisch P, Ulbricht W, Grüning B (1990) Tenside Surf Det 27:366–379
4. Hill RM, He M, Lin Z, Davis HT, Scriven LE (1993) Langmuir 9: 2789–2798
5. He M, Hill RM, Lin Z, Scriven LE, Davis HT (1993) J Phys Chem 97: 8820–8834
6. He M, Lin Z, Scriven LE, Davis HT (1994) J Phys Chem 98:6148–6157
7. Hill RM, He M, Davis HT, Scriven LE (1994) Langmuir 10:1724–1734
8. Stürmer A, Thunig C, Hoffman H, Grüning B (1994) Tenside Surf Det 31: 90–98
9. Messier A, Schorsch G, Rouviere J, Tenebre L (1989) Progr Colloid Polym Sci 79:249
10. Steytler DC, Sargeant DL, Robinson BH, Eastoe J, Heenan RK (1994) Langmuir 10:2213–2218
11. Silwet L77 is a commercial ethoxylated trisiloxane surfactant from Union Carbide Corp
12. Kahlweit M, Strey R (1985) Angew Chem 97:655–669
13. Bourrel M, Schechter RS (eds) (1988) Microemulsions and related systems. Marcel Dekker, New York
14. Kahlweit M, Strey R, Firman P (1986) J Phys Chem 90:671–677
15. Shinoda K, Kaneko TJ (1988) Dispersion Sci Technol 9:555

16. Kahlweit M, Busse G, Faulhaber B (1995) Langmuir 11:1576–1583
17. Kahlweit M, Busse G, Faulhaber B (1995) Langmuir 11:3382–3387
18. Kahlweit M, Busse G, Faulhaber B, Eibl H (1995) Langmuir 11:4185–4187
19. Kahlweit M, Busse G, Faulhaber B (1996) Langmuir 12:861–862
20. Wagner R, Richter L, Wersig R, Schmaucks G, Weiland B, Weissmüller J, Reiners J (1996) Appl Organomet Chem 10:421–435
21. Wagner R, Richter L, Weiland B, Reiners J, Weissmüller J (1996) Appl Organomet Chem 10:437–450
22. Wagner R, Richter L, Wu Y, Weiland B, Weissmüller J, Reiners J, Hengge E, Kleewein A (1997) Appl Organomet Chem (in press)
23. Wagner R, Richter L, Wu Y, Weissmüller J, Reiners J, Kleewein A, Hengge E (1997) Appl Organomet Chem (in press)

Progr Colloid Polym Sci (1998) 111:113–116
© Steinkopff Verlag 1998

T. Wolff
D. Nees

Influence of solubilized stilbazolium salts in *trans* and *cis* configuration on the percolation equilibrium in AOT-isooctane–water microemulsions: Light induced switching of conductivity

Prof. Dr. T. Wolff (✉)
Technische Universität Dresden
Institut für Physikalische Chemie
und Elektrochemie
D-01062 Dresden
Germany

Dr. D. Nees*
Universität Siegen
Physikalische Chemie
Siegen
Germany

*Present address
Surfactant Science Group
School of Chemistry
University of Hull
Hull HU6 7LS
United Kingdom

Abstract Solubilized *trans* and *cis* isomers of *N*-methyl and *N*-hexadecyl hydroxystilbazolium bromide had specific influences on the percolation temperature (as revealed by the onset of conductivity) in microemulsions composed of AOT, isooctane, and water: in the methyl derivative both *cis*- and *trans*-HSB induce shifts of the temperature at which conductivity sets on, the *cis* form being more efficient. In the hexadecyl compound a conductivity onset at increased temperature induced by the *trans* form was partially reverted when the *cis* isomer was present. Since *trans* → and *cis* stilbazolium salts can be interconverted photochemically, conductivity may be switched isothermally by exposing samples to light. Some competition between *trans* → *cis* photoisomerization and photodimerization was found depending on concentration ratios in the systems.

Key words Conductivity – microemulsion – percolation – photoisomerization

Introduction

Microemulsions composed of surfactant, water and oil are thermodynamically stable systems which are macroscopically monophasic [1]. They exist in two types: in water rich systems, the so-called oil-in-water – o/w – microemulsions form consisting of oil droplets in a continuous water phase; oil rich systems exhibit water droplets in a continuous oil phase (water-in-oil – w/o – microemulsion), cf. Fig. 5 in [2] or more generally Figs. 9–15 in [1]. In both cases, surfactants are accumulated in the interfacial regions mediating the dispersed state. When starting with a w/o microemulsion containing an ionic surfactant the water-to-oil ratio is gradually increased (at an appropriate temperature) the system may convert to an o/w microemulsion passing through a bicontinuous region with an average interfacial surface curvature equal to zero. At a certain ratio called percolation point the system changes from almost non-conductive (in the w/o case) to conductive

since water becomes a continuous phase. As percolation points depend on temperature, systems at water-to-oil ratios close to the percolation point can be switched from conductive to non-conductive (and vice versa) upon varying the temperature.

Isothermally percolation can be induced by applying electric fields [3] or by the addition of certain solubilizates that affect percolation temperatures specifically [4]. Recently we reported photochemical switching of conductivity in microemulsions brought about by solubilized acridizinium bromide and its photodimerization [5]: since monomer and dimer induce different percolation temperatures the conductivity could be varied isothermally upon exposure of samples to light. As one of the possible origins of the effect it was discussed that the divalent cations as formed upon photodimerization of acridizinium monocations may vary the ionic strength in the system. We will show below that this is not a necessary prerequisite for photochemical switching of conductivity in microemulsions, since monomolecular reactions such as the *trans-cis* photoisomerization of hydroxystilbazolium salts (Scheme 1) are well capable of inducing conductivity changes.

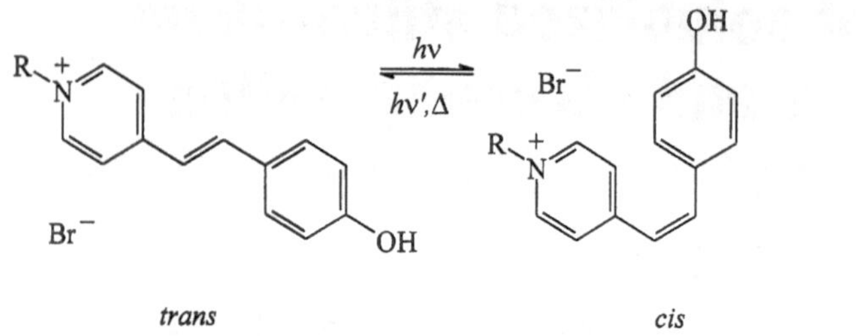

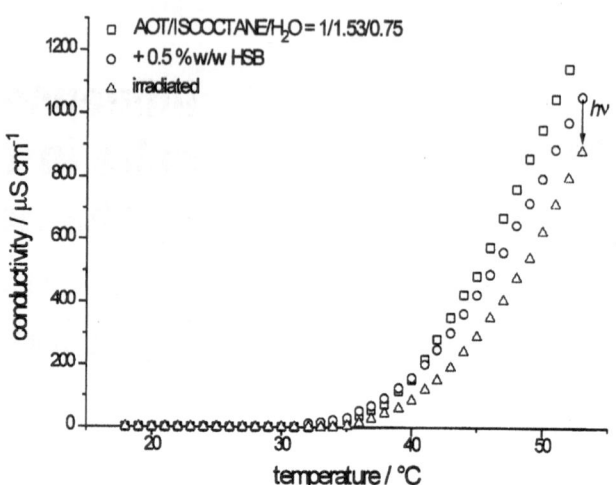

Scheme 1 Photoisomerization of hydroxystilbazolium (HSB) salts

HSB: R = CH₃   C16-HSB: R = n-C₁₆H₃₃

Fig. 1 Conductivity vs. temperature diagrams for the system AOT:isooctane:water = 1:1.53:0.75 (weight ratios, μE I) in the absence and presence of 0.5% (by weight) of HSB, and after irradiation, i.e., photochemical *trans* → *cis* isomerization

Results

Figure 1 displays the conductivity of a microemulsion composed of AOT, isooctane, and water (1:1.53:0.75 ratio by weight = μE I) as a function of temperature. Up to 31 °C the system is (practically) non-conductive. At higher temperatures increasing conductivity indicates that percolation has taken place. When to this mixture 0.5% (by weight) of *trans*-HSB (see Scheme 1) is added, no significant change in the temperature of the onset of conductivity can be found. However, the conductivity increase at higher temperatures is less steep. After exposure to Pyrex filtered light of a mercury lamp (i.e., after photoisomerizing *trans*-HSB to *cis*-HSB to a great extent, cf. [6–9]) the onset of conductivity is shifted by some degrees to higher temperatures, and at a given temperature the conductivity is generally below that of the non-irradiated system.

When more HSB is added (1%, see Fig. 2), the described influences become more pronounced: the curving of the conductivity vs. temperature plot is reduced in the presence of *trans*-HSB and the onset of conductivity shifts to lower temperatures. Irradiation reverts the latter effect, and leads, again, to a generally reduced conductivity. A hypsochromic shift of the long wavelength absorption peak of HSB (measured without dilution in a thin layer between quartz plates) indicates the formation of parallel H-aggregates – cf. [10–12] – of HSB in this system.

Upon changing the composition of the microemulsion to a mixture in which the capability of taking up water is almost exhausted, i.e., AOT/isooctane/water = 1:3:2 (= μE II), we observe a distorted conductivity vs. temperature plot (Fig. 3). Such distortions generally appear when phase transitions occur. Above 37 °C – when range III in Fig. 3 is entered – the upper two-phase region is reached. Here measured conductivity values can be reproduced only qualitatively (and interesting modifications of the

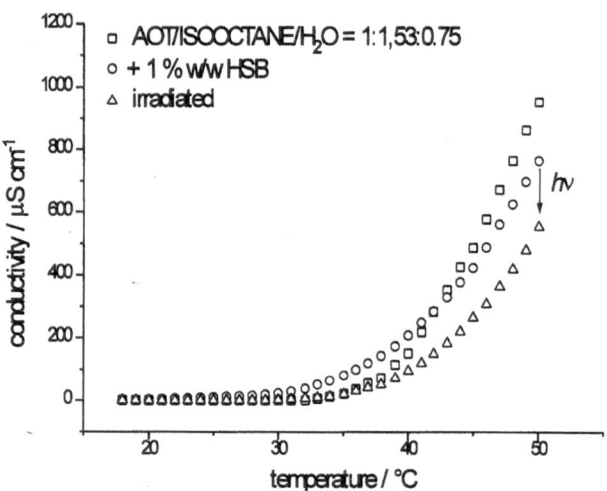

Fig. 2 Conductivity vs. temperature diagrams for the system AOT:isooctane:water = 1:1.53:0.75 (weight ratios, μE I) in the absence and presence of 1% (by weight) of HSB, and after photochemical *trans* → *cis* isomerization

flow behavior occur as indicated in the figure, which shall not be a matter of this investigation).

In the presence of 0.5% HSB these features vanish as shown in Fig. 4. *Trans*-HSB already shifts the onset of conductivity to a 7 °C higher temperature, irradiation (*trans–cis* isomerization) causes a further shift by 3–4°. Phase transitions do not occur up to 60 °C; the conductivities extend to considerably higher values as compared to μE I (cf. Figs. 1 and 2), as a consequence of the increased water content. Almost complete thermal back isomerization (see Scheme 1) can be achieved upon storing the

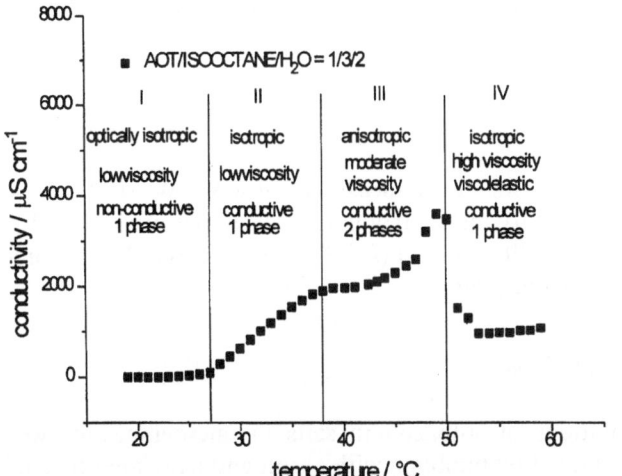

Fig. 3 Conductivity vs. temperature diagram for the system AOT:isooctane:water = 1:3:2 (weight ratios, μE II)

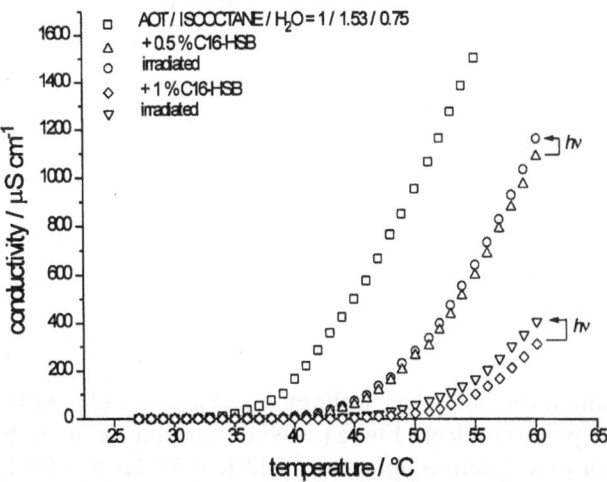

Fig. 5 Conductivity vs. temperature diagrams for the system AOT:isooctane:water = 1:1.53:0.75 (μE I) in the absence and presence of 0.5 and 1% (by weight) of C_{16}-HSB, and after photochemical *trans → cis* isomerization

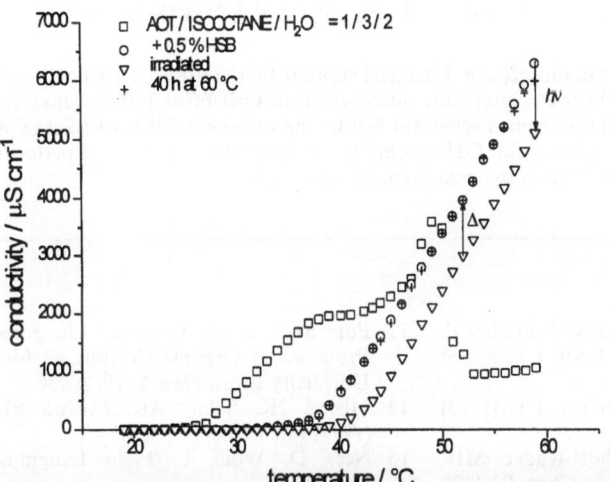

Fig. 4 Conductivity vs. temperature diagrams for the system AOT:isooctane:water = 1:3:2 (μE II) in the absence and presence of 0.5% (by weight) of HSB, after photochemical *trans → cis* isomerization, and after subsequent thermal back isomerization

Discussion

Scheme 1 indicates that re-isomerization from the *cis*-form to the *trans*-form can also be achieved photochemically using another (shorter) wavelength hv'. In the systems studied here this was not tried since AOT bearing carbonyl goups will absorb most of this light thereby inhibiting the re-isomerization and giving rise to side reactions originating from the photolysis of AOT.

According to our previous study in lyotropic liquid crystalline systems [13] HSB destabilizes flat surfaces such as in lamellar LLC phases as revealed by lowered transition temperatures from the lamellar to the isotropic phase. In turn we may expect a stabilization of the bent surfaces of the water droplets in our microemulsion studied here, leading to the higher percolation temperatures observed. This interpretation is corroborated by the observation of features due to the formation of parallel H-aggregates, cf. [10–12], in the UV/VIS-spectra of HSB in w/o microemulsions. These H-aggregates may hold the surface bent, extending the w/o range at the expense of the bicontinuous and o/w ranges. The H-aggregates also were found responsible for a thermally irreversible photodimerization of HSB as a side reaction to isomerization. It is thus in keeping that in μE I (containing smaller water droplets with stronger bent surfaces) this side reaction is more effective as in μE II, leading to the observed incomplete thermal back reaction in μE I.

Opposite influences of the irradiation of HSB and C16-HSB, respectively, on the macroscopic phase behavior of surfactant systems was found in AOT-LLC systems at certain compositions [14]. This was explained by the

sample at 60 °C for 40 h. In the μE I systems the thermal back reaction was incomplete.

Effects brought about by a hydroxystilbazolium salt bearing a C_{16}-chain, C16-HSB (cf. Scheme 1), are shown in Fig. 5 for μE I. The addition of 0.5% C16-HSB shifts the onset of conductivity to a 7 °C higher temperature, 1% C16-HSB causes a stronger shift. The irradiation of C16-HSB has only a little influence on the conductivity; the direction of this little influence, however, is opposite to the observations in HSB containing systems, i.e. the conductivity increases after irradiation.

greater hydrophobicity of the C_{16}-chain leading to a preferred incorporation of this chain in the AOT layer at the water–isooctane interface. Similarly, in the microemulsion studied here a solubilization site (or partition between sites) differing from that of the quite water soluble HSB may have lead to the opposite behavior.

Experimental

Materials

Sodium bis-2-ethylhexylsulfosuccinate (aerosol OT, AOT) was purchased from Fluka (>99%). Isooctane and methanol were spectroscopic grade (Aldrich). Water was triply distilled. 1-Methyl-4-[2-(4-hydroxyphenyl)-vinyl]-pyridinium bromide (HSB) and 1-n-hexadecyl-4-[2-(4-hydroxyphenyl)-vinyl]-pyridinium bromide (C16-HSB) were available from previous investigations [13, 14].

Samples

For the preparation of samples solutions of AOT in water at the desired mass ratio were mixed with the necessary quantity of isooctane. The microemulsion then formed upon stirring. The stilbazolium salts easily dissolved in the microemulsion.

Apparatus

Conductivities were measured on a WTW Multilab 540 conductivity meter. Irradiations were performed employing glass filtered light ($\lambda > 310$ nm) of a 100 W high pressure mercury lamp.

Irradiations

Solutions of stilbazolium salts in microemulsions were deaerated by bubbling with argon and irradiated through the gas–liquid interface. Reaction progress was monitored UV/VIS spectroscopically after appropriate dilution of samples by methanol [13]. Dilution was necessary in order (i) to reduce optical densities and (ii) to destroy microemulsions which can cause considerable scattering. Irradiations were processed until constancy of UV/VIS spectra.

Acknowledgement Financial support by the Fonds der Chemischen Industrie is gratefully acknowledged. One of us (DN) thanks the "Aktionsgemeinschaft zur Förderung wissenschaftlicher Projekte" at the Universität-GH-Siegen for providing the possibility of performing some of the measurements reported.

References

1. Dörfler H-D (1994) Grenzflächen- und Kolloidchemie. VCH, Weinheim
2. Chen S-H, Chan S-L, Strey R (1990) J Chem Phys 93:1907
3. Schlicht L, Spilgies J-H, Runge F, Lipgens S, Boye S, Schübel D, Ilgenfritz G (1996) Biophys Chem 58:39
4. Garcia-Rio L, Leis JR, Mejuto JC, Pena ME, Iglesias E (1994) Langmuir 10:1676
5. Nees D, Cichos U, Wolff T (1996) Ber Bunsenges Phys Chem 100:1372
6. Steiner U, Abdel-Kader MH, Fischer P, Kramer HEA (1978) J Am Chem Soc 100:3190
7. Abdel-Kader MH, Steiner UE (1983) J Chem Ed 60:160
8. Abdel-Halim ST, Abdel-Kader MH, Steiner UE (1988) J Phys Chem 92:2808
9. Ikeda N, Mataga N, Steiner U, Abdel-Kader MH (1983) Chem Phys Letters 95:66
10. Herz AH (1974) Photogr Sci Eng 18:323
11. Pope M, Swenberg CE (1982) Electronic Processes in Organic Crystals. Oxford University Press, New York, p 436
12. Mishra JK, Sahay AK, Mishra BK (1991) Ind J Chem 30A:886
13. Nees D, Wolff T (1996) Langmuir 12:4960
14. Nees D, Blenkle M, Koschade A, Wolff T, Baglioni P, Dei L (1996) Progr Colloid Polym Sci 101:75

Progr Colloid Polym Sci (1998) 111:117–126
© Steinkopff Verlag 1998

Th. Engels
W. von Rybinski
P. Schmiedel

Structure and dynamics of surfactant-based foams

Th. Engels (✉) · W. von Rybinski
P. Schmiedel
Henkel KGaA
Henkelstraße 67
D-40191 Düsseldorf
Germany

Abstract In nearly all areas in which surfactant containing systems are used, the tailoring of the foam performance of the products according to specific demands or needs during the application plays a major role. In detergency or cleansing processes, systems with rather low or no foaming abilities are requested. On the other hand, in the cosmetics industry the foam performance of body and hair care products have to be optimized so that the consumer will subjectively experience a most positive optical and sensorial impression during the application.

From the colloid chemistry or physics point of view, parameters like surface tension, surface elasticity and surface viscosity, etc. are responsible for the adjustment of the desired foam properties of the systems.

For dynamic processes, e.g. the generation of foam, not only equili-brium data, but kinetic parameters like the dynamic surface tension have to be regarded.

This paper will summarize the results of studies of the dynamic surface tension of different surfac-tants. A correlation has been found between the dynamic surface tension and results from a newly developed foam test, which is in excellent agreement with foam properties in application tests. The importance of the study will be demonstrated by two different types of product developments:

– a new detergent with optimized foam properties
– hair shampoo application and correlation to the subjective consumer impression.

Key words Foam – detergency – cosmetics – dynamic surface tension – applications

Introduction

In colloidal science phenomena like "Foaming" and Defoaming", together with related studies on foam lamellae, have gained a high degree of interest in many research groups worldwide. They deal with the many-fold aspects that can be related to this area. As a consequence, numerous publications exist (for a first review see [1–3]).

Although the basic principles that govern the foaming and defoaming action have been thoroughly studied in detail and are believed to be well understood, major guide-lines for the optimization of chemical systems with respect to their foaming or defoaming performance have only scarcely been published. As a consequence, in industrial product development much efforts have to be spent on systematic screening studies.

This publication deals with the topic "Foam" from a chemical industry point of view and especially covers the following aspects: correlation of foam kinetics with dynamic surface activity and consequences for the product development of detergents and hair care shampoos.

Experimental

Chemicals

Alcohol ethoxylates C_xE_y and alkylsulfonates C_xSO_3 with a purity of almost 100% (according to GC-measurements) have been purchased from Nikkol Chemicals. The linear C_{11-13}-alkylbenzenesulfonate (LAS or ABS) can be purchased from different suppliers. The studied C_{12-18}-fatty alcohol polyglycol ether with an average ethoxylation degree of 7, $C_{12,14}$-polyglycoside with an average polymerization degree of 1.4, $C_{12,14}$-fatty alcohol polyglycol ethersulfate with an average ethoxylation degree of 2 and a cocamidopropyl betaine (according to CTFA-nomenclature) were technical-grade surfactants of the HENKEL KGaA.

Dynamic surface tension

The dynamic surface tension data of the test solutions have been measured with the Lauda MPT 1 bubble pressure tensiometer. With commercial equipment, measurements of bubble surface ages ranging from approx. ten seconds to ten milliseconds are possible. A modified set-up at the *MPI für Kolloid- und Grenzflächenforschung* in Berlin-Teltow allowed for a further reduction of the bubble surface age down to 1 ms.

Rotor-test

The Rotor-test has been especially designed for a quick and reliable assessment of the foam kinetics of test solutions with and without the addition of defoaming agents [4] (see Fig. 1). The solution is thermostated in a cylindrical 2 l vessel with a water jacket. A stirrer with a rotating disk at the bottom end of the stirring rod is immersed. The disk has been modified in order to maximize the amount of foam generated. The revolutions per minute (rpm) of the stirrer can be set very precisely and controlled via an optical display. Usually, the stirrer is operated at 900, 1100 or 1300 rpm. During foam generation the foam height is recorded every 10 s. For the measurement itself the Rotor is stopped so that the foam column which is distorted by rotating and centrifugal forces can level out. The reading will only take about 3–5 s and does not affect neither foam height nor foam structure significantly. Usually, after 3 min the foam generation is complete. With this equipment a maximum of about 1200 ml of foam can be generated.

SSF-test (stress-stability of foam)

With the SSF-test the mechanical stability of foams can be measured (see Fig. 2). The foam of the test solution is generated in a cylindrical vessel with a thermostated water jacket by means of nitrogen gas passing through a very fine and even metal sieve. After 2 min the gas flow is stopped and an aluminum plate with a definite weight (usually 50 g) is placed on top of the foam column. The plate compresses the foam and passes down to the bottom of

Fig. 1 Rotor-test

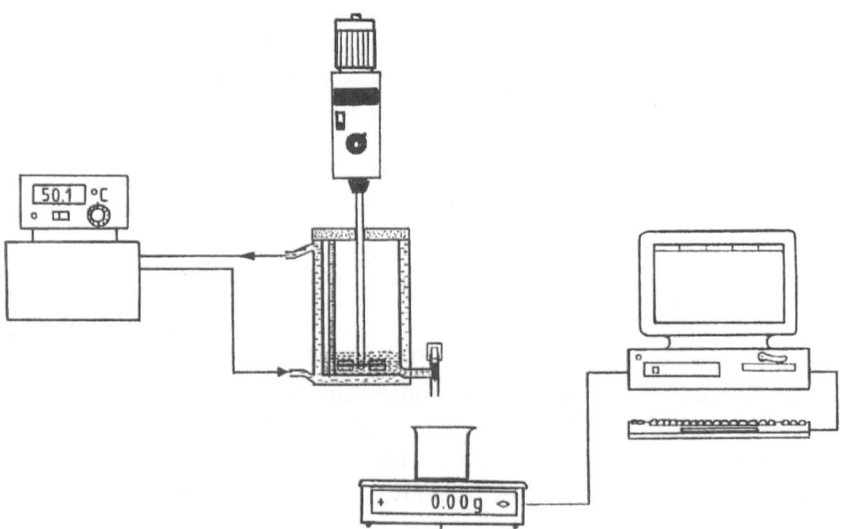

Progr Colloid Polym Sci (1998) 111:117–126
© Steinkopff Verlag 1998

Fig. 2 SSF-test (stress-stability of foam)

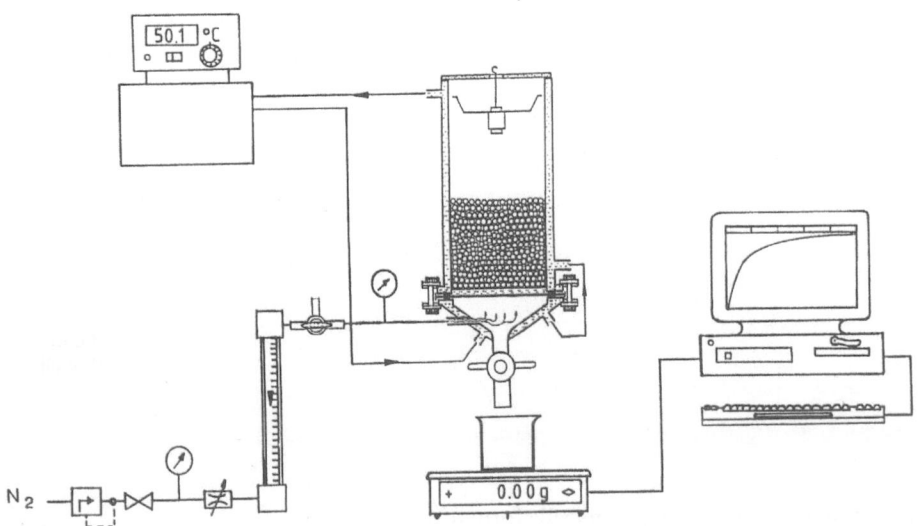

the vessel destroying the foam. The height or position of the plate as a function of time is a measure of the resisting force or the foam stability. Optionally defoamers may be added to the test solution prior to or after foam generation.

Half-head-test

The half-head test is a one-to-one comparison under realistic conditions of, e.g. a new hair shampoo development against a reference product. The two products are applied on each of the two hemispheres of the head of a test person. Usually, parameters like the foam kinetics (flash foam), the feel of the foam as well as the feel of the wet and dry hair, the visual appearance and the rinse-off characteristics of the foam are evaluated by a group of five experts.

Correlation of foam kinetics with dynamic surface activity

In general, the performance of a product can be described and quantified by physical chemical parameters. This has been shown for interfacial activity and wetting behavior of household products as well as technical applications [5, 6]. In addition to this, the optimization of the foam properties is one of the key objectives during the product development of surfactant-based products like shampoos, detergents, manual dishwashing agents, etc. For the optimization of foam parameters like volume, stability or kinetics, profound knowledge about the interrelation of experimental properties with more fundamental para-

meters is essential for an effective product optimization. Rosen and Solash [7] studied the relationship between initial foam height, H, as measured by the Ross–Miles test, and such factors as surfactant concentration, critical micelle concentration, c.m.c., of the solute, surface tension, σ, of the solution, surface area, A, of the foam, and the work involved in the production of the foam surface. The initial foam height was remarkably constant in character for the systems studied and there was a linear relationship between H and A. For five of the six solutes used, the work, dE, of producing the foam surface, dA, fell within a narrow range at concentrations in the neighborhood of the c.m.c., indicating an inverse relationship between A and σ ($dE = dA \cdot \sigma$). For these solutes, H approx. equals $(1000/\sigma) - 3.6$. The maximum in H occurred in the vicinity of the c.m.c. In a later study on six commercial anionic-zwitterionic or anionic-nonionic binary mixtures and a sodium dodecylbenzenesulfonate (LAS)-soap mixture Rosen and Zhu [8] investigated the relationship between synergism in Ross–Miles foaming and the existence of different types of synergism in equations derived from static parameters like concentrations, c.m.c's, mixing ratios, surface tension reductions etc. of the binary surfactant mixtures. Synergism in foaming effectiveness, measured by initial foam heights, appeared to be related to synergism in surface tension reduction effectiveness, but not to synergism in surface tension reduction efficiency or in mixed micelle formation. The LAS-soap system showed negative synergism in foaming effectiveness, correlated with negative synergism in surface tension reduction effectiveness. There appeared to be no correlation between synergism in foaming efficiency and synergism in either surface tension

Fig. 3 Major correlations
between experimental foam
kinetics as well as stability and
fundamental dynamic
interfacial parameters

Foaming of *Surfactant-Water* Systems
Correlation of experimental data with fundamental parameters

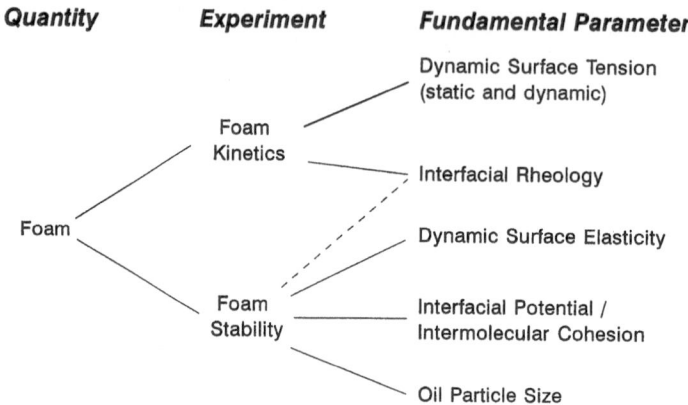

reduction efficiency or mixed micelle formation. There also appeared to be no unambiguous relationship between foam stability, measured by the ratio of the 5 min to the initial foam height, and the average area per surfactant molecule at the aqueous solution/air interface.

So far all these investigations concentrated on static parameters and were more or less unsatisfactory concerning the establishment of a correlation between foam kinetics and fundamental interfacial parameters. In a series of papers Varadaraj et al. [9–13] compared the fundamental interfacial properties of linear and branched sulfate and ethoxy sulfate surfactants derived from Guerbet alcohols. Based on a new method of analysis for the treatment of dynamic surface tension data suggested by Rosen et al. [14, 15] Varadaraj demonstrated that Ross–Miles "static" foam stabilities correlated with $R_{1/2}$, the rate of dynamic surface tension reduction. It is stated that according to Marangoni et al. [16], when the rate of surface tension reduction, $R_{1/2}$, is very high, the surface tension gradient between old and new areas of surface rapidly disappears leading to foam collapse.

It was the intention of the present study to perform additional investigations with emphasis on the "dynamic" aspects during foam generation (i.e. foam kinetics) and stabilization. In this respect Fig. 3 represents the assumed major correlations between experimental "dynamic" foam kinetics as well as stability and fundamental parameters concerning "dynamic" interfacial and solution properties. Though very important, the mechanical influence of the test equipment on foam generation and destruction has not been included in the picture. Likewise the interfacial rheology was assumed to be more or less independent on the systems studied.

During foam production with the Rotor-test the air bubbles generated under the influence of the rotational

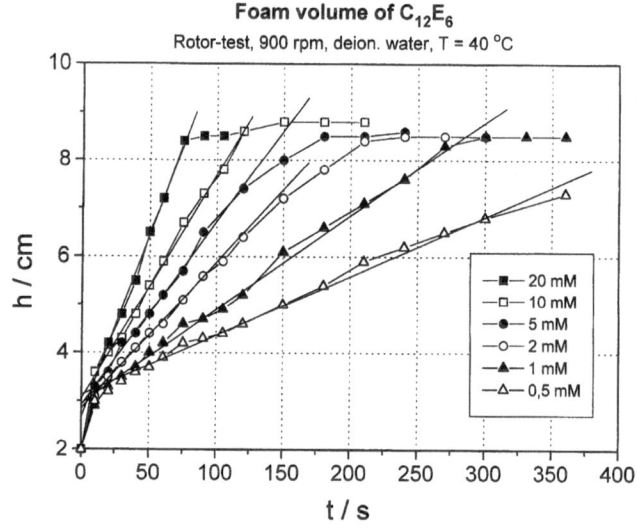

Fig. 4 Time-dependent foam heights of $C_{12}E_6$ at 40 °C as function of the surfactant concentration

mechanics of the dissolver disc have to be stabilized against coalescence via adsorption of surfactants at the air–solution interface. This process is believed to be very similar to the surfactant adsorption at a growing gas bubble during a dynamic surface tension study using a bubble pressure tensiometer.

For the correlation of the Rotor-test foam kinetics data with dynamic surface tensions model surfactants like sodium dodecylsulfonate (high foam) and hexaethylene glycol-dodecyl ether (poor foam) have been studied systematically. The dodecylsulfonate was preferred to dodecylsulfate in order to avoid hydrolysis problems, which may influence the surface tension data [17].

Figure 4 shows the time-dependent foam heights of $C_{12}E_6$ at 40 °C as a function of the surfactant concentra-

Progr Colloid Polym Sci (1998) 111:117–126
© Steinkopff Verlag 1998

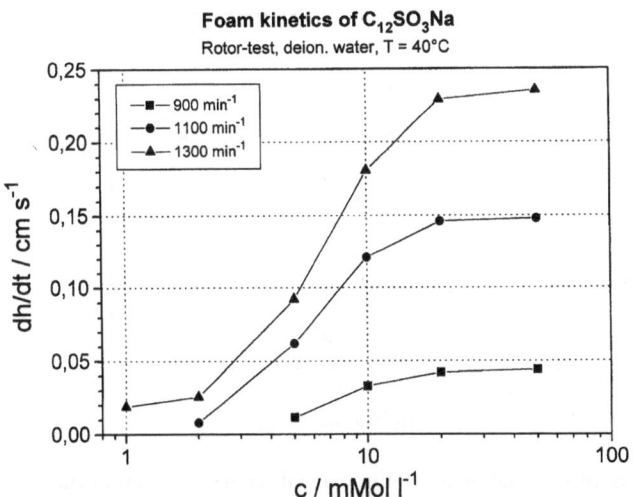

Fig. 5 Foam kinetics of $C_{12}SO_3Na$ as function of concentration and revolutions per minute of the rotor at 40 °C (Krafft point of $C_{12}SO_3Na = 33$ °C; c.m.c.: 11 mM)

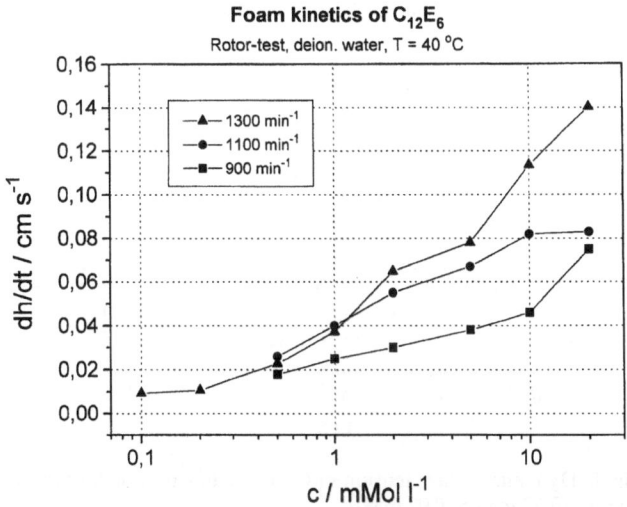

Fig. 6 Foam kinetics of $C_{12}E_6$ as function of concentration and revolutions per minute of the rotor at 40 °C (c.m.c.: 0.07 mM)

tion using deionized water (Rotor-test operating at 900 rpm). The bulk volume of the test solution itself adds two centimeters to the ordinate value. Thus for the determination of the foam height from the diagram two centimeters have to be subtracted. As a function of time the foam height increases almost linearly up to nearly 9 cm for all the concentrations measured. The foam height data level out at a specific concentration dependent point in time indicating that the solution is not able to incorporate air any more. The slope of the linear part of the foam height data vs. time curve calculated by linear regression represents the foam kinetics of the solution. Figures 5 and 6 show the kinetics of $C_{12}E_6$ and $C_{12}SO_3Na$ as a function of concentration and revolutions per minute of the rotor at 40 °C (Krafft point of $C_{12}SO_3Na = 33$ °C). Whereas the ionic $C_{12}SO_3Na$ exhibits a steep increase in foam kinetics when approaching its critical micelle concentration (c.m.c. = 11 mM) followed by a level out of the kinetics data, the nonionic $C_{12}E_6$ does not foam below its c.m.c. (= 0.07 mM) and the foam kinetics only slowly increases above the c.m.c. which might result from a different mechanism governing the foam kinetics.

The dynamic surface tensions of the two surfactants are displayed in Figs. 7 and 8. In the case of $C_{12}SO_3Na$ dynamic surface tensions are generally high at concentrations well below the c.m.c., nearly independent from the surface age at almost all concentrations and a function of the overall concentration. At concentrations, c, above the c.m.c there is no big difference in the dynamic surface pressures dependent on surface age and concentration. These data suggest that in the case of $C_{12}SO_3Na$ the surfactant adsorption at the air–solution interface may be

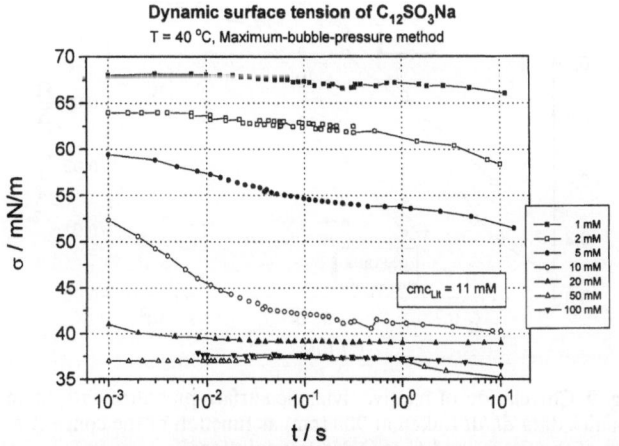

Fig. 7 Dynamic surface tension of $C_{12}SO_3Na$ as function of concentration at 40 °C (c.m.c.: 11 mM)

mainly dominated by the monomers rather than the micelles. On the other hand for nearly all concentrations of $C_{12}E_6$, measured only above the c.m.c., a considerable dependence of the dynamic surface tension on the bubble surface age and the concentration can be detected. This indicates that the micelle concentration is the major parameter concerning surface activity.

For the purpose of comparison with foam data the relative dynamic surface pressures, Π/Π_{eq}, were calculated for a certain surface age, t, from the dynamic and equilibrium surface tension data according to the following equation:

$$\Pi(t)/\Pi_{eq} = (\sigma_{water} - \sigma(t))/\sigma_{water} - \sigma_{eq}\,,$$

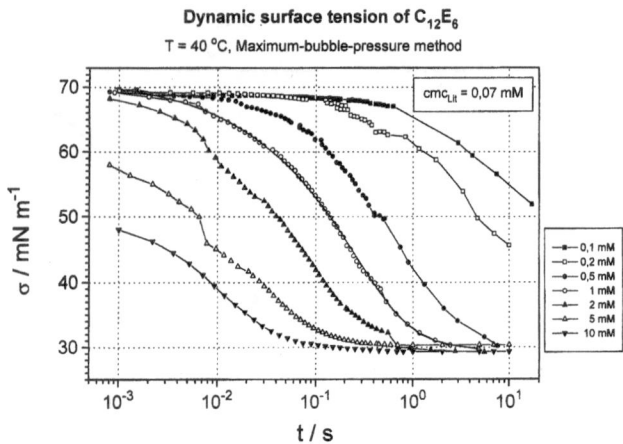

Fig. 8 Dynamic surface tension of $C_{12}E_6$ as function of concentration at 40 °C (c.m.c.: 0.07 mM)

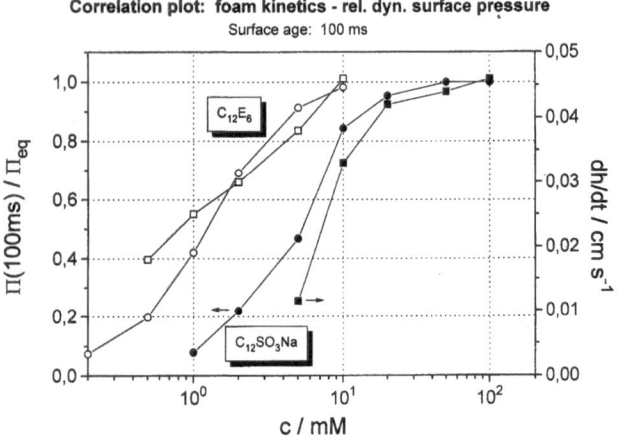

Fig. 9 Correlation of relative dynamic surface pressures with foam kinetics data dh/dt (taken at 900 rpm) as function of the concentration of the $C_{12}E_6$ and $C_{12}SO_3Na$ test solutions

where t is the surface age of the bubble, σ_{eq} the equilibrium surface tension and σ_{water} the equilibrium surface tension of water ($=69$ mN/m at 40 °C).

At a surface age, t, of approx. 100 ms the relative dynamic surface pressures were in line with the foam kinetics data dh/dt as a function of the concentration of the $C_{12}E_6$ and $C_{12}SO_3Na$ test solutions (see Fig. 9). This means that experimental foam kinetics data can be described successfully in terms of relative dynamic surface pressures at specific surface ages which are characteristic for the time scale of the foam generation process. Current investigations concerning the influence of variations in the alkyl chain length of the surfactant and the salt dependence have been summarized in Fig. 10. Increasing the alkyl chain length at a fixed molar concentration or increasing the salt concentration leads to a reduction of the

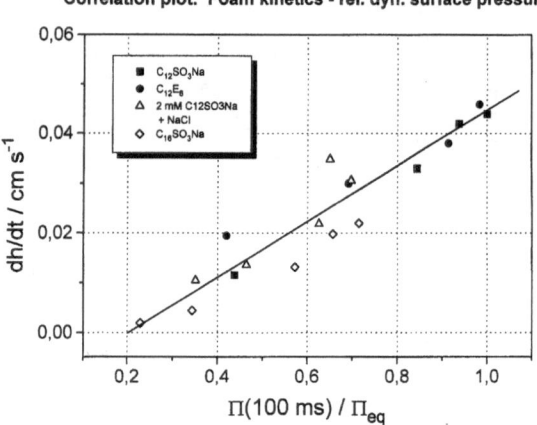

Fig. 10 Correlation of relative dynamic surface pressures with foam kinetics data dh/dt (taken at 900 rpm) as function of type of surfactant, alkyl chain length and salt concentration of the test solution

c.m.c. and the micellar kinetics which as a consequence has an influence on the relative dynamic surface pressure. The presence of micelles at the expense of monomeric surfactants as well as the higher molecular weight of the surfactants with longer alkyl chains reduces the rate at which monomeric surfactants can be incorporated into the freshly generated foam surfaces for stabilization.

As a matter of fact the experimental data displayed in Fig. 10 suggest that in general a fairly satisfying correlation exists for the studied systems between foam kinetics and relative dynamic surface pressures stressing the importance of the dynamic aspects during foam generation and stabilization. However, it is not claimed that the relative dynamic surface pressure is the one and only parameter governing the foam kinetics. After all we finally believe that the correlation can be further improved if additional parameters will be taken into account especially when systems containing surfactant mixtures are taken into consideration.

Foam characteristics and detergent concepts

The correlation between foam characteristics and physical chemical parameters like the dynamic surface tension has consequences for the development of complex formulations. For the development of detergents and the optimization of the washing process the Sinner's wash cycle addresses the parameters of major influence. According to it the chemistry, the temperature, time and the mechanics dominate the washing performance [18]. With respect to the last point the mechanical energy of an automatic drum-type washing machine contributes to more than 25% to total detergency. This is why up to now detergents

Progr Colloid Polym Sci (1998) 111:117–126
© Steinkopff Verlag 1998

with almost no foam have been developed because the presence of foam dampens the mechanical impact of the rotating drum on the textiles and thus reduces the washing performance.

On the other hand it is well known that mechanical stress may lead to considerable textile damage, color fading and deformation of textile shape. As a consequence a detergent which produces an extremely fine foam with high mechanical stability should reduce the mechanical stress which is highly desirable especially for fine textiles made of silk or viscose. During the washing process the foam then sheathes and protects the textiles against damage leading to an improved performance with regard to color and shape retention.

A newly developed detergent demonstrates these aspects. Its extremely fine and stable foam results from an optimized surfactant combination of linear alkyl-benzenesulfonate (LAS or ABS), C_{12-18}-fatty alcohol polyglycol ether with an average ethoxylation degree of 7 and $C_{12,14}$-polyglycoside with an average polymerization degree of 1.4.

Solid mineral particles like soda, hydrophobic or hydrophilic silica or zeolites are important components in powder detergents. The presence of these particles in the foam lamellae can significantly destabilize or stabilize the foam depending on their size, shape, chemistry and hydrophilic/hydrophobic nature [19–23]. In Fig. 11 the positive influence of the addition of hydrophilic silica on the foam stability of the new product at 40 °C and 60 °C is demonstrated using the SSF-test for a foam that has additionally been loaded with 30 mg/l sebum representing a typical textile soil.

Different fine textile detergents have been compared at 40 °C using the SSF-test and applying different kinds and amounts of textile soils (see Fig. 12). The addition of 4 g/l

Foam stabilities of Light-duty detergents
T = 40 °C; 160 mg CaO/l; recommended dosage of detergent

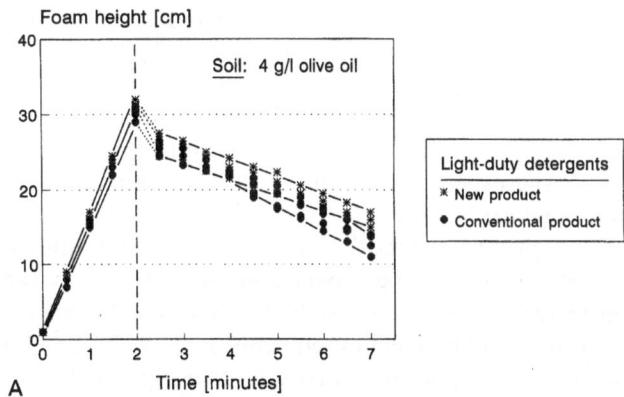

A

Foam stabilities of Light-duty detergents
T = 40 °C; 160 mg CaO/l; recommended dosage of detergent

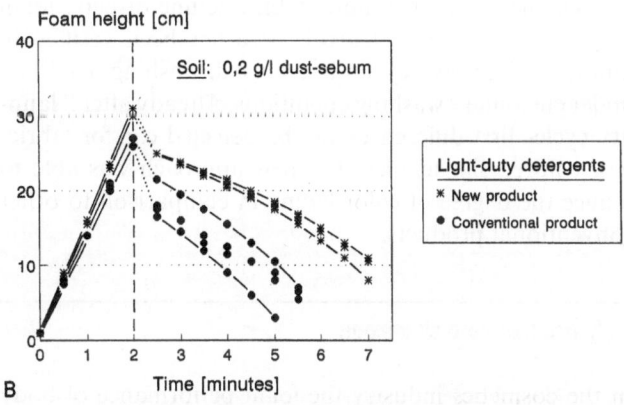

B

Foam stabilities of Light-duty detergents
T = 40 °C; 160 mg CaO/l; recommended dosage of detergent

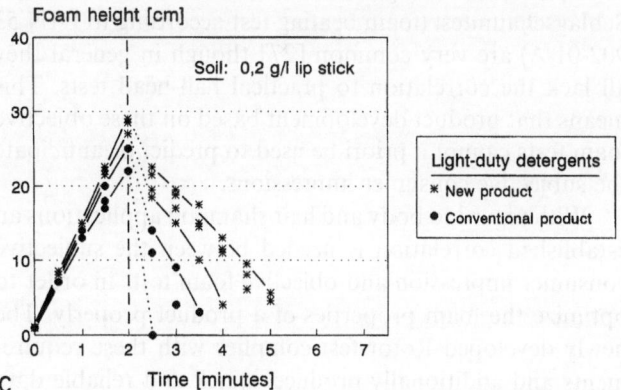

C

Foam stabilities of Light-duty detergents
T = 40 °C and 60 °C; 160 mg CaO/l; recom. detergent dosage

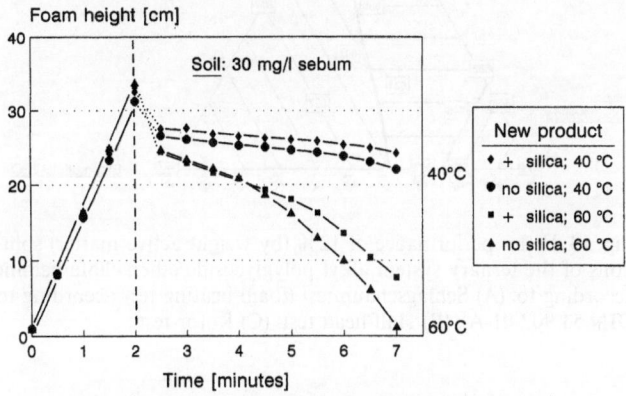

Fig. 11 Foam stability of the new product (SSF-test)

Fig. 12 Foam stability of fine textile detergents (SSF-test): (A) soil: 4 g/l olive oil; (B) soil: 0.2 g/l dust-sebum; (C) soil: 0.2 g/l lip stick

olive oil as model soil to the test solutions shows only minor advantages for the new product with, in general, rather high foam stabilities. The scatter of the foam data between different measurements shown in Fig. 12A is only small proving the good reproducibility of the SSF-test.

Generally, soils which are composed of oily liquid and crystalline solid particles [24–26] are better defoamers than oily components like olive oil alone. As a consequence the experiments with 0.2 g/l dust-sebum soil exhibit significantly lower foam stabilities (Fig. 12B) together with a high scatter of the data in the case of a conventional product whereas the new foam concept is verified by the considerably higher foam stabilities and smaller scatter between data of different experiments. Finally, if 0.2 g/l lip stick (again a liquid (oily wax)-solid (red pigments) composition; see Fig. 12C) is taken as soil the lowest foam stabilities are encountered and as with the preceding two soils, again the new product provides the more stable foam.

Whether or not the more stable and fine-disperse foam will result in improved textile care has been verified in multiple laundering cycles in the usual washing machines under customary washing conditions. Already after 2 laundry cycles first differences can be detected e.g. for fabric-new colored silk textiles. The new approach was able to reduce the degree of color fading in comparison to other conventional products.

Body and hair care shampoos

In the cosmetics industry the foam performance of body and hair care products have to be optimized so that the consumer will subjectively experience a most positive optical and sensorial impression during application.

Up to now among others the Ross–Miles test and the Schlagschaumtest (foam beating test according to DIN 53 902-01-A) are very common [27] though in general they all lack the correlation to practical half-head tests. This means that product development based on these objective foam tests cannot a priori be used to predict or anticipate the subjective consumer impressions.

With regard to body and hair shampoo applications an established correlation is needed between the subjective consumer impression and objective foam tests in order to optimize the foam properties of a product properly. The newly developed Rotor-test complies with these requirements and additionally produces quick and reliable data about the foam kinetics of products or solutions with a high degree of differentiation. The Rotor-test is widely applicable with respect to concentration, water hardness and temperature (0–approx. 70 °C) of the test solution as well as chemical type and amount of soils.

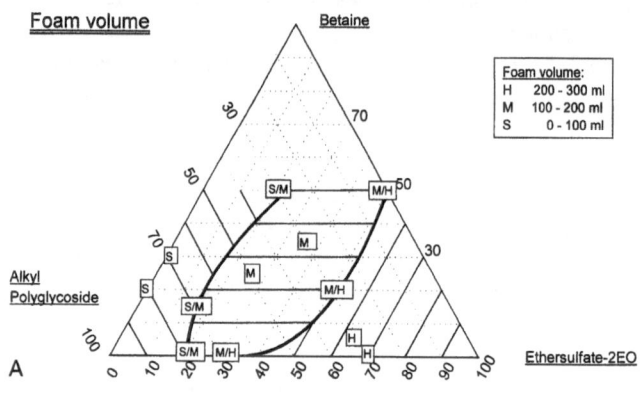

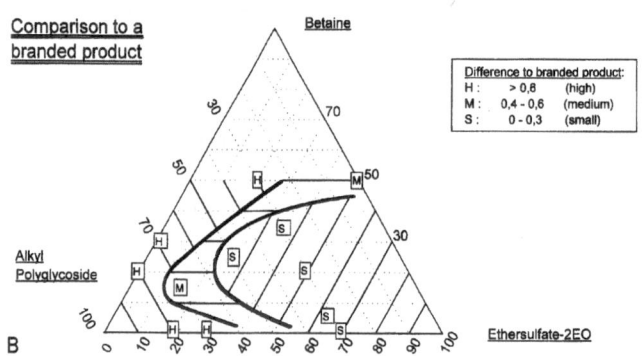

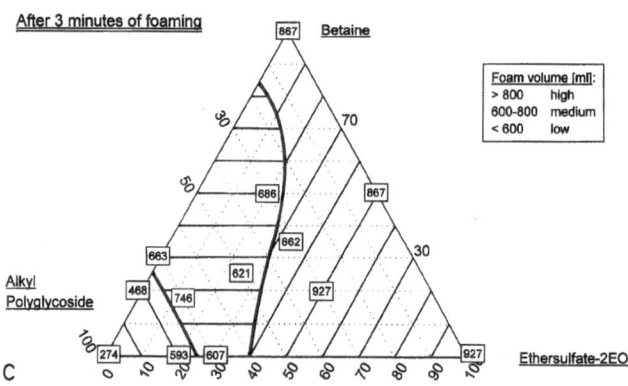

Fig. 13 Foam performance of 12% (by weight active matter) solutions of the ternary system alkyl polyglycoside/ethersulfate/betaine according to: (A) Schlagschaumtest (foam beating test according to DIN 53 902-01-A); (B) Half-head test; (C) Rotor-test

Progr Colloid Polym Sci (1998) 111:117–126
© Steinkopff Verlag 1998

The optimization of the foam performance of alkyl polyglycosides – a new class of mild and readily biodegradable surfactants made from renewable resources such as coconut oil and starch and thus satisfying the requirements for a sustainable development [28] – may act as an example.

In Fig. 13 the foam performances of 12% (by weight active matter) solutions of the ternary system alkyl polyglycoside/ethersulfate/betaine (at 40 °C with 150 mg/l CaO water hardness (=15°dH) and addition of 0.1 g/l sebum) are compared according to experiments using the Schlagschaumtest (foam beating test according to DIN 53 902-01-A), half-head test and Rotor-test. In the Schlagschaumtest (Fig. 13A) the foam volume is directly related to the ethersulfate content. Starting in the ethersulfate-rich corner of the phase triangle the foam volume almost constantly decreases on reduction of the ethersulfate content down to zero until the binary system alkyl polyglycoside – betaine is reached passing through a quite extensive area with average foam volumes. When these findings are compared to the results of the half-head test (Fig. 13B) it becomes obvious that the subjective half-head test results are more polarized. The "foam" is either good or bad. Only a very small area of mediocre foam performance can be found. Additionally the half-head tests tends to produce evidence for an allegedly higher foam performance of the test product. The area with the best foam results in the ethersulfate-rich corner extending at the expense of the alkyl polyglycoside/betaine-rich side when compared to the results of the Schlagschaumtest.

The foam performance according to the Rotor-test (Fig. 13C) though different from the half-head test from its first visual appearance, exhibits some essential concurrences. First of all the Rotor-test too manifests a large area with good foam performance on the ethersulfate-rich side. Secondly, the area with minor foam quality is limited to the alkyl polyglycoside-rich corner of the phase triangle. Finally no linear dependence of the foam performance on the ethersulfate content of the formulation exists as with

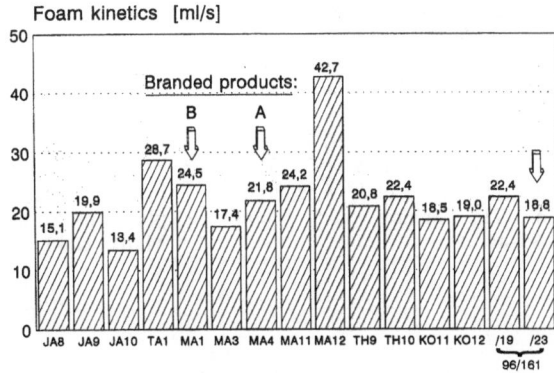

Foam kinetics using 20 g/l product
T=40°C, 150 mg CaO/l, 0,1 g/l sebum

Fig. 14 Foam kinetics of shampoo market products at concentration of use

the Schlagschaumtest but rather an irregular relationship between foam performance and phase triangle composition is brought to view.

The high degree of differentiation of the Rotor-test compared to the half-head test is shown in Fig. 14. Using the half-head test the foam kinetics of market product A cannot be distinguished from market product B. With the Rotor-test even very small differences can be uncovered which are indispensable for an effective product development.

The above examples clearly demonstrate the importance of suitable physical chemical parameters for the foam characterization. A specific foam test developed on these results enables the characterization and optimization of foam properties.

Acknowledgements We would like to thank Dr R. Miller, Max-Planck Institute for Colloid and Interface Science (Berlin), for his assistance in dynamic surface tension measurements in the short-time range.

References

1. Prud'homme RK, Khan SA (eds) (1996) Foams – Theory, Measurements, and Applications, Surfactant Science Series 57. Marcel Dekker, New York
2. Garrett PR (ed) (1993) Defoaming – Theory and Industrial Applications, Surfactant Science Series 45. Marcel Dekker, New York
3. Rosen MJ (1989) Surfactants and Interfacial Phenomena, 2nd ed, Ch 7. Wiley, New York
4. Engels Th, Kahre J, Tesmann H, to be published
5. Nickel D, Speckmann HD, von Rybinski W (1995) Tenside Surf Det 32:470
6. Engels T, Förster T, Mathis R, von Rybinski W (1991) Abstracts of the 7th ICSCS – Vol I, Part 2, 265
7. Rosen MJ, Solash J (1969) J Amer Oil Chem Soc 46:399
8. Rosen MJ, Zhu ZH (1988) J Amer Oil Chem Soc 65:663
9. Varadaraj R, Bock J, Valint P, Zushma S, Brons N (1990) J Colloid Interface Sci 140(1):31
10. Varadaraj R, Bock J, Valint P, Zushma S, Thomas R (1991) J Phys Chem 95:1671
11. Varadaraj R, Bock J, Valint P, Zushma S, Brons R (1991) J Phys Chem 95:1677
12. Varadaraj R, Bock J, Valint P, Zushma S, Brons N (1991) J Phys Chem 95:1679

13. Varadaraj R, Bock J Valint P, Zushma S (1991) J Phys Chem 95:1682
14. Hua X-Y, Rosen MJ (1988) J Colloid Interface Sci 124:652
15. Hua X-Y, Rosen MJ (1991) J Colloid Interface Sci 141:180
16. Gibbs JW (1931) Collected Works of J. Willard Gibbs. p 300, Longmans, Green, New York
17. Mysels KJ (1990) Colloid Surf 43:241
18. Puchta R, Grünewälder W (1973) Textilpflege, Waschen und Chemischreinigen. VIg. Schiele und Schön Berlin, 35–44

19. Garrett PR (1993) in Defoaming – Theory and Industrial Applications, Surfactant Science Series 45. Marcel Dekker, New York, pp 1–117
20. Dippenaar A (1982) Int J Mineral Process 9:1–22
21. Livshitz AK, Dudenkov SV (1954) Tsvet Metally 30(1):14
22. Livshitz AK (1965) Dudenkov SV, Proc 7th IMPC New York, p 367
23. Ferch H, Leonhardt W (1993) in Defoaming – Theory and Industrial Applications, Surfactant Science Series 45. Marcel Dekker, New York, pp 221–268

24. Aveyard R, Cooper P, Fletcher PDI, Rutherford CE (1993) Langmuir 9:604
25. Aveyard R, Binks P, Fletcher PDL Rutherford CE (1994) J Disp Sci Technol 15:251
26. Aveyard R, Binks P, Fletcher PDL, Peck TG, Rutherford CE (1994) Colloid Interface Sci 48:93
27. Domingo X, Fiquet L, Meijer H (1992) Tens Surf Det 29:16
28. Hill K, von Rybinski W, Stoll G (1997) Alkyl Polyglycosides. VCH Verlagsgesellschaft, Weinheim

Progr Colloid Polym Sci (1998) 111:127–134
© Steinkopff Verlag 1998

MACROMOLECULES

P. Fischer
H. Finkelmann

Lyotropic liquid-crystalline elastomers

P. Fischer* · Prof. Dr. H. Finkelmann (✉)
Institut für Makromolekulare Chemie
der Universität Freiburg
Sonnenstr. 5
D-79104 Freiburg
Germany
E-mail: finkelma@ruf.uni-freiburg.de

*Present address:
Université Catholique de Louvain
Unité Cinétique et Macromolecules
Place L. Pasteur 1
B-1348 Louvain-la-Neuve
Belgium

Abstract The linkage of nonionic amphiphiles via their hydrophobic ends to the monomer units of a polymer backbone leads to broad hexagonal (H_1) and lamellar (L_α) phase regimes in binary mixture with water. Cross-linking of these linear polymers yields rubber-like samples with elastomeric properties, which do not dissolve, but swell with a certain amount of water to form lyotropic mesophases. Mechanical deformation of samples in the mesophase causes a reversible alignment of the hexagonal or lamellar domains. Stressing of the gels during the synthesis of the elastomers enables to lock-in polymer anisotropy. As a result, spontaneously well-aligned LC-samples are produced by swelling with water. The type of alignment depends on the symmetry of the mesophase and the mechanical field. Due to the scaling between macroscopic dimensions and local anisotropy, changes of micellar shape and mesophase structure are visualized as changes of the sample length. This is measured by the hygroelastic method, a combination of simultaneous measurements of water sorption and sample length. It can be shown that the phase transformation between isotropic (L_2) and L_α-phase, which is driven by the change of partial vapor pressure of H_2O at constant temperature, is accompanied by a discontinuous lengthening of the network in the direction of the stress applied during synthesis. Due to its highly anisotropic swelling behavior, the sample length remains nearly constant within the L_α-phase, although water is absorbed and the volume increases.

Key words Amphiphilic polymers – elastomers – lyotropic liquid-crystalline phases – mechanical field – hygroelastic effect – stimuli-responsive hydrogels

Introduction

A trend in modern polymer science consists of exploring what has been termed as tandem molecular interactions [1]. Combining specific organized components within a macromolecule yields new materials, which do not only sample the properties of their constituents, but also show new ones owing to the "tandem interactions" – the synergism of these qualities. Examples of such new materials are LC-ionomers [2], block copolymers with LC-blocks [3, 4] or end-capped by ionic species [5], thermoplastic LC-block [6] and permanently cross-linked LC-elastomers [7].

The range of these cooperative phenomena can even be broadened using additional physical constraints as imposed, e.g. by interfaces or mechanical fields, which may cause different structure formation, orientation or alignment processes. Prominent examples are shear effects on liquid-crystalline [8] or block copolymers [9, 10] interface effects on polymer-dispersed liquid crystals [11], surface adsorption of alternately charged polyelectrolyte layers [12] or geometrical constraints on the behavior of thin films from block copolymers [13] and polymer blends [14].

In this paper we present an example of such a tandem molecular interaction which to date has only hardly been investigated. The first component are amphiphiles, which display numerous aggregation phenomena induced by hydrophobic interactions. The second one are weakly cross-linked polymer networks, which exhibit rubber-like mechanical properties. We will describe the interplay between amphiphilic and elastomeric behavior on the lyotropic mesophases of these materials and the effect of mechanical fields on their liquid-crystalline organization.

Motivation

In solution, depending on concentration and temperature, the micelles formed by monomeric or polymeric surfactants can transform to lyotropic liquid crystals. Especially, the lamellar phase (L_α) composed of regularly stacked sheet-like micelles and the hexagonal phase (H_1) built from hexagonally arranged rod-like aggregates, are often observed.

Polysoaps that consist of amphiphilic side-chains linked to a polymer backbone [15] already show tandem interactions due to the superposition of their micellar organization and polymer-specific properties:

- The aggregation number N of a spherical micelle is limited by its finite surface area and the requirement of space-filling which limits the radius. If the degree of polymerization of a polysoap exceeds N, the micelle has to grow in one or two dimensions yielding an anisotropic rod-like or sheet-like micellar shape. Compared to the monomeric analogues, this is often manifested in the phase behavior: the H_1 and L_α-phases are favored, whereas the I-phases (bcc or fcc-packed spherical micelles) are suppressed [16].

- The geometrical constraints dictated by an anisometric micellar shape act as a template, which influences the polymer conformation. A hydrophobic polymer backbone that is linked to the hydrophobic ends of the amphiphilic side-chains has to remain mainly in the hydrophobic interior of the micelle. As sketched in Fig. 1 this will directly lead to a significant polymer anisotropy, which is in contrast to the isotropic coil conformation that is normally encountered in solution. For a planar micelle the packing of the polymer chains is strongly confined in the direction normal to the interface. As a consequence, an oblate polymer conformation results. For rodlike micelles the random walk is even more constrained in two dimensions. A prolate overall polymer shape with the main axis coinciding with the rod axis is thus expected.

The cross-linking of polysoaps introduces properties, which are well known from conventional elastomers: the materials show form stability, an equilibrium degree of swelling in solvents and the phenomenon of entropy elasticity, which is the basis for their rubber-like behavior. Mechanical deformation of the networks leads to anisotropic chain conformations. As entropy favors random coiling with similar expansion in all Cartesian coordinates, the polymer anisotropy of the network strands will relax to a spherical conformation upon cessation of the mechanical force.

The tandem interactions between elastomeric and micellar qualities should be exemplified in two important aspects, which are the motivation for our investigations on these materials:

(1) The limited degree of swelling and the constraints on supramolecular organization by the cross-linked polymer strands should influence the lyotropic phase behavior of amphiphilic side-chain polymer networks. This will be described in detail in the following sections.

(2) By mechanical deformation of the networks, polymer anisotropy can be easily induced. A crucial question is,

Fig. 1 Polymer conformations in dependence of different micelle geometries. The coordinate systems show schematically the restriction of chain extension in different directions

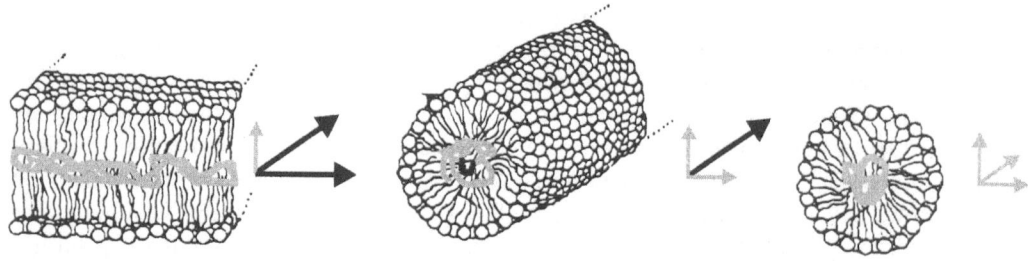

Progr Colloid Polym Sci (1998) 111:127–134
© Steinkopff Verlag 1998

whether this interferes with the anisotropy of the backbone conformations caused by the anisotropic micellar shape in lyotropic mesophases. We will show that this interaction will lead to alignment phenomena of the mesophase structure, which can be exploited for the synthesis of the so-called "lyotropic liquid single-crystal elastomers". Their macroscopically anisotropic character can be accessed by studying macroscopic properties, which will be exemplified in the section on the "hygroelastic" behavior.

Materials

For the synthesis of lyotropic liquid-crystalline elastomers two different pathways have been used. The first of these was the preparation of methacrylates esterified with amphiphilic alcohols followed by subsequent radical copolymerization with hydroxyethylmethacrylate. The resulting polymers can be characterized prior to cross-linking, which was achieved by the addition reaction of the hydroxy moieties with (methylene diphenyl diisocyanate) (MDI) [17].

The second way consists of a polymer-analogous hydrosilylation reaction. Amphiphilic precursors end-capped with an 1-alkene moiety react with polymethylhydrogensiloxane (PMHS). The polymer strands are interconnected by the presence of amphiphilic bis(1-alkene) compounds.

As precursor molecules nonionic amphiphiles are preferred as they avoid complex polyelectrolyte behavior of the resulting polymer. To date only methyl-oligoethyleneoxy-alkyl ethers have been used, which are functionalized for polymer synthesis at the hydrophobic end. Due to the availability of cheap α,ω-functionalized synthons the hydrophobic part normally consists of an undecylene $((CH_2)_{11})$-moiety. By variation of the chain length of the ethyleneoxy part the phase behavior can be tailor-made.

Hexamers and longer units induce hexagonal phases, smaller ones favor lamellar phases, but give also rise to very low critical solution temperatures (LCST) [18].

The composition of the networks described in this paper is sketched in Scheme 1.

The amphiphilic side-chain precursors are a tetra (ethyleneoxy)methyl-(-undecenyl-)ether (1a) and a hexa (ethyleneoxy)methyl-(ω-undecenyl-α,α-D$_2$)ether (1b). The deuteron labeling of the hexaethyleneoxy compound enables investigations on the phase behavior, state of order and dynamics of these systems by ^{2}H-NMR-spectroscopy, which will be presented in some forthcoming papers. Cross-linking is achieved by hydrosilylation reaction with a bifunctional bis(ω-undecenyl)-dodeca (ethyleneoxy)ether (2) in a ratio of polyoxymethylsilylene/1/2 = 1/0.85/0.075.

Phase behavior

As already stated, the limitation of molecular mobility in amphiphilic side-chain polymers and the trend towards anisotropic micellar shapes normally leads to an increase of concentration and temperature ranges of the H_1 and L_α-mesophases compared to the monomeric analogues [16]. In this section we will focus on the question whether the lyotropic phase behavior is further modified by crosslinking the polymer backbones. A considerable effect on the mesophase stability should not be surprising as conformational freedom of the chains should be significantly reduced due to their mutual connection by cross-linking. On the other hand, the covalent connections between different micelles should decrease the dynamics of the molecules and aggregates within the mesophase. Thermal fluctuations of the micelles, which give rise to the fragmentation of lyotropic aggregates, should be damped and thus the mesophases should become more stable.

Scheme 1 Chemical composition of the investigated networks

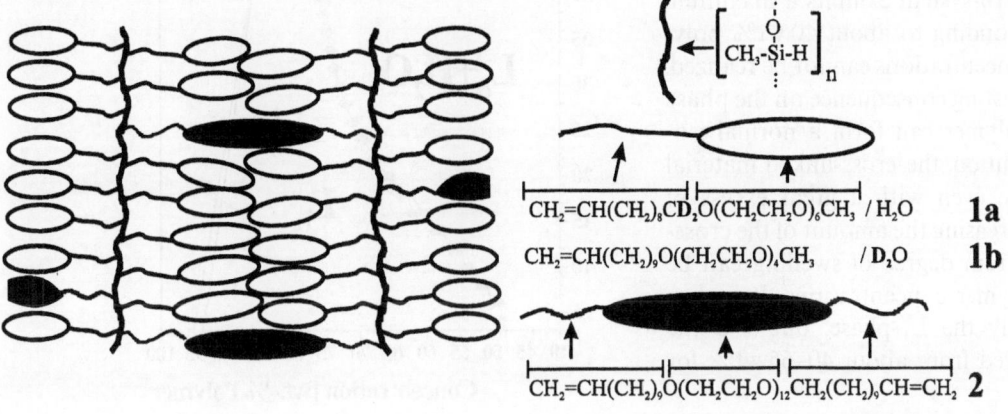

$$CH_2=CH(CH_2)_8CD_2O(CH_2CH_2O)_6CH_3 \; / \, H_2O \quad \textbf{1a}$$

$$CH_2=CH(CH_2)_9O(CH_2CH_2O)_4CH_3 \quad / \, D_2O \quad \textbf{1b}$$

$$CH_2=CH(CH_2)_9O(CH_2CH_2O)_{12}CH_2(CH_2)_9CH=CH_2 \quad \textbf{2}$$

To investigate the effect of cross-linking on the phase behavior, two important requirements have to be met: first the networks have to be carefully deswollen to remove the sol fraction. A second important aspect is that the chemical composition of the cross-linking agent may change the hydrophobic–hydrophilic balance and might induce changes in the phase behavior.

Bis(ω-undecenyl)-dodeca(ethyleneoxy)ether (2), the cross-linker, which was synthesized to avoid this effect, is nearly similar to a head group dimer of the hexaoxyethylene system (1b). Both the contour length and the similar amphiphilic character should ensure that the cross-links do only minimally disturb the local membrane structure for "intermicellar" connections. On contrast, the dodeca-oxyethylene part which is not compatible with a paraffine environment should force unfavorable interactions for "intramicellar" cross-links, which do not bridge over a hydrophilic region.

We find especially by ^{2}H-NMR-specroscopical results that the phase behavior is not significantly changed in the regime of high polymer concentration. The maximum clearing temperature of (1b) of about 70 °C was not changed within the limits of the experimental error even for a system with a composition of PMHS/1b/2 = 1/0.6/0.2. This is in contrast to the results of Löffler on methacrylate networks [17] and to previous results on the same systems, where the mesophases are destabilized with increasing amount of a 1,4-bis(undeceneoxy)benzene cross-linking agent. Thus, the topology and hydrophilic–hydrophobic balance of the cross-linker are a crucial factor for the phase behavior.

Contrary to these results, in the high concentration regime, a remarkable effect of cross-linking on the phase behavior is observed at the water-rich side of the phase diagram. This is attributed to the equilibrium degree of swelling of the network as mentioned above. Because of the given PMHS-prepolymers (degree of polymerization ≈ 70), we are not able to synthesize networks of sufficient mechanical stability, if the concentration of cross-linker is below molar ratios of PMHS:1b:2 = 1: 0.93:0.035. Therefore, the 1b-system exhibits a maximum degree of swelling corresponding to about 40 wt% polymer and lower polymer concentrations cannot be realized. This has a direct and interesting consequence on the phase behavior. Whereas the polymer can form a normal isotropic phase (L$_1$) upon dilution, the cross-linked material remains in the mesophase even with a great excess of water. Furthermore, by increasing the amount of the cross-linking agent, the equilibrium degree of swelling can be shifted even to higher polymer concentrations. It is thus possible to avoid not only the L$_1$-phase, but also the H$_1$-phase, which is observed from about 40–65 wt% for the linear polymer.

The equilibrium degree of swelling therefore constitutes a demixing line in the phase diagram. In contrast to monomeric amphiphiles, the whole demixing line can be readily recorded in a *single* experiment by simply measuring the extension of sample length (λ_y) and width (λ_x) as a function of temperature for a polydomain sample immersed in water. Provided the swelling is isotropic ($\lambda_x = \lambda_y \equiv \lambda$), the volume degree of swelling Q and the corresponding weight fraction of polymer ω_2 can be calculated by

$$\omega_2 = \frac{\rho_2}{(Q-1)\rho_1 + \rho_2} = \rho_2/[(\lambda^3 - 1)\rho_1 + \rho_2] , \qquad (1)$$

where ρ_1 and ρ_2 denote the densities of water and polymer. In Fig. 2 the phase diagram and calculated demixing line for an elastomer with a composition of PMHS/1a/2 = 1/0.85/0.075 is shown. The data were recorded with a computer-controlled apparatus composed of a thermostat, a water bath and a video camera with a setup similar to the one used for the hygroelastic measurements.

Owing to the marked temperature dependence of the hydrophilicity of the ethylene glycol units the polymer weight fraction varies from about 50 to 95 wt% for a temperature variation from 5 to 60 °C. This is similar to the behavior of the linear polymer 1a, which has a broad two-phase regime with a LCST of 10 °C. The equilibrium line of swelling for the network runs nearly parallel to the demixing line of the polymer above 10 °C. Below this

Fig. 2 Phase diagram calculated from swelling equilibrium measurements of a network with the composition PMHS:1a:2 = 1:0.85:0.075

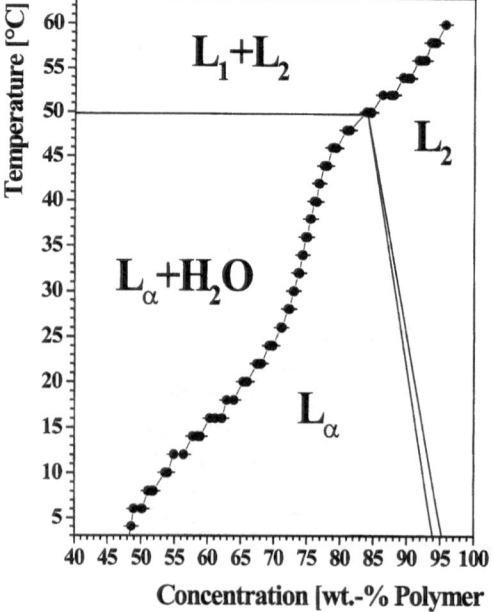

temperature the polymer shows a pure L_α-phase up to 20 wt% polymer, whereas the network cannot swell to concentrations below 50 wt%.

Effects of mechanical fields and synthesis of lyotropic single-crystal elastomers

To answer the question, whether the polymer anisotropy of the network chains induced by mechanical deformation interacts with the anisotropy due to the templating by the micelle shape, a network from **1b** was swollen up to ca. 70 wt% with water and compressed to about $\lambda = 0.8$. X-ray diffraction patterns and ^{2}H-NMR-spectra [19] of the sample were taken before and after compression. In Fig. 3 the small angle part of the diffraction patterns is shown with the intensities of the first-order layer reflections of the L_α-phase. Before compression the sample displays a concentric diffraction ring with a homogeneous azimuthal intensity distribution. This indicates a polydomain structure, i.e. the lamellar stacking of the micelles is regular only within small regions. Neighboring regimes exhibit a different orientation of the director, which is defined as the normal to the micellar interface. Upon compression of the sample, the azimuthal intensity is focused into two separate maxima parallel to the force direction. Thus, a realignment of domains occurs which results in a monodomain structure with the director aligned parallel to the compression axis throughout the whole sample.

Similar effects have been observed for networks in the hexagonal phase after stretching [20] or by uniaxially constrained swelling in a cylindrical tube [17].

The alignment effect needs not necessarily be a consequence of the induced network anisotropy. Transitions from a polydomain to a monodomain structure also occur for monomeric or polymeric lyomesophases that are subjected to shear flow [21]. To exclude the possibility that

shear of the micelles causes the orientation, the polymer anisotropy has to be generated before the mesophase develops. In practice this can be easily done by applying a trick similar to the synthesis of liquid single-crystal elastomers (LSCEs) [22]. During the preparation of the networks in a nonselective solvent, the material can be mechanically deformed by stretching or compressing after the gel point has been reached. The cross-links, which are formed after the sample is loaded, lock in the network anisotropy and prevent relaxation when the mechanical force is removed. If such a network swells with water into the lyotropic mesophase the "inherent mechanical field" indeed affords macroscopically aligned mesophases.

Depending on the symmetry of the lyomesophase and the field applied during the synthesis, different types of alignment are observed. Uniaxial compression along the z-axis ($\lambda_x = \lambda_y > 1 > \lambda_z$, with $\lambda_i = l_i/l_{i,0}$ for $i = x, y, z$ as the change of sample dimension in the respective Cartesian coordinate) yields a monodomain in the L_α-phase, aligned parallel to the deformation axis.

For networks stretched during synthesis ($\lambda_x = \lambda_y < 1 < \lambda_z$) a different orientation is observed which can be elucidated from the X-ray patterns of a L_α-network shown in Fig. 4. Parallel to the elongation axis a homogeneous azimuthal intensity distribution is observed, while the sample is well aligned in the plane perpendicular to it. This can be explained by a so-called planar distribution. All domains are ordered with their director in the plane perpendicular to the force axis causing sharp diffraction crescents, but their orientation within this plane is random yielding powder-like diffraction rings.

For the H_1-phase different types of alignment are observed. The phase behavior of networks built from **1b** allows a phase transition from the L_α- to the H_1-phase at 40–60 wt% by lowering the temperature from 40 to 0 °C. As proved by ^{2}H-NMR spectra, the hexagonal phase epitaxially grows from an aligned lamellar one with the rod axis perpendicular to the former layer normal [18], a

Fig. 3 Small-angle X-ray diffraction images and models of alignment for a network from **1b**, (a) before, (b) after compression in the lamellar phase (at ca. 70 wt% polymer)

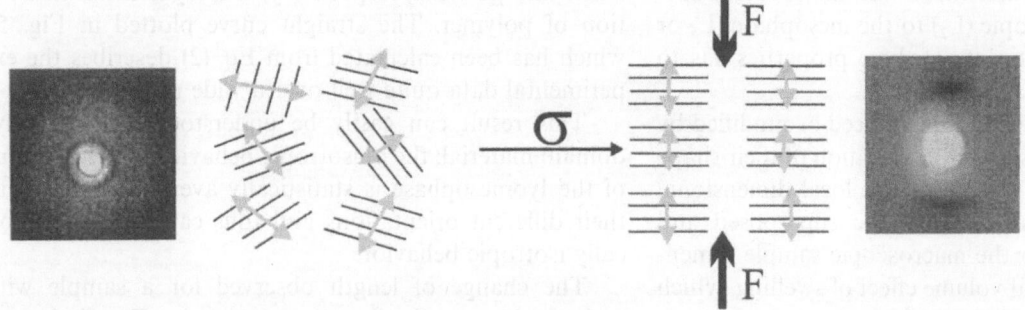

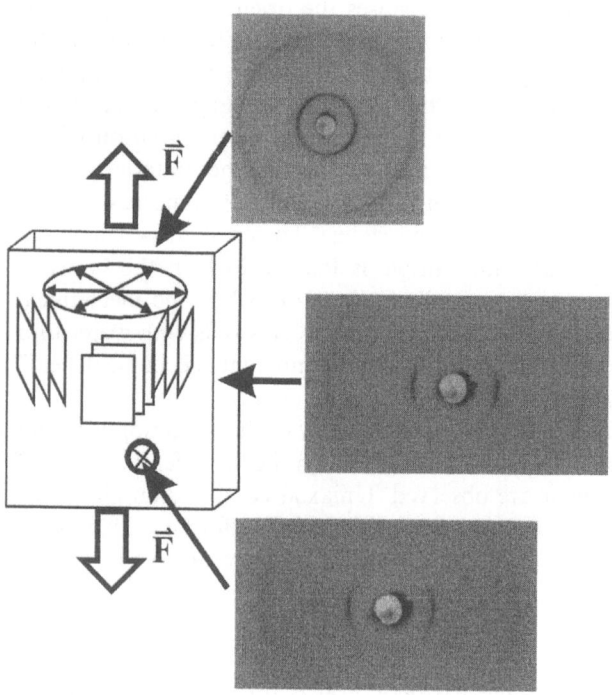

Fig. 4 Small-angle X-ray diffraction images of a network from **1b**, which has been uniaxially stretched during synthesis and swollen with water without mechanical stress. The diffraction patterns observed from different directions and the resulting model of domain orientation are depicted

behavior that is similar to single crystals grown in the magnetic field from monomer amphiphiles [23, 24],

It has to be noted that an extension of $\lambda \approx 1.2$–1.3 during synthesis is sufficient to get well-aligned mesophases. This is in contrast to the considerably higher drawing ratios that have to be applied for normal polymers in order to observe a significant anisotropy of their properties.

The hygroelastic effect

The macroscopic alignment of the mesophase throughout the elastomer should have two important effects: (i) the anisotropic physical properties should be reflected in macroscopical observables, and (ii) at the first-order transformation from an isotropic (L_2) to the mesophase (L_α or H_1) a significant discontinuity of these properties has to occur.

Polymer anisotropy will be introduced or modified by the formation of micelles and by a variation of their shape and size. In an oriented network these local dimensional changes on a molecular level will be superposed and should be transferred to the macroscopic sample dimensions. Beyond the normal volume effect of swelling, which occurs due to a variation of temperature and concentra-

tion, changes of the length or width of the elastomer are thus directly coupled to the underlying aggregation processes and yield information on these.

These processes can most easily be observed by altering the polymer concentration at a constant temperature. To achieve this, we constructed a computer-controlled apparatus [18]. It combines a setup for the measurement of the polymer sorption by varying the relative humidity of a home-built hygrostat with a device to access the sample length similar to thermoelastic measurements. We refer to this combination of methods, which is new to our knowledge, as hygroelastic measurements. The sample is flushed with a nitrogen stream containing water of variable partial vapor pressure. The sorption of water and thus the polymer concentration is recorded in situ by weighing the sample with a microbalance. The dimensions of the network are determined with a CCD-camera. Details of this setup will be published in a forthcoming publication.

An illustrative example of the potential of the hygroelastic measurements for the investigations on lyotropic LSCEs is given in Fig. 5. The relative sample length (length at a certain concentration divided by the length of the dry elastomer, $\lambda_z = l/l_0$) is shown for a polydomain network synthesized without mechanical stress and for an aligned sample with a planar layer distribution. The long axis of this sample coincides with the axis of mechanical stress applied during synthesis. At first glance, the two curves demonstrate that there are distinct dimensional effects that are only caused by the mesophase alignment because the samples are identical in their chemical composition. In Fig. 5a, the data for the polydomain show a continuous and monotonous behavior throughout the whole concentration regime independent of the different changes of micellar organization. The curve is very similar to the one expected for normal isotropic swelling ($\lambda_x = \lambda_y = \lambda_z \equiv \lambda$) by simply assuming additivity of partial volumina. Under these conditions the relative length is given by

$$\lambda = \sqrt[3]{\frac{V}{V_0}} = \sqrt[3]{1 + \frac{m_1/\rho_1}{m_{20}/\rho_2}} = \sqrt[3]{1 + \left(\frac{1}{\omega_2} - 1\right)\frac{\rho_2}{\rho_1}}, \qquad (2)$$

where m_1, m_{20} (ρ_1, ρ_2) denote the masses (densities) of water and polymer, respectively, and ω_2 is the mass fraction of polymer. The straight curve plotted in Fig. 5a which has been calculated from Eq. (2) describes the experimental data quite well over a wide range.

This result can easily be understood for the polydomain material: the anisotropic behavior of the domains of the lyomesophase is statistically averaged out due to their different orientations and thus causes macroscopically isotropic behavior.

The change of length observed for a sample with a planar director distribution is depicted in Fig. 5b. It does

Progr Colloid Polym Sci (1998) 111:127–134
© Steinkopff Verlag 1998

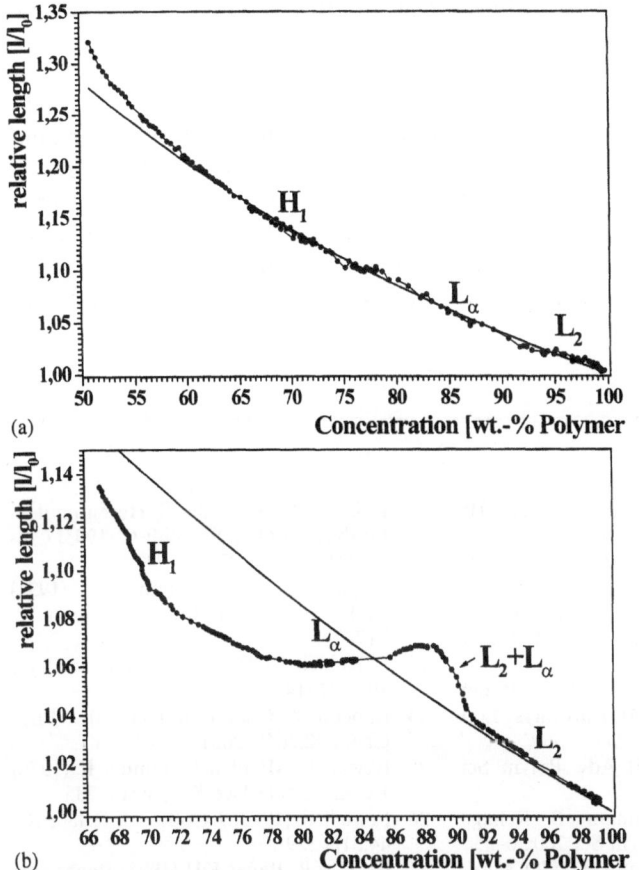

(a)

(b)

Fig. 5 Hygroelastic curves for two samples with a composition of PMHS:**1b**:2 = 1:0.85:0.075: (a) polydomain network (b) network with a planar domain distribution. The solid line has been calculated for $\rho_1 = 1.0$ and $\rho_2 = 1.12$ according to Eq. (2)

not deviate from the calculated curve (2) in the range from 100 to 91 wt%, which is attributed to the isotropic L_2-phase. From about 91 to 89 wt% a steep increase of λ_z is observed during the phase transformation L_2–L_α, which is followed by a plateau reaching to approximately 78 wt%. Then a continuous increase of the slope occurs leading to a steepness of the curve, which is considerably larger than the one calculated for isotropic swelling.

In the L_2-phase probably only small aggregates exist having no uniform orientation and thus nearly isotropic swelling behavior. At the phase transformation to the L_α-phase the micellar size increases to form the more or less planar bilayers, which constitute this mesophase. The simultaneous increase of polymer anisotropy and molecular alignment is manifested as a jump of sample length at this first-order phase transformation. The behavior resembles to some extent the behavior of thermotropic LSCEs, which are built from cross-linked side-chain polymers bearing mesogens. At the temperature-driven phase transformation LC–i they show length

changes, which vary for different systems in the range from $\Delta\lambda \approx 0$–60% [7].

Within the lamellar phase, the bilayer structure of alternating hydrophobic and hydrophilic regions will lead to an anisotropic incorporation of water. Only the hydrophilic ethyleneoxy parts will swell significantly, thus the bilayer distance d increases. For monomeric amphiphiles the swelling behavior of a L_α-phase can often be characterized by the relation $d = \delta\phi^{-1}$ with δ as the bilayer thickness and ϕ as the volume fraction of amphiphile [25, 26] If the planar bilayer structure remains intact, only the distance between the layers, and not their longitudinal extension should increase. For the given planar alignment, the scaling of molecular and macroscopic dimensions consequently leads in a first approximation towards a plateau of sample length within the lamellar regime.

It has to be noted that the cited swelling of the lamellar phase has been observed for ionic amphiphiles where the bilayers are separated by an intermediate water layer. The incorporation of water results in swelling of this water layer. For our materials with their voluminous non-ionic headgroups, in the given concentration regime opposing bilayers will be at least in contact or even interdigitate and the water molecules predominantly will be grouped among the headgroups and not between opposing oligo(ethyleneoxy) chains.

The long rod-like micelles of the H_1-phase demand a higher polymer anisotropy than the bilayers of the L_α-phase as already explained above. As a consequence, during the H_1–L_α-transformation the sample length should increase significantly as exemplified in Fig. 5b. The results of ^{2}H-NMR-investigations of this process with the **1b**-system have been interpreted with a continuous, second-order-like phase transformation [18]. The continuous increase of slope of Fig. 5b in the regime below 78 wt% polymer can also be understood by a broad biphasic regime or a second-order behavior.

Conclusions

The tandem interactions of mechanical fields with the entropy elasticity of polymer networks and the organization of amphiphiles to micelles and lyotropic liquid crystals yield a class of new materials, the lyotropic liquid single crystal elastomers.

Besides their anisotropic swelling behavior, these materials give rise to other macroscopically anisotropic properties, such as, e.g. birefringence or anisotropic diffusion and permeation. The structural similarity of the L_α-phase to biomembranes and more or less ordered arrays of biomembranes like the human skin makes them interesting in a biomedical context [27].

With the novel hygroelastic method we were able to show that changes in the amphiphilic self-organization are directly reflected in the macroscopic sample dimensions due to a scaling between molecular and macroscopic properties. The phase transformation from the isotropic to the L_α-phase causes a length jump of $\Delta\lambda \approx 3\%$ as observed by hygroelastic measurements. By a proper molecular design of anisotropic hydrogels, this discontinuous increase of sample length should be significantly augmentable. Within the field of mechanochemical conversion, it could be applied for smart materials, e.g. cheap mechanical

switches, which react on the variation of parameters like temperature, pH or relative humidity. Such stimuli-responsive hydrogels have attracted considerable research interest during the past [28]. These materials are normally based on the isotropic volume change during demixing. As has been pointed out by de Gennes et al. in a recent series of theoretical papers [29, 30], this concept suffers from some shortcomings. Their suggestions to use anisotropic gels for innovations within this area are somewhat similar to the ideas that motivated our experimental work.

References

1. Zentel R, Galli G, Ober CK (eds) (1996) Macromol Symp 107:1–304
2. Wiesemann A, Zentel R, Lieser G (1995) Acta Polymer 46:25–36
3. Adams J, Gronski W (1989) Makromol Chem Rapid Commun 10:553–562
4. Walther M, Finkelmann H (1996) Progr Polym Sci 21(5):951–979
5. Schädler V, Wiesner U (1997) Macromolecules 30(21):6698–6701
6. Sänger J, Gronski W, Maas S, Stühn B, Heck B (1996) Macromolecules 30(22):6783–6787
7. Disch S, Schmidt C, Finkelmann H (1996) In: Salamone JC (ed) Polymeric Materials Encyclopedia, Vol 5. CRC Press, Boca Raton, FL, pp 3794–3801
8. Grabowski D A, Schmidt C (1994) Macromolecules 27:2632–2634
9. Wiesner U (1997) Macromol Chem Phys 198:3319–3352
10. Chen ZR, Kornfield JA, Smith SD, Grothaus JD, Satkowski MM (1997) Science 277(5330):1248–1253
11. Bouteiller L, Lebarny P (1996) Liq Cryst 21(2):157–174

12. Decher G (1996) In: Sauvage JP, Hosseini MW (eds) Comprehensive Surpramolecular Chemistry, Vol 9. Pergamon, Oxford, pp 507–528
13. Spatz JP, Roescher A, Sheiko S, Krausch G, Möller M (1995) Adv Mater 7(8):731–735
14. Krausch G, Mlynek J, Straub W, Brenn R, Marko JF (1994) Europhys Lett 28(5):323–328
15. Laschewsky A (1995) Adv Polym Sci 124:1–76
16. Finkelmann H, Lühmann B, Rehage G (1982) Colloid Polym Sci 260:56–65
17. Löffler R, H. Finkelmann H (1990) Makromol Chem Rapid Commun 11:321–328
18. Fischer P (1997) Dissertation. Universität Freiburg
19. Fischer P, Schmidt C, Finkelmann H (1995) Macromol Rapid Commun 16:435–447
20. Albrecht H (1997) Dissertation, Universität Freiburg

21. Lukaschek M, Müller S, Hasenhindl A, Grabowski D A, Schmidt C (1995) Colloid Polym Sci 274:1–2747
22. Küpfer J, Finkelmann H (1991) Makromol Chem Rapid Commun 12:717–726
23. P. Kékicheff (1991) Mol Cryst Liq Cryst 198:131–144
24. Rançon Y, Charvolin J (1988) J Phys Chem 92:2646–2651
25. Ekwall P, Mandell L, Fontell K (1970) J Colloid Interface Sci 33:215–235
26. Fontell K (1973) J Colloid Interface Sci 44:318–329
27. Hermes R, Bauer KH (1995) Pharmazie 50(7):481–486
28. Dusek K (ed) (1993) Adv Polymer Sci 110:1–269
29. de Gennes PG, Hébert M, Kant R (1997) Macromol Symp 113:39–49
30. Hébert M, Kant R, de Gennes PG (1997) Journal de Physique I 7(7):909–919

Progr Colloid Polym Sci (1998) 111:135–143
© Steinkopff Verlag 1998

MACROMOLECULES

E. Killmann
D. Bauer
A. Fuchs
O. Portenlänger
R. Rehmet
O. Rustemeier

Adsorption of polyelectrolytes on colloidal particles – Electrostatic interactions and stability behaviour

E. Killmann (✉) · D. Bauer · A. Fuchs
O. Portenlänger · R. Rehmet
O. Rustenmeier
Institut für Technische Chemie
der Technischen Universität München
Lichtenbergstraße 4
D-85748 Garching
Germany

Abstract Adsorption isotherms of neutral poly(ethyleneoxide) (PEO), cationic poly-L-lysine (PLL) and homo- and copolymers of diallyl-dimethyl-ammoniumchloride (DADMAC) and N-methyl-N-vinyl-acetamide (NMVA) on colloidal silica and latex particles were determined from the concentration in the supernatant solution. Layer thickness and flocculation of the particles are measured by photon correlation spectrometry. Adsorbed amount, layer thickness and stability of the suspensions are influenced by ionic strength and pH, molar mass and charge densities of the polymers and of the surface. Correlations between flocculation and adsorption parameters are accomplished. With neutral PEO the stability is determined by the layer thickness adjusted by molar mass and coverage; thicknesses $\geq 4\,\text{nm}$ stabilize up to high electrolyte concentrations. With high positively charged PE layers (pH ≤ 7) stable latex suspensions are guaranteed and large electrokinetic influences on the diffusion coefficient depending on the PE/electrolyte ratio are observed. With increasing ionic strength, latex covered with PE layers of low molar mass flocculates, while for that covered with PE layers of high molar mass stabilization occurs. Adsorption of PDADMAC and P(DADMAC-co-NMVA) result in spontaneous flocculation if a certain amount of PE depending on the charge density according to a definite coverage is added. Flocculation at increasing ionic strength can be controlled by the adsorption of PDADMAC depending on the molar mass. At full coverage the suspensions are stabilized if the thickness of the layer is large enough, dependent on ionic strength, charge density and molar mass of the PE.

Key words Latex – silica – colloid stability – polymer and polyeletrolyte adsorption

Introduction

The regulation of the stability of dispersed particles by adsorbed polymer and polyelectrolyte layers is of high scientific and technological significance [1]. Polyelectrolytes (PE) find widespread application in many indus-trial processes and in numerous products of our daily life [2]. An important property of PE is their tendency to adsorb on solid surfaces. This may be due to either Coulombic or other polar or non-polar forces.

The adsorptive behavior of PE on charged surfaces, the structure of the adsorbed layer and the adsorbed amount is determined by the molecular and electrochemical structure of the PE, the charge density and molar mass of the macroions, the surface charge density and by the ionic strength of the solution. The electrostatic interactions control dominantly the conformation of the adsorbed macroions, the segment distribution and the thickness of the charged polymer layer. pH and ionic strength are of important influence [3]. The adsorptive behavior of PE causes the modification of solid surfaces and interfaces by PE-based coatings and adhesives [4].

The conformation of the adsorbed PE and the charge of the PE-covered particles have great influence on the stability and the flocculation behavior of suspensions. Because of these particular properties PEs are used either as stabilizers, e.g. of dye dispersions or as flocculants, e.g. in waste water treatment or in the paper producing industry. But usually the PEs are added empirically. So we try to get some rules for the dosage to obtain either stable suspensions or optimal flocculation.

Experimental

Adsorption isotherms of neutral polyethylene oxide, PEO, of cationic Poly-L-lysine, PLL, of poly-diallyl-dimethyl-ammoniumchloride, PDADMAC, of poly-N-methyl-N-vinyl-acetamide, PNMVA, and of copolymers of DADMAC with NMVA, P(DADMAC-co-NMVA) on colloidal, negatively charged silica and polystyrene latex particles are obtained by depletion measurements, for PEO by turbidity, for PE by polyelectrolyte titration of the supernatant solution. Hydrodynamic layer thicknesses are measured by diffusion with photon correlation spectrometry (PCS). Additional complicating electrokinetic effects are discussed. Linear PDADMAC of different molar masses have a large charge density because every monomer carries a positive charge. To vary the charge density of the polyelectrolytes DADMAC is statistically copolymerized with NMVA.

The substrate precipitated silica (own preparation) is suspended by ultrasonification. Because of the dissociation of the silanol groups in aqueous suspension, silica is negatively charged. The charge density depends on the pH of the suspension. Colloidal polystyrene latex samples of negative surface charge (prepared by Dr. Jaeger, Fraunhofer Institut für Angewandte Polymerforschung, Teltow, and from Dow Chemical Corporation, USA) and of positive surface charge (Dr. Jaeger) were used without any pretreatment.

Details of the silica and latex samples and the molecular characterization parameters for the polymers and polyelectrolytes are published for PEO in [5] for PDADMAC, PNMVA and P(DADMAC-co-NMVA) in [6–8] and for PLL in [9].

Procedures

Adsorption isotherms were measured at room temperature (23 ± 2 °C) using the depletion method. The detailed procedure for the different systems are reported in [5–7]. The concentrations of the supernatant cationic PE solutions are determined by polyelectrolyte titration with the anionic potassium polyvinylsulfate and the cationic o-toluidine blue as an indicator at low salt concentrations up to 1×10^{-2} mol/l NaCl only.

At higher salt concentrations dialysis of the PE solution was performed to reduce the content of salt, before polyelectrolyte titration [7]. With this procedure the PE concentration could be determined up to a high ionic strength of 1 mol/l NaCl with high accuracy. The concentration of the neutral PNMVA was determined by UV-absorption [6, 7], and the concentration of PEO by a turbidity method [5, 11].

The zeta potential measurements were accomplished in a Malvern Zetamaster.

The hydrodynamic layer thicknesses of the adsorbed layers were determined by measuring the diffusion coefficients of uncovered and covered particles by PCS at 25 °C [5, 12]. The hydrodynamic layer thickness was obtained by applying the Einstein–Stokes equation $a = kT/(6\pi\eta D)$ (k is the Boltzmann constant, T the temperature in K and η the dynamic viscosity of the suspension) to the diffusion coefficients D of the bare and modified particles and subtracting the radii, a. Thicknesses could only be determined under stable conditions without aggregation, confirmed by the angle and the time dependence of the diffusion coefficients. These conditions are directly certified by scanning electron micrographs of the bare and fully PE covered silica particles. Without adsorbed PE the suspensions are only stable at low ionic strength, while at higher salt concentrations aggregation is observed. If the particles are fully covered with PE, stable suspensions are obtained. So hydrodynamic layer thicknesses can be measured in the plateau region of the adsorption isotherms.

PCS is also used to follow the coagulation and flocculation of the colloidal silica and polystyrene latex particles, bare and covered with adsorbed layers, by the decreasing diffusion coefficient of the aggregating particles [9, 10]. Aggregation is initiated by direct dosage of different amounts of PE ranging from low coverage to saturation as well as by addition of NaCl to adsorbed layer covered particles under stable conditions. The rate constants of aggregation are evaluated by the second-order Smoluchowski theory with the assumption of spherical aggregated particles and volume proportional light scattering amplitude [10].

Progr Colloid Polym Sci (1998) 111:135–143
© Steinkopff Verlag 1998

Representative samples in different states of aggregation, at different coverages with PE and with different ionic strengths were selected to produce micrographs by raster electron microscopy (REM). The dispersions were filtered at a reduced pressure of approximately 900 mbar through membrane filters (0.1 μm). The deposed particles were washed carefully with bidistilled water to remove the salt. These particles were vapor deposited with gold and imaged by REM.

Experimental results and discussion

Polyethylene oxide layers

The adsorption isotherms on silica and latex surfaces show high affinity character. The adsorbed plateau amounts, A, in H_2O increase with the molar mass of PEO from 0.65 to 1.3 mg/m^2 on latex which is somewhat steeper than on silica where the increase is from 0.5 to 0.7 mg/m^2 in the molar mass region 1×10^4 to 1×10^6 g/mol. An increase of A is also observed with increasing NaCl concentration [14]. The hydrodynamic thicknesses of the adsorption layers, δ, measured by PCS are plotted in Fig. 1 versus the corresponding adsorbed amounts A for both adsorbents. Steep increases of the thicknesses in the plateau region are found on latex (Fig. 1) and silica (not shown here). The maximum plateau thicknesses are strongly dependent on molar mass with somewhat weaker dependence on silica than on latex [5, 13].

The stability of the covered latex and silica particles is determined by the layer thickness given by different molar masses or by different coverages of PEO. Layers of small thickness reduce the critical electrolyte concentration for flocculation. Adsorbed layers of large thickness $d \geq 4$ nm

stabilize the suspension even at high electrolyte concentrations demonstrating the steric influence of the tails of the adsorbed PEO macromolecules [10, 14].

Poly-L-lysine layers

The charge density of the poly-L-lysine (PLL) chains is dependent on pH. The adsorption isotherms of PLL on latex show high affinity character. At small ionic strength ($c_{NaBr} \leq 1 \times 10^{-4}$ M) the adsorbed amount is very small (0.25 mg/m^2) and independent of the molar mass. The adsorbed amount increases distinctly at higher electrolyte concentration and becomes strongly increasing with the molar mass. At plateau adsorption of PLL the negative surface charge of the latex becomes overcompensated to positive charge [9].

The PCS measurements of the diffusion coefficients of the suspended particles become difficult because of the superimposing by the electroviscous effect as a consequence of the opposite polarization of the electrostatic double layer leading to smaller diffusion coefficients (Fig. 2). With plateau adsorption of highly charged PLL the diffusion coefficient of the covered positively charged latex particles in water decreases more and more with increasing PLL concentration (not shown here) and with molar mass of PLL (Fig. 2) independent of the particle concentration. By addition of salt ($c_{NaBr} > 1 \times 10^{-3}$ M) and also by adsorption of neutral PEO layers of large thickness (> 22 nm) this effect is eliminated [5]. The reason for this effect may be due to electrostatic interactions. In analogy to observations of Förster et al. [15] for dissolved PE we conclude that the diffusion of the highly charged latex

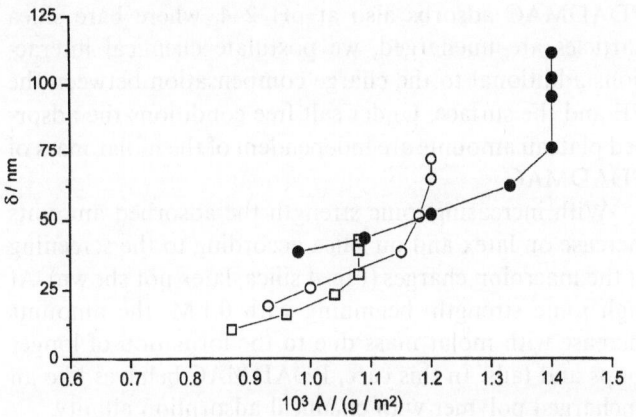

Fig. 1 Dependence of the adsorbed layer thickness δ on the adsorbed amount A of PEO on latex 233 in H_2O at 25 °C. ●: $M_w = 900\,000$, ○: $M_w = 325\,000$, □: $M_w = 160\,000$

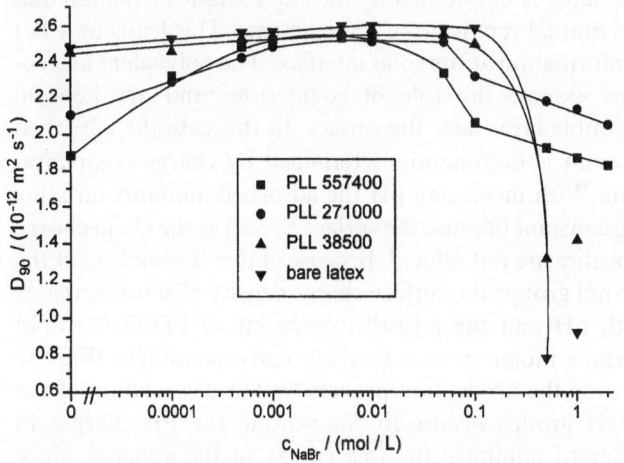

Fig. 2 Dependence of the diffusion coefficient D_{90} on log c_{NaBr}; latex (negative) covered with PLL of different molar mass; $c_{latex} = 2.5 \times 10^{-6}$ g/ml; pH = 6–7

138
E. Killmann et al.
Stability of colloidal particles at polyelectrolyte adsorption

particles is influenced by interactions with domains or clusters formed in the solution. More substantiating arguments are given in [9]. At an NaBr concentration of about 0.002–0.01 M the diffusion coefficients have the same value independent of molar mass resulting in a layer thickness of only 3–5 nm, much smaller than the layer thickness of the neutral PEO [9].

With increasing electrolyte concentration $c_{NaBr} > 0.1$ M the surface charges as well as the segment charges are screened and their mutual repulsions are reduced. The improved segment approach and the decreasing solvent quality favors the PLL adsorption. This results in a three times higher adsorbed amount and larger layer thickness increasing with c_{NaBr} and molar mass of PLL. Latex particles covered with PLL of low molar mass $M_w < 100\,000$ flocculate, particles covered with PLL of high molar mass $M_w \geq 100\,000$ become more and more stabilized because of the steric repulsion of the PLL layers of large thickness (Fig. 2). Flocculation rate constants are also evaluated by the Smoluchowski theory.

Electromicrographs measured by REM (not shown here) demonstrate directly the remaining single particles covered with PLL of high molar mass up to 2 M NaBr. Contrary to this behavior bare particles and particles covered with PLL of low molar mass become clearly aggregated [9]. The existence of non-aggregated particles at low ionic strength proves that the decrease of the diffusion coefficient under these conditions is not caused by particle aggregation but by cluster or fluctuating domain formation.

Layers of PDADMAC, P(DADMAC-co-NMVA) and PNMVA

The adsorption isotherms of the cationic PDADMAC, P(DADMAC-co-NMVA) and the neutral PNMVA are of high affinity character. At low ionic strength the adsorption of the PE on the oppositely charged surfaces of silica and latex is dominated by the electrostatic attraction and the mutual repulsion of the segments. This leads to a flat conformation at the solid interface. The polyvalent macroions assume the role of counterions and are kept in a double layer near the surface. In this case the adsorbed amount is dominantly determined by charge compensation. With increasing pH the adsorbed amounts on latex are constant because the surface as well as the chain charge densities are not altered. Because of the dissociation of the silanol groups the surface charge density of silica increases with pH and the adsorbed amounts of PDADMAC of various molar masses increase correspondingly (Fig. 3). During the adsorption process further dissociation of the SiOH groups occurs to compensate the PE charges in order to minimize the free energy of the systems. Since

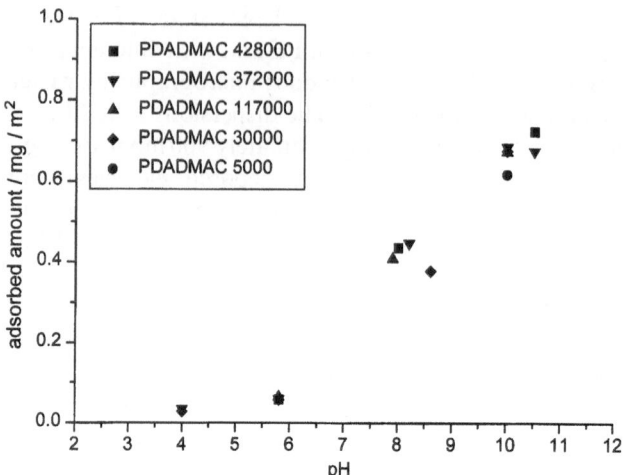

Fig. 3 Adsorbed amounts of PDADMAC of different molar masses on silica as a function of pH without added salt

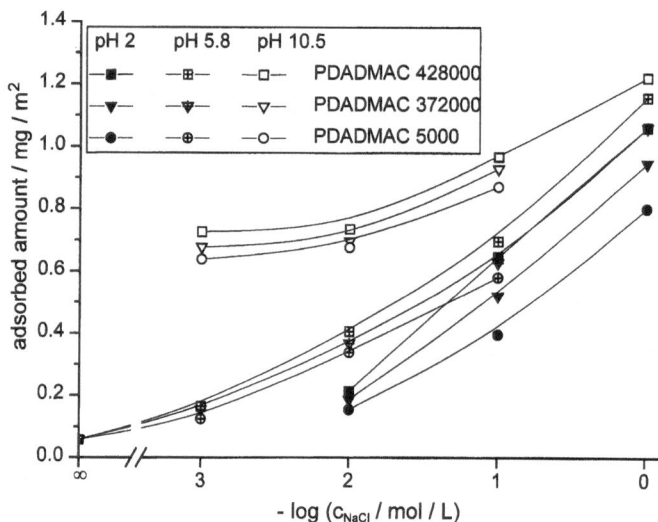

Fig. 4 Adsorbed plateau amounts of PDADMAC on silica as a function of pH and NaCl concentration

PDADMAC adsorbs also at pH 2–4, where bare silica particles are uncharged, we postulate chemical interactions additional to the charge compensation between the PE and the surface. Under salt-free conditions the adsorbed plateau amounts are independent of the molar mass of PDADMAC.

With increasing ionic strength the adsorbed amounts increase on latex and on silica according to the screening of the macroion charges (Fig. 4 silica, latex not shown). At high ionic strength, beginning with 0.1 M, the amounts increase with molar mass due to the formation of longer loops and tails. In this case, PDADMAC behaves like an uncharged polymer with chemical adsorption affinity.

Progr Colloid Polym Sci (1998) 111:135–143
© Steinkopff Verlag 1998

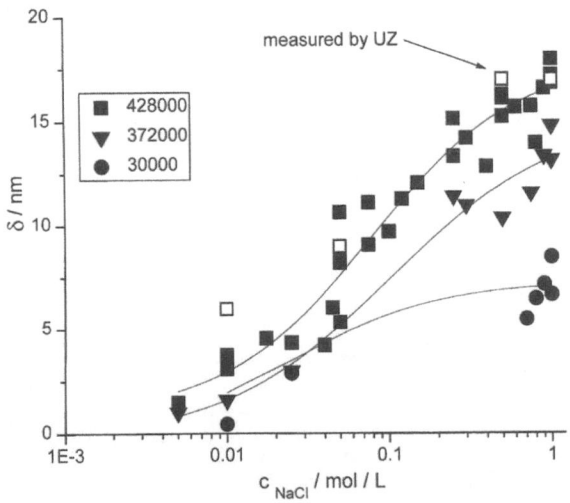

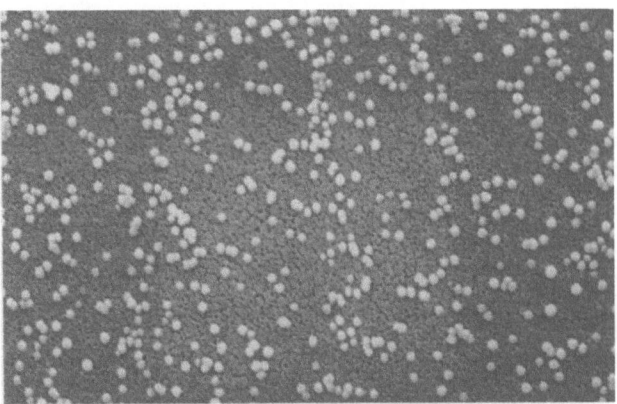

Fig. 6 Electromicrograph of bare silica in water without salt, pH 5.8

Fig. 5 Hydrodynamic thicknesses of PDADMAC layers of different molar mass on silica as a function of NaCl concentration

Experimental evidence for this explanation is given by the hydrodynamic layer thickness in the plateau of the isotherms on silicas. The thickness increases because of the screening of the segments. At high salt concentration the thicknesses increase with the molar mass because of the formation of more extended tails (Fig. 5). The plot of the hydrodynamic thicknesses of PDADMAC of the molar masses 428 000, 372 000 and 30 000 g/mol on silica as a function of the adsorbed amount (not shown here) demonstrates: At low adsorbed amounts the increase of the layer thickness with adsorbed amount is independent from the molar mass of the macroions. With increasing adsorbed amount longer tails are formed, the thickness increases and becomes larger with increasing molar mass.

The preceding interpretation could be certified by direct raster electronmicrographs of the bare and covered particles under different conditions (Figs. 6–9). These micrographs show stable suspensions for particles fully covered with PDADMAC up to high salt concentrations whereas the suspensions of bare particles are only stable at low salt concentrations but show aggregates at high ionic strength.

The adsorbed amount depends strongly on the surface charge of latex. Even at very high ionic strength under extreme screening conditions the cationic PDADMAC does not adsorb on positively charged latex (Fig. 10). This refers to an extremely weak chemical attraction between the PE segments and the latex surface and a strong contribution of the electrostatic interactions even if the charges are screened extensively.

From the adsorption isotherms of the copolymers of DADMAC and NMVA and of the neutral PNMVA the dependence on the polymer charge density can be

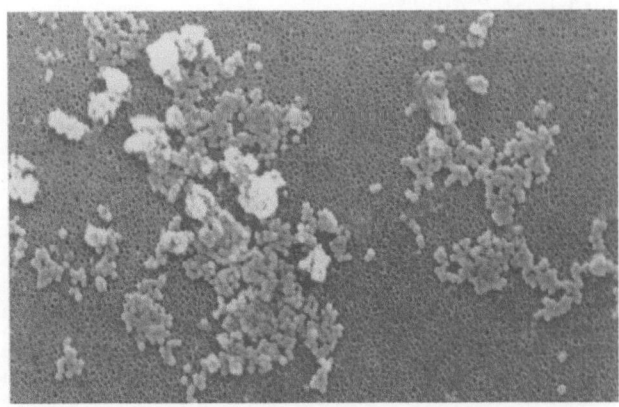

Fig. 7 Electromicrograph of bare silica in 1 M NaCl at pH 5.8

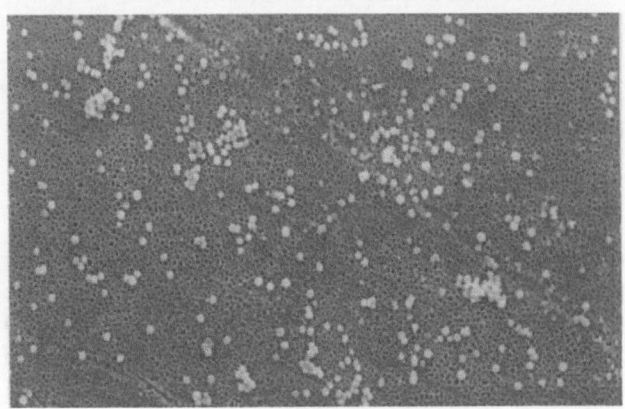

Fig. 8 Electromicrograph of silica fully covered with PDADMAC 428 000 in 0.01 M NaCl at pH 5.8

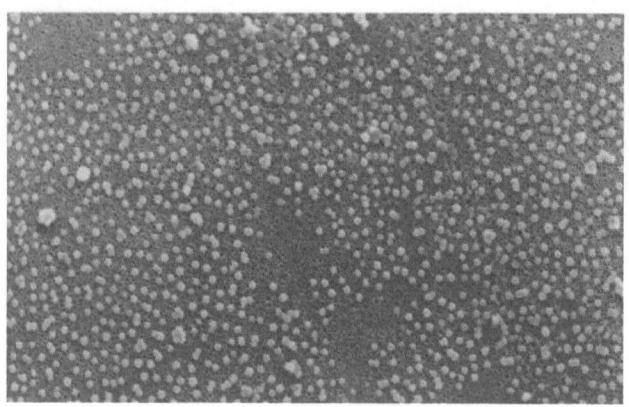

Fig. 9 Electromicrograph of silica fully covered with PDADMAC 428 000 in 1 M NaCl at pH 5.8

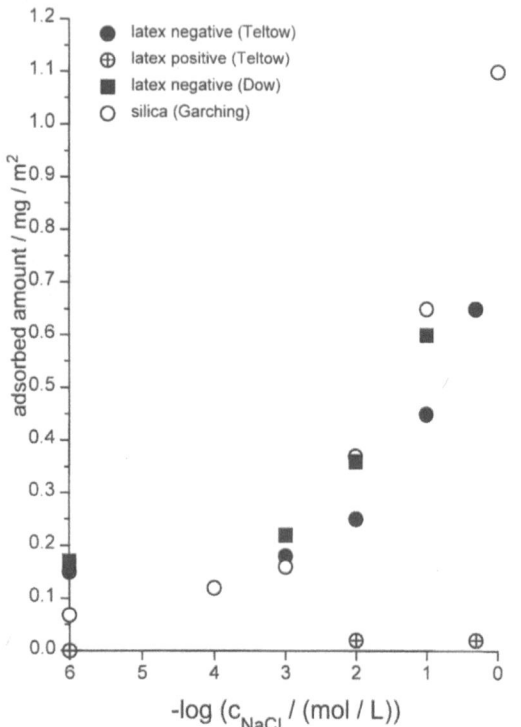

Fig. 10 Adsorbed plateau amounts of PDADMAC 372 000 on polystyrene latex and silica as a function of c_{NaCl} in water

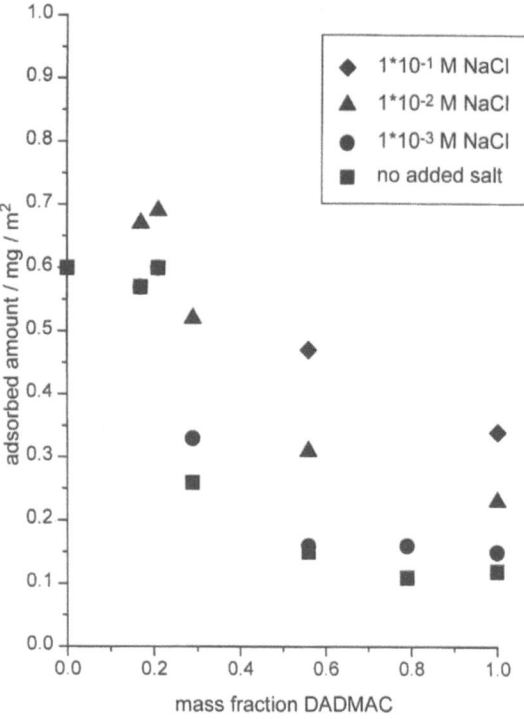

Fig. 11 Adsorbed plateau amounts of PDADMAC, P(DADMAC-co-NMVA) and PNMVA on latex as a function of mass fraction DADMAC at different NaCl concentrations

evaluated. On latex as well as on silica (not shown here) an increase of the adsorbed plateau amount with the decrease of the polyelectrolyte charge density is observed (Fig. 11) [2, 3]. With decreasing charge density the stiffness of the polymer chain reduces and loops and tails can be formed more easily resulting in higher adsorbed amounts. This tendency is pronounced especially in suspensions with no salt addition. If the electrolyte concentration is increased

two different regimes are recognized: The adsorbed amounts of highly charged polyelectrolytes (mass fraction of DADMAC, 0.3–1.0) show a distinct increase with the salt concentration, whereas the adsorbed amounts of weakly charged copolymers (mass fraction, 0.0–0.3) show small dependence on salt concentration, because already without added salt the adsorbed amount is large due to the formation of loops and tails. In order to indicate the two regimes more clearly, a plot of the relative adsorption data for latex (not shown here) where the adsorbed amounts are divided by the adsorbed amounts of the same polymer without added salt is instructive [7]. The relative changes of the adsorbed amounts after the addition of salt are comparable for all PEs of different charge densities. If the compensation of the surface charges by polyelectrolyte charges dominates the adsorption mechanism exclusively, the number of adsorbed charged segments would be the same for all homo- and copolyelectrolytes. The measured data do not follow this dependence exactly [7]. This could be explained with the counterion condensation on the polymer chain, reducing the effective charge density in the adsorbed state also.

The adsorbed amounts of the P(DADMAC-co-NMVA) on silica and latex (not shown here) increase at every pH with decreasing charge density of the copolymers

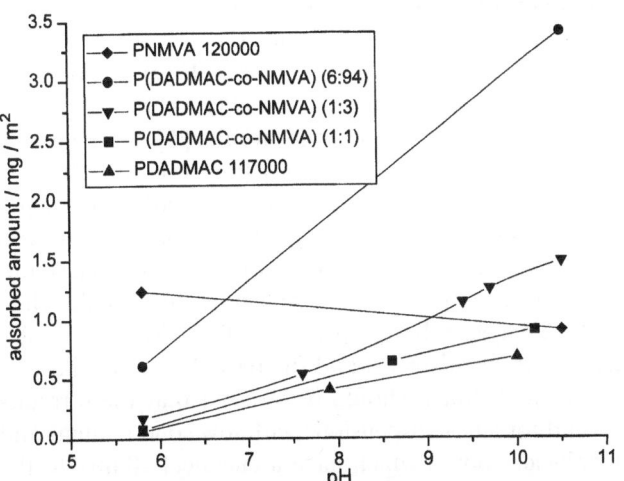

Fig. 12 Adsorbed amounts of PDADMAC, P(DADMAC-co-NMVA) and PNMVA on silica as function of pH without added salt

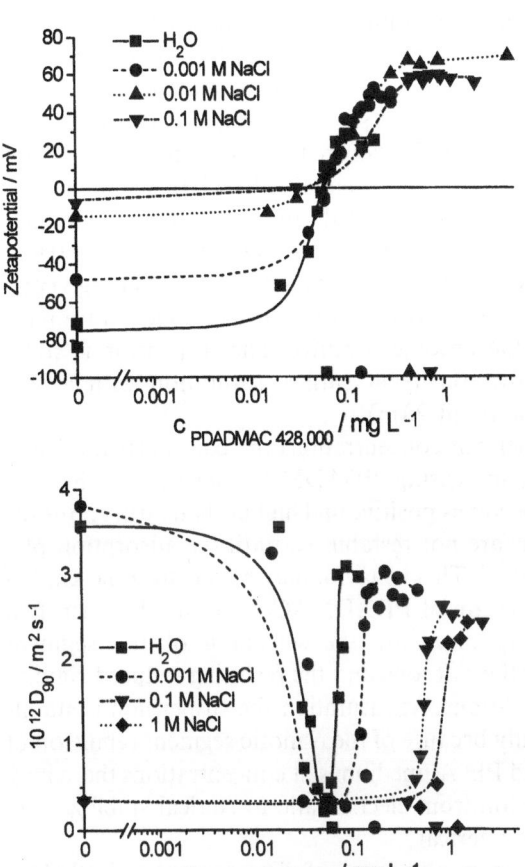

Fig. 13 Zeta potential and diffusion coefficient of silica as a function of the coverage with PDADMAC 428 000

(Fig. 12) [6]. The adsorption is determined by an equilibrium between a low electrostatic repulsion between the macroions and a large electrostatic attraction between the PE and the surface leading to the same number of adsorbed charges of the PE and to a corresponding increase of the adsorbed amounts with decreasing PE charge density at pH 5.8. The adsorbed amounts of PDADMAC and of the P(DADMAC-co-NMVA) increase with pH whereas the values of the neutral PNMVA decreases.

The adsorbed amounts of every copolymer such as the PDADMAC homopolymers on silica increase with the ionic strength as shown in Fig. 10, because of the screening of the PE charges and the formation of loops and tails [6]. At high ionic strength the adsorption is dominated by the chemical affinity between polyelectrolyte segments and silica surface. The adsorbed amounts of PNMVA independent of the ionic strength and the almost same values for all charged and uncharged polymers at the highest ionic strength ($c_{NaCl} = 1$ mol/l) lead to the conclusion that the NMVA and DADMAC segments have almost the same chemical adsorption energies.

The exceptional behavior found at pH 10.5 with decreasing adsorbed amount of PNMVA and of the copolymers with large content of NMVA supports the conclusion of a small chemical affinity between NMVA and silica at high pH. The salt ions are able to displace the polymer segments.

The different adsorption behavior's and especially the various structures of the adsorbed layers of PDADMAC and of P(DADMAC-co-NMVA) have great influence on the stability and flocculation behavior of the suspensions. As an experimental parameter for these investigations the diffusion coefficient measured with photon correlation

spectroscopy is used. A high diffusion coefficient means less or non-aggregated particles, a low diffusion coefficient is obtained with large aggregates. To get information about the flocculation mechanisms the zeta potentials are compared with the diffusion coefficients of the silica particles as a function of the added concentration of PDADMAC at different salt concentrations. In Fig. 13 both parameters for silica covered with PDADMAC of the molar mass 428 000 g/mol at pH 5.8 are plotted. The marks at the abscissa axis indicate these concentrations at which the adsorption plateaus are reached. In water without any added salt the bare silica particles have a zeta potential of about − 80 mV. With increasing salt concentration the negative value of the zeta potential decreases because of the screening of the surface charge. The adsorption of the polycation also leads to a decreasing zeta potential value. At a PDADMAC concentration of about 0.05 mg/l, which is just below the saturation in water, we observe an elctrokinetic charge compensation. At the adsorption plateau the zeta potential is positive and does not increase further.

Considering the diffusion coefficients we observe flocculation of the bare silica particles at salt concentrations higher than 0.01 M. Then the zeta potential is reduced to −20 mV because of the screening of the surface charge by the salt ions. The PE covered particles flocculate at low ionic strength at zeta potential values larger than 20 mV. In water without any added salt, aggregation is obtained at about 50% of saturation at a zeta potential of −50 mV. We assume that this is due to a mosaic-like charge compensation. With increasing PDADMAC concentration the zeta potential becomes positive. The suspension is stabilized electrostatically at surface saturation with a zeta potential of about 20 mV.

At higher salt concentrations the bare particles aggregate. With increasing PDADMAC adsorption the zeta potential becomes positive and higher than 20 mV. But the suspensions are not restabilized until the adsorption plateau is reached. This behavior may be due to some kind of bridge formation of PDADMAC molecules between two particles supported by the screening of the segment charges by the salt ions and the resulting reduced rigidity of the polycations. At saturation the suspension is stabilized sterically because of the osmotic segment repulsion of the screened PE. At medium salt concentrations there may be a transition from electrostatic to sterical stabilization, called electrosterical.

The zeta potential curves of silica covered with PDADMAC with the low molar mass of 5000 g/mol are similar to those obtained with the adsorption of PDADMAC of the high molar mass of 428 000 g/mol. This is expected, because there is little difference between the adsorbed amounts of these two PDADMACs up to a salt concentration of 0.1 molar NaCl. Nevertheless PDADMAC 5000 causes a different flocculation behaviour of the silica suspension. At low ionic strength a smaller flocculation region is observed. Aggregation occurs only if the surface charge is screened by the adsorbed PDADMAC and if the zeta potential value is smaller than 20 mV. So there is no mosaic flocculation with the small polycation. At surface saturation the suspension becomes stable because of the electrostatic repulsion between the polycation covered particles. At PDADMAC concentrations below saturation leading at different salt concentrations to higher zeta potentials of about +20 to +50 mV we observe some flocculation which can be explained by bridging. The suspensions are stabilized sterically when the particles are fully covered with PDADMAC 5000 as with PDADMAC of the high molar mass.

In contrast to PDADMAC 428 000 PDADMAC 5000 does not stabilize silica at any polyelectrolyte concentration at 1 M NaCl. This is refered to the smaller layer thickness which cannot stabilize the suspension sterically.

PDADMACs of the other molar masses show a compromise between the behavior of PDADMAC of the high molar mass and the low molar mass. Even at other pH values and with P(DADMAC-co-NMVA) of low charge density we have obtained similar results [6]. Summarizing the results of state and kinetics (not yet published) of flocculation of silica fastest flocculation with large aggregates is obtained when the surface charge of the particles is screened either by salt ions or by the adsorbed macroions leading to an amount of the zeta potential smaller than 20 mV. To get stable suspensions, PE with high molar mass has to be adsorbed and the particle surface must be fully covered. But it should be stressed that these results are valid for silica suspensions with low solid content and for polyelectrolytes which have a chemical affinity to the substrates.

The flocculation behavior of PDADMAC covered latex particles corresponds in principle with the characteristics reported for silica particles. In pure water flocculation occurs if the zeta potential of the PE covered particles becomes close to zero, PDADMAC of low and high molecular weights show the same behavior as demonstrated for PDADMAC 428000 (Fig. 14). At higher coverage with PE the negative particle charge and the zeta potential get

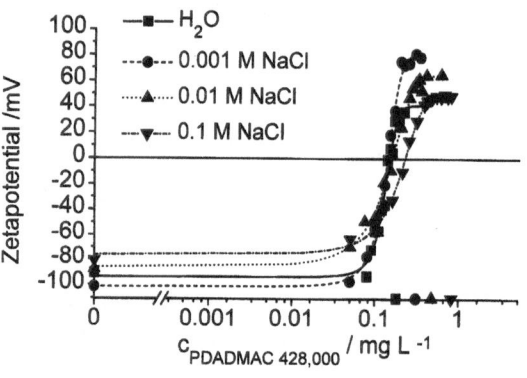

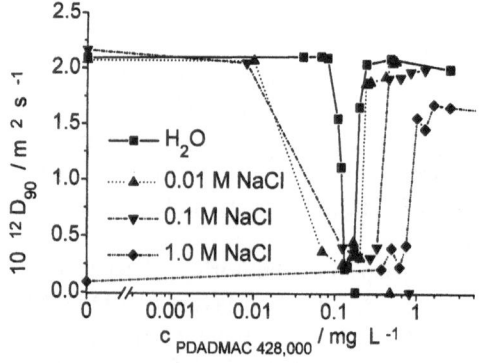

Fig. 14 Zeta potential and diffusion coefficient of latex as a function of the coverage with PDADMAC 428 000

Progr Colloid Polym Sci (1998) 111:135–143
© Steinkopff Verlag 1998

overcompensated to positive zeta potentials of ca. 30 mV. Electrostatic repulsion of the colloidal particles is responsible for the stability of those suspensions.

At increasing ionic strength flocculation occurs at coverages lower than the plateau values. At ionic strength of 1 M no stabilization can be achieved by plateau adsorption of PDADMAC with molar mass 5000. In contrast adsorption of PDADMAC with higher molar masses e.g. 428 000 and of the copolymers P(DADMAC-co-NMVA) with molar mass $M_w \approx 100 000$ results in the stabilization of the lattices up to NaCl concentrations of 1 M.

Conclusions

Neutral polyethylene oxides, PEO, adsorb on silica and latex surfaces with flat layers at low coverage and a distinct increase of the layer thickness near the plateau adsorption. In comparison to bare particles flocculation is promoted after the addition of electrolyte by the adsorption of PEO layers with hydrodynamic thicknesses lower than 4 nm. With PEO layers of larger thicknesses steric stabilization of the suspensions is initiated.

The adsorption of cationic poly-L-lysine, PLL, on negatively and positively charged latex particles is dependent on pH. Screening of the layer by increasing electrolyte concentration eliminates the influence of fluctuating PLL domains in the solution on the diffusion coefficients and results in large adsorbed amounts and layer thicknesses. At low molar mass of PLL the suspension flocculates and at high molar masses ($M_w \geq 100 000$) electrosteric stabilization occurs.

Cationic poly (diallyl-dimethyl-ammoniumchloride), PDADMAC, and copolymers of DADMAC with N-vinyl-acetamide form very flat surface layers on silica and latex at low ionic strength because of segment–segment repulsion. With decreasing charge density or by charge screening, loops and tails are formed leading to higher adsorbed amounts and their increase with molar mass. Chemical surface affinity of the polyelectrolyte segments enhances the pure electrosorption of the PE. Counterion condensation also takes place on the adsorbed PE. Possible reasons of the incomplete charge compensation are the distance of the PE segments to the charged surface and the counterion condensation. Charge reversal of the negative substrates to positively charged particles is obtained at high coverages of the polyelectrolytes.

Flocculation of the negative silica and latex particles can be achieved by adsorbing the cationic PE in a definite coverage region as well as by the increase of the ionic strength. Plateau adsorption of high molar mass PE stabilize the suspension up to a high ionic strength.

References

1. Napper DH (1977) J Colloid Interface Sci 58:390–406
2. Dautzenberg H, Jaeger W, Kötz J, Philip B, Seidel Ch, Stscherbina D (1994) Polyelectrolytes: Formation, Characterization and Application. Carl Hanser Verlag, Munich
3. Fleer GJ, Cohen Stuart MA, Scheutjens JMHM, Cosgrove T, Vincent B (1993) Polymers at Interfaces, 1st ed. Chapman & Hall, London
4. Bárány S, Baran AA, Solomentseva I, Velichanskaya L (1994) In: Schmitz KS (ed) Macroion Characterization.

American Chemical Society, Washington DC, pp 406–420
5. Killmann E, Sapuntzjis P (1994) Colloids Surf A 86:229–238
6. Bauer D, Killmann E, Jaeger W (1997) Prog Colloid Polym Sci, accepted for publication
7. Rehmet R, Killmann E (1997) Colloid Surf A, accepted
8. Bauer D, Buchhammer H-M, Fuchs A, Jaeger W, Killmann E, Lunkwitz K, Rehmet R, Schwarz S, to be published
9. Rustemeier O, Killmann E (1997) J Colloid Interface Sci 190:360–370

10. Killmann E, Adolph H (1995) Colloid Polym Sci 273:1071–1079
11. Killmann E, Maier H, Kaniut P, Gütling N (1985) Colloids Surf 15:261
12. Killmann E, Sapuntzjis P, Maier H (1992) Makromol Chem Macromol Symp 61:42
13. Killmann E, Maier H, Baker JA (1988) Colloids Surf 31:51
14. Killmann E, Portenlänger O (1997) Dissertation TU München, unpublished results
15. Förster S, Schmidt M, Antonietti M (1990) Polymer 31:781

Progr Colloid Polym Sci (1998) 111:144–150
© Steinkopff Verlag 1998

MACROMOLECULES

B.U. Kluß
R. Zimehl

Some aspects of colloidal polymers: Solvent structure originated interactions

B.U. Kluß · Dr. R. Zimehl (✉)
Institute of Inorganic Chemistry
University of Kiel
Olshausenstr. 40-60
D-24098 Kiel
Germany

Abstract The coagulation of some aqueous latex dispersions has been studied. Type-1-dispersions were made by emulsifier free emulsion polymerization of styrene and stabilized by anionic or cationic groups at the surface of the latex particles. Because of charge stabilization a sufficiently large amount of electrolyte forced the particles to coagulate by double-layer compression. The value of the critical coagulation concentration for monovalent counter ions is intimately connected to the water structure and to the kind of particle surface groups. The interaction of ions with the boundary region developed between the particle and the aqueous dispersion medium is also influenced by temperature and by the addition of organic molecules which can interfere with the microstructure of the solid–solution interface. The impact of a direct change of the particle surface characteristics on stability is illustrated for several type-2-latices which were synthesized by seed and feed copolymerization of styrene and glycidyl methacrylate. Particles with different surface structure were derived from the same stock dispersion by the nucleophilic addition of D-glucamine and taurine to the reactive epoxy group at the particle surface. Differences in the critical coagulation concentration may then be attributed to changes in the structure of the latex surface. The alteration of the boundary region separating the latex particles from the dispersion medium is well reflected by kinetic and electrophoretic experiments.

Key words Emulsion polymers – reactive latices – coagulation – specific ion effects – adsorption – water structure

Introduction

Latices seemed to be almost ideal colloids for the investigation of dispersion stability. The latex dispersion is formed from hydrophobic and smooth spherical polymer particles immersed in an aqueous environment. Charged fragments from the initiator molecules accumulate at the particle water interface and the latex dispersion is electro- statically stabilized. Thus, the stability of latex dispersions against the addition of electrolytes should be well matched with the DLVO theory which has been derived for lyophobic colloids and is based on a pure electrostatic model [1, 24]. But even hydrophobic docosane particles [3] and differently charged polystyrene latices [4–7] are far away from this ideal. The stability of the latices is largely influenced by the structure of the boundary region between hydrophobic polymer particles and the dispersion

Progr Colloid Polym Sci (1998) 111:144–150
© Steinkopff Verlag 1998

medium [6–9]. Especially for latex dispersions used in technical application the situation is even more complicated: Emulsifier molecules, hydrophilic initiator fragments and charged groups with their supporting polymer chains make the surface of the latex particles "bristly" or "hairy" [4, 6, 10]. This may provide some additional stabilizing mechanisms even when the ionic strength in the dispersion is high enough to screen the charges on the surface of the particles. The polar or ionic groups protruding from the particle surface may promote the formation of a hydrophilic diffuse solvation shell surrounding the particles. Thus, a more appropriate model for the stability of hydrophilic latex dispersions should involve solvation as an extra stability factor.

Experimental

Monomers: Styrene was supplied by Hüls AG, Marl, Germany, glycidyl methacrylate (GMA) was delivered by Fluka, Germany. The monomers were used as received.
Initiators: Azo-bis-isobutyramidine dihydrochloride (ABA·2HCl) and azo-bis-N,N'-dimethylene-isobutyramidine dihydrochloride (ADMBA·2HCl) was kindly gifted by Wako Chemicals, Pure Chemical Industries Ltd., Japan, and 4.4'-azo-bis-4-cyanovaleric acid (ACVA) was delivered by Sigma, Germany.
Chemicals: Deionized water was used throughout the experiments. All other chemicals were of p.a. purity grade and used without further purification.
Preparation of latices: The latex dispersions were prepared by different emulsion polymerization techniques [4, 7, 11]:
Type-1-latices: The electrostatically stabilized particles (latex PSS(−), PSC(−), PSA1(+) and PSA2(+)) were pre-

pared by emulsifier free emulsion polymerization at 90 °C [4, 5] with the initiators indicated in Table 1.
Type-2-latices: The dispersions PSG(−), PSHG(−) were prepared by emulsion copolymerization of styrene with glycidyl methacrylate (GMA) and the initiators indicated in Table 1 [12].
Modification of type-2-latices: The desired amount of well-cleaned stock dispersion was diluted in deionized water containing small amounts of EMK-30 stabilizer. The mixture was heated to 80 °C and nitrogen was bubbled through the dispersion. The desired amount of modifier (and catalysts, if necessary) dissolved in a small amount of water was added. The total volume of the dispersion was 250–800 ml. The dispersion was held at 80 °C for an elaborated interval, cooled down and filtered [12].
Characterization of the dispersions: The mean diameter and the surface charge density of latex particles were determined by photon correlation spectroscopy and by a streaming current technique [11, 12]. The coagulation experiments were performed on diluted latex dispersions (volume fraction of the lattices $\varphi_{eff} \approx 0.01\%$). The critical coagulation concentration was either evaluated by test tube experiments (visual inspection of the coagulating dispersions after 24 and 48 h) [3–6] or by kinetic determination of the stability factor W (measuring the increase in turbidity with time at the beginning of coagulation). Full details of the determination methods are referred to in [11].

Results and discussion

To analyze water structure originated effects on dispersion stability one has to account for different contributions,

Table 1 Polystyrene latices prepared for coagulation experiments

No.	Initiator	Comonomer	Particle diameter in nm	Surface charge in $\mu C\,m^{-2}$
1	KPS	—	102	−8[b]
2	ACVA·2KOH	—	98	−12[b]
3	ADMBA·2HCl	—	98	+16[b]
4	ABA·2HCl	—	120	+19[b]
5	KPS/S$_2$O$_5^{2-}$	GMA	91[a]	−3[c]
6	H$_2$O$_2$/ASCA	GMA	67[a]	−4[c]

[a] Determined by TEM.
[b] Determined by pH ≈ 7.
[c] Determined at pH ≈ 11.
ABA·2HCl = azo-bis-N,N'-diisobutyramidine-dihydrochloride.
ADMBA·2HCl = azo-bis-N,N'-dimethylenisobutyramidine-dihydrochloride.
KPS = potassium peroxodisulfate.
ACVA = 4,4'-azo-bis-4-cyanovaleric acid.
H$_2$O$_2$/ASCA = hydrogen peroxide/ascorbic acid.
GMA = glycidylmethacrylate.

namely structural changes in the close vicinity of the hydrophobic particle surface [13–15] and interactions between dissolved molecules or ions with the modified water structure at the particle–water interface [16]. The alteration of the water structure in the dispersion medium well beyond the particle surface must also be considered [9, 17, 18]. A suitable relative measure for the stability of a dispersion can be easily derived from test tube experiments. The critical salt concentration c_k at which a floc phase is formed is determined by visual inspection. In Fig. 1 the c_k values of potassium chloride derived from test tube experiments are displayed for the modified type-2-latex particles (Table 2). The starting point on the left-hand side of Fig. 1 is the c_k value of a type-2-latex dispersion which has been synthesized by homopolymerization of styrene only. The dashed reference lines indicate the region of critical coagulation concentrations for electrostatically stabilized type-1-dispersions [4–7, 9]. By surface modification of the latex particles the stability of the dispersion is generally increased. The increase in the c_k of potassium chloride follows the order PS < PSG < PSGG < PSGT and PSH < PSHGT < PSG < PSHGG. The increase in the critical coagulation concentration is more pronounced for the particles synthesized by modification of the latex PSH. The increase in c_k is extreme for the latex PSHGG. The charge density for some latices was evaluated from streaming current (SCD) measurements. At pH $\approx$ 11 only relatively small differences between the stock particles and the modified latices were measured. The charge of the particles is not significantly influenced by the modification and the increase in stability is not linked to the charge density of the polymer particles. From SCD titration and the measurement of the electrokinetic mobility at different pH values an IEP for the amphoteric PSHGG latices was estimated to be at pH – IEP $\approx$ 6.5. At pH values $\ll$ pH$_{IEP}$ the particles are positively charged, and at pH $\gg$ pH$_{IEP}$ the particles carry negative charges. It has been already shown in separate papers [4, 7, 19] that the nature of the counterion, among other parameters, has a perceptible influence on colloid stability. Very often specific ion sequences can be found and the arrangement of the ions is specifically for a certain colloid [11]. The appearance of ion sequences (so-called Hofmeister series in protein chemistry) is related to water structure-mediated interactions in polyelectrolyte solutions [20, 21]. To get to the bottom of water structure–mediated interactions in the PSHGG water interface the critical coagulation concentration of some monovalent electrolytes for latex PSHGG were measured by the kinetic method. The c_k at different pH-values for the amphoteric polystyrene latices are presented in Fig. 2. For pH $\approx$ 3, i.e. well below the IEP of pH $\approx$ 6.5, the latex particles are positively charged due to the protonation of the glucamine residue at the particle surface. The addition

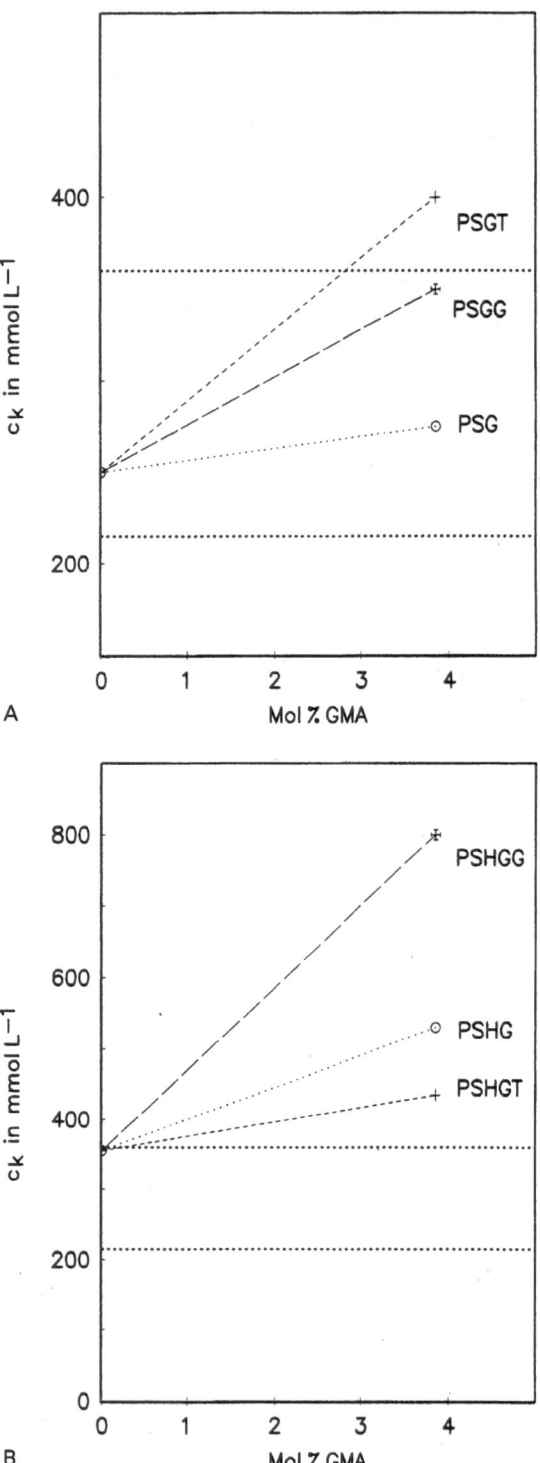

Fig. 1 Critical coagulation concentrations of potassium chloride at pH $\approx$ 11 for modified type-2-dispersions: (A) synthesized from stock latex PS, (B) synthesized from stock latex PSH

of electrolyte compresses the diffuse double layer and the surface charge is screened by the anions of the added salt. Indeed there is a quite distinct ranking of the measured

Table 2 Properties of the modified polystyrene particles

No.	Latex designation	Stock latex	Nucleopohile	Particle diameter in nm[a]	Surface charge in $\mu C\,m^{-2b)}$
5	PSG	—	—	91	− 3.0
7	PSGG	PSG	D-glycamine	86	− 5.1
8	PSGT	PSG	taurin	88	− 5.9
6	PSHG	—	—	67	− 4.1
9	PSHGT	PSHG	D-glucamine	78	− 4.5
10	PSHGG	PSHG	taurin	61	− 4.0

[a] Determined by TEM.
[b] Determined at pH $\approx$ 11.

c_k values. The c_k increase in the order $F^- \approx I^- < Br^- \ll Cl^-$. At pH $\approx$ 11 the screening of the surface charges is determined by the cations. The critical coagulation concentration is exceptionally high for sodium and potassium chloride and the sequence in c_k is $Cs^+ \ll Li^+ \ll Na^+ \approx K^+$.

The pattern of the critical coagulation concentration for different monovalent electrolytes and electrostatically stabilized anionic and cationic type-1-latices is presented in Fig. 3. The magnitude of c_k for the polystyrene latices in Fig. 3 is different but the arrangement of the c_k values of different alkali chlorides and potassium halides clearly demonstrate the influence of water structure–mediated interactions even on the stability of the type-1-polystyrene dispersions. The B coefficient of the Doyle–Jones equation has been chosen as an appropriate parameter for the water ion interaction [24]. Structure-ordering ions generally have positive B coefficients and structure-disordering ions are identified by negative values of B. It can be clearly seen from the graphs that the critical coagulation concentration was generally reduced with increasing structure breaking tendency (negative values of the ionic B coefficient) of the counterions. The pattern of the critical coagulation concentration of different monovalent electrolytes for the amphoteric latices (Fig. 2) is superficially similar to the pattern observed for conventionally charge stabilized polystyrene dispersions (Fig. 3).

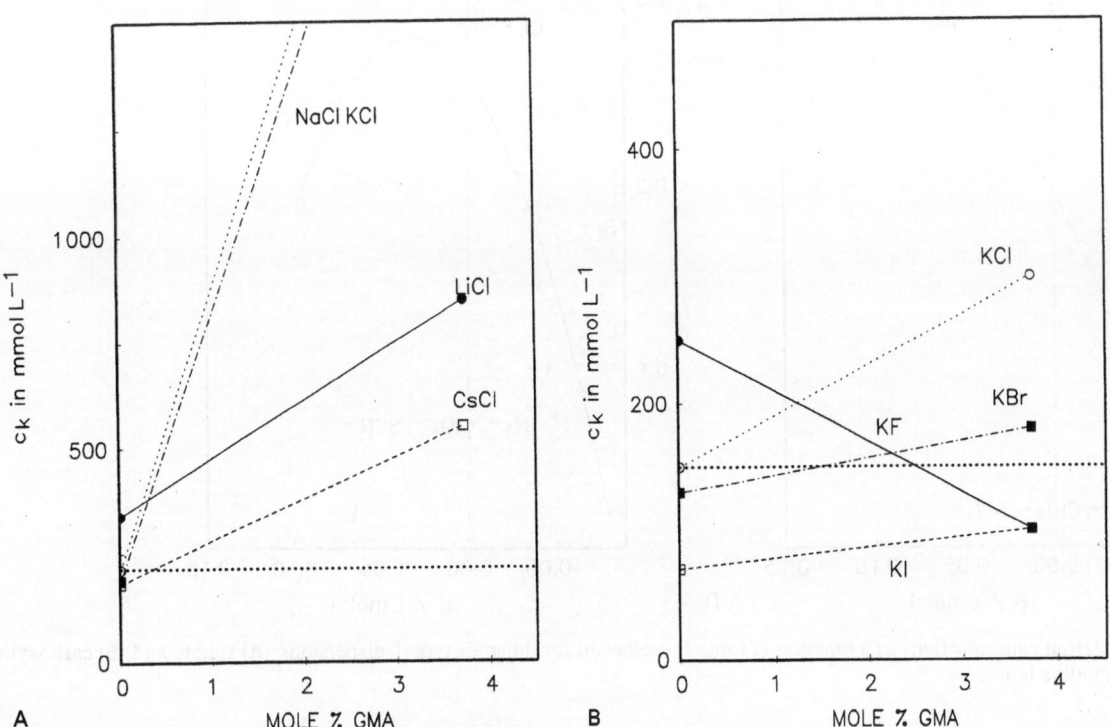

Fig. 2 Critical coagulation concentrations of different electrolytes for amphoteric type-2-latex PSHGG: (A) anionic polymer particles at pH $\approx$ 3, (B) cationic polymer particles at pH $\approx$ 11

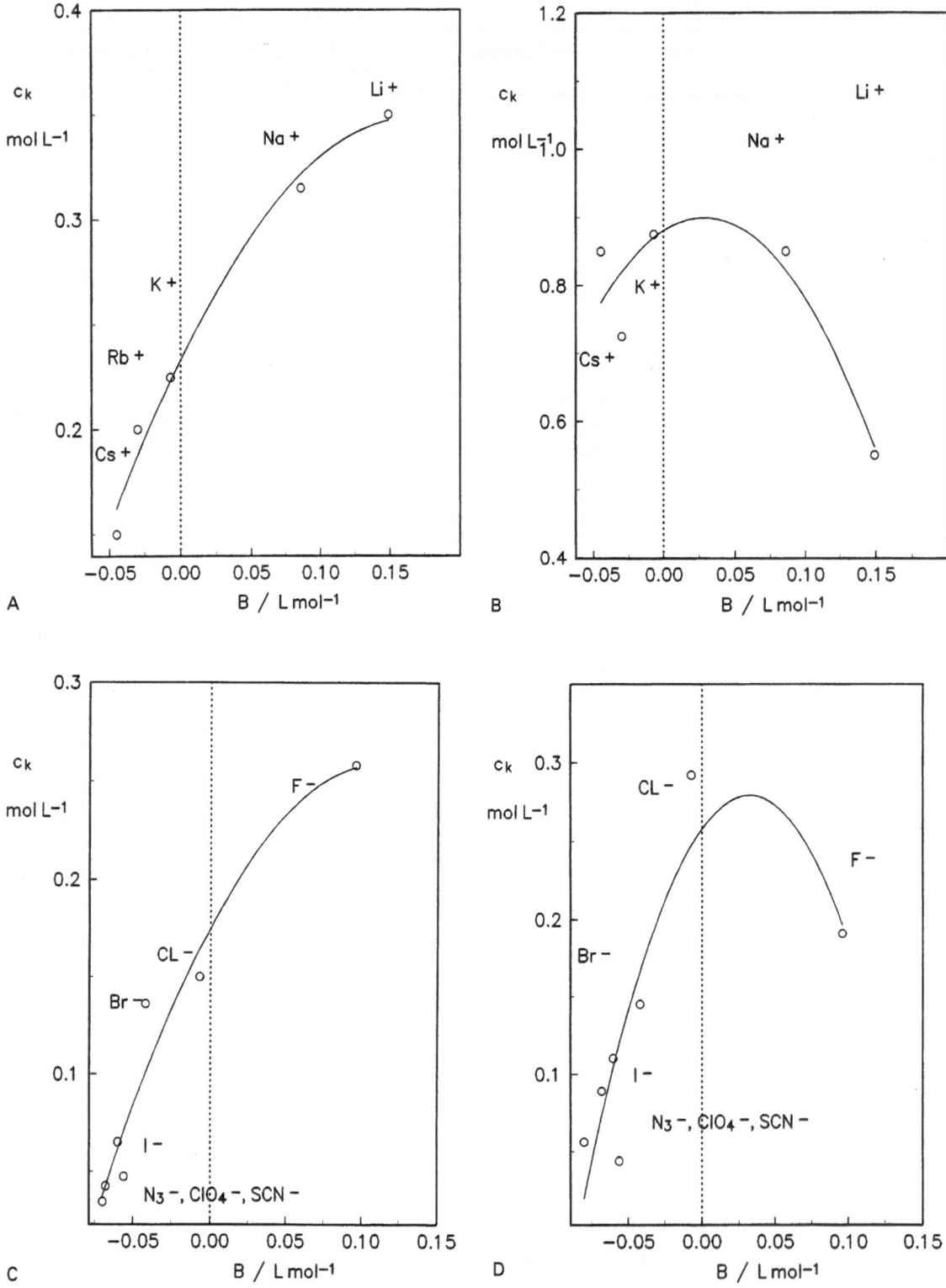

Fig. 3. Critical coagulation concentrations as a function of ionic B coefficient for different type-1-dispersions: (A) sulfate and (B) carboxylic latices, (C) and (D) amidine latices

Progr Colloid Polym Sci (1998) 111:144–150
© Steinkopff Verlag 1998

Discussion

Based on our results, structure mediated hydration forces involved in the coagulation of latex dispersions cannot be discussed without several speculations. For the type-1-dispersions the ionic end groups accumulate at the surface of the particles and one might expect that the surface properties of the final latex particles are mainly determined by the functional groups of the initiator. From the charge density determination (Tables 1 and 2) the surface area covered by the ionic groups can be estimated. Most part of the particle surfaces is still uncovered by ionic groups and remains basically hydrophobic. For the modified copolymer particles of latex PSHGG we can speculate from charge density determination and elementary analysis [12] that the surface of the particles is covered with sulfonate groups and poly-(GMA-co-D-glucamine) chains and that more or less densely packed patches of hydrated moieties may exist at the particle water interface. The systematic variation of the critical coagulation concentration with the nature of the counterion at otherwise fixed conditions is often interpreted in terms of a specific ion adsorption. In a pure electrostatic model low coagulation concentration is equivalent to well screened surface charges. The counterions accumulate at the particle surface and relatively less ions will remain in the diffuse part of the electrical double layer. The potential drop in the close vicinity of the particle surface is large. Then only a relatively low potential at the beginning of the diffuse double layer guarantees dispersion stability [11, 19]. The measured decrease in the critical coagulation concentration for the sulfate latex from $Li^+ \rightarrow Cs^+$ and for the amidine latices from $F^- \rightarrow SCN^-$ may be interpreted by a generic Stern layer adsorption: Small cations and small anions attract each other very strongly because of the high electrostatic field strength that they produce (electrostatic attraction), big cations and big anions bind by hydrophobic bonding [23]. Ion selectivity is linked to the functional groups at the particle surface and is determined by the electrostatic field strength at the fixed charged site. But all of the species involved in the interaction are surrounded by water molecules and most of them remain fully hydrated at least in the first steps of interaction. Accordingly there is a competition between the fixed charges, the mobile ions and the water dipoles. Fixed ionic sites with low field strength cannot attract cations away from the water dipoles [24]. The smallest cation, Li^+, is attracted most strongly by water molecules and the relatively hydrophobic surfaces of the sulfate latices will have a much more diffuse hydration shell when interacting with Li^+ than let us say with Cs^+, which is exactly the sequence that we observed in the coagulation experiments. Similar mechanisms may be operative for the coagulation of cationic latices by anions, although their behavior is somewhat less predictable with this simple measure. The strongly hydrated polyelectrolyte chains at the particles surface of the copolymer latices interact selectively with Na^+ and K^+ ions. Especially at low surface charge of the latex particles, the hydrated polyelectrolyte chains at the particle surface will stabilize more strongly with Na^+ and K^+ than with Li^+ or Cs^+. The c_k is increased in the presence of D-glucamine (Figs. 2 and 3) and the enhancement of the stability is extremely evident for the critical coagulation concentration derived from rate constant data for the first steps of the coagulation [8, 9, 12]. For the same reason irregularities in the anion sequence were obtained.

Under the premise of specific adsorption and the formation of a water originated surface structure the critical coagulation concentration should be very sensitive to temperature [25]. An increase in temperature may decrease the value of the critical coagulation concentration because of an "evaporation" of the "condensed" layer of water bound to counterions and to the polymer chains at the particle surface. Indeed with increasing temperature the c_k decrease and at 70 °C we could not detect significant ion sequences [9]. Thus, we conclude that the thermal desorption process is nearly complete at ≈70 °C [12].

The properties of the structural layer will not only depend on temperature but also on charge density. The double-layer field strength tends to orient the water dipoles and this could mean that at high potential the water structure is partially disrupted. There is vast literature for the structural layer of water molecules at the surface of oxide and silver halide particles [21]. The surface entropy calculated by Lyklema for the silver iodide/water system increases with increasing negative potential of the colloidal particles, which was interpreted as an increase in configurational disorder [21, 25]. This picture is supported by NMR experiments on polyvinyltoluene particles. At room temperature the correlation time for molecular motion of water molecules in the dispersion was longer than for bulk water and the correlation time increased with decreasing surface charge [26]. The structuring of water molecules might provide an additional repulsive potential particularly noticeable when the electrostatic potential becomes very small.

Beyond the water structuring at the boundary region of the latex particles and the levelling by temperature, several other mechanism have to be considered. Organic molecules like urea, n-alkyl ureas, acetamide, acetone or 1-propanol can increase or reduce the salt sensitivity of a latex dispersion [7]. The composition of the dispersion medium can be used as a moderator of the particle–particle interaction. Especially in a solvent mixture, adsorption layer formation occurs on the surface of the dispersed particles [13, 27]. The composition of the adsorption layer generally differs from both the initial and the equilibrium compositions of the liquid mixture due to molecular interaction at the solid–liquid interface [14, 15].

The most straightforeward way of determining the extra stability term due to hydrated hairy layers would be to measure the force acting between two latex spheres. The interaction forces between $2\,\mu m$ latex particles has been investigated by atomic force microscopy [28], and by using another ultrasensitive force measurement method, colloidal particle scattering (CPS), it has recently been demonstrated that a layer of immobilized water molecules in the hairy layer effectively lowers the van der Waals interactions, thus increasing the relative importance of electrostatic repulsion [29]. But in aqueous media, not only the macroscopic changes in parameters like the "Hamaker constant" or the dielectric increment influence the dispersion stability. More subtle effects, such as competitive adsorption between water and dissolved organic molecules or the displacement of specifically adsorbed ions may cause unexpected phenomena [30, 31]. Moreover, hydration effects may interfere with the ion distribution at the solid–liquid interface. Indeed several investigators argue that the water structure near the particle interface is strongly modified and that a salt deficient zone surrounding polymer chains can introduce repulsive image charge forces [32–34]. These new results seem to give further evidence that the usual double layer model might not be sufficient in the case of latex dispersions.

Conclusion

Based on the foregoing discussion we will come to the conclusion that the coagulation of latex particles can only occur if two contributions are sufficiently suppressed:

- Screening of surface charges (i.e. compression of the diffuse double layer) and
- disruption of structured solvent shells (i.e. compression of the diffuse solvation field).

The structured water can be regarded as composed of a number of regions of different molecular mobility compared to bulk water. The solvation energy barrier for a given disperse system can presumably disappear at high enough temperatures and can be moderated by ions and organic compounds. We do not exclude the possibility of effects other than changes in the water structures playing a role, but at the moment this is the model we postulated to explain the coagulation data.

Acknowledgements The authors are very grateful to Hüls AG for financial support and to Prof. I. Dékány, as a part of this idea was borne during a long term stay in his laboratory at Szeged, Hungary. We also thank Dr. J. Wieboldt (PolymerLatex GmbH & CoKG) and Dr. J. Priewe (Schering AG) for their assistance and helpful discussions and Prof. G. Lagaly for encouragement.

References

1. Verwey EJW, Overbeek JThG (1948) Theory of the stability of lyophobic colloids. Elsevier, Amsterdam, New York
2. Hiemenz PC (1986) Principles of colloid and surface chemistry, 2nd ed. Marcel Dekker, New York, p 721
3. Dunstan DE (1993) J Chem Soc Faraday Trans 89:521
4. Zimehl R, Lagaly G (1986) Progr Colloid Polymer Sci 72:28
5. Zimehl R, Wieboldt J (1989) TIZ Int Powder Mag 113:629
6. Zimehl R, Lagaly G, Ahrens J (1990) Colloid Polymer Sci 268:924
7. Zimehl R, Lagaly G (1987) Colloids Surf 22:225
8. Zimehl R, Priewe J (1996) Progr Colloid Polym Sci 101:116
9. Zimehl R (1997) Colloids and Surfaces, accepted
10. Wieboldt J, Zimehl R, Ahrens J, Lagaly G, Progr Colloid Polym Sci, submitted for publication
11. Lagaly G, Schulz O, Zimehl R (1997) Dispersionen und Emulsionen. Steinkopff, Darmstadt
12. Kluß BU (1992) Thesis. University of Kiel
13. Dékány I (1993) Pure Appl Chem 65:901
14. Zimehl R, Dékány I, Lagaly G (1992) Colloid Polym Sci 270:68
15. Zimehl R, Ruffmann B, Vózár A, Dékány I (1996) Progr Colloid Polym Sci 101:166
16. Mielke M, Zimehl R (1997) Progr Colloid Polym Sci, submitted for publication
17. Marosi T, Dékány I, Lagaly G (1994) Colloid Polymer Sci 272:1136
18. Leberman R, Soper AK (1995) Nature 378:364
19. Stern O (1924) Z Elektrochem 30:508
20. Lyklema J (1991) Fundamentals of interface and colloid science, Vol I, Fundamentals, Ch 5. Academic Press, London
21. Lyklema J (1995) Fundamentals of interface and colloid science, Vol II: Solid–Liquid Interfaces, Ch 3. Academic Press, London
22. Nightingale ER (1966) in Conway BE, Barradas RG (eds) Chemical Physics of Ionic Solutions 7:87, Wiley, New York
23. Lyklema J (1983) in Parfitt GD, Rochester CH (eds) Adsorption from solution at the solid/liquid interface, Ch 5. Academic Press, London
24. Eisenman G (1965) in Reilley CN (ed) Advances in Analytical Chemistry and Instrumentation, Vol 4. Wiley-Interscience, New York
25. Lyklema J (1966) Discussions Faraday Soc 42:81
26. Johnson GA, Lecchini SMA, Smith EG, Clifford J, Pethica BA (1966) Discussions Faraday Soc 42:120
27. Vincent B, Király Z, Emmett S, Beaver A (1990) Colloids Surf 49:121
28. Li YQ, Tao NJ, Pan J, Garcia AA, Lindsay SM (1993) Langmuir 9:637
29. Wu X, van de Ven TGM (1996) Langmuir 12:3859
30. Ottewill RH, Vincent B (1972) J Chem Soc Faraday Trans 68:1533
31. Vincent B (1992) Adv Colloid Interface Sci 42:279
32. Brooks DE (1973) J Colloid Interface Sci 43:687
33. Florin E, Kjellander R, Eriksson JC (1984) J Chem Soc Faraday Trans 80:2889
34. Bahadur P, Pandya K, Almgren M, Li P, Stilbs P (1993) Colloid Polym Sci 27:657

Progr Colloid Polym Sci (1998) 111:151–157
© Steinkopff Verlag 1998

ASSOCIATION COLLOIDS

O.A. El Seoud
R.C. Bazito
G.K. Barlow

A proton NMR study on the structure of interfacial water of aqueous micelles: Effects of the structure of the surfactant

O.A. El Seoud (✉) · R.C. Bazito
Instituto de Química
Universidade de São Paulo
C.P. 26077
05599-970 São Paulo, S.P., Brazil
E-mail: elseoud@iq.usp.br

G.K. Barlow
Department of Chemistry
The University of York
Heslington, York YO1 5DD
United Kingdom

Abstract Fractionation factors, φ_M, of interfacial water at surfaces of anionic, cationic, and zwitterionic micelles were determined from the dependence of the ^1H NMR chemical shift of H_2O on the isotopic composition of the solvent (H_2O plus D_2O). Except for one surfactant, 3-dodecylamido-1-(N,N-dimethyl)-propane betaine, values of φ_M are greater than unity, showing that interfacial water is more structured than bulk water, whose fractionation factor = 1, by definition. Interfacial water structuring increases with increasing bulk of the headgroup of cationic surfactants, but is unaffected by increasing the length of the hydrophobic group of anionic surfactants (sodium tetradecylsulfate, versus sodium dodecylsulfate), or by the presence of two oxyethylene units at the interface (sodium dodecyl-dioxyethylenesulfate, versus sodium dodecylsulfate). The betaine surfactant does not affect the structure of interfacial water because of strong intermolecular interactions between the head ions (i.e., $-N^+(CH_3)_2-CH_2-CO_2^-$) of neighboring molecules in the micelles. Values of φ_M are: sodium tetradecylsulfate, 1.06; sodium dodecyldioxyethylenesulfate, 1.06; dodecyl-dimethyl(n-butyl)-ammonium bromide, 1.08; dodecyltri-n-propylammonium bromide, 1.11; dodecyldimethyl(2-hydroxyethyl)ammonium bromide, 1.06; and 3-dodecylamido-1-(N,N-dimethyl)propane betaine, ≈ 1.

Key words Interfacial water structuring – aqueous micelle – surfactants

Introduction

Interest in studying the properties of water at interfaces of organized assemblies stems from the fact that interfacial water is involved in hydration of the surfactant headgroup and, when surfactants act as catalysts for chemical reactions, in solvation of reactants, transition states, and in proton transfers [1–3]. Extrinsic lipophilic or water-insoluble reporter molecules have been used to probe the microscopic properties (e.g. viscosity, polarity) of interfacial water of organized assemblies, and to evaluate the extent of hydrocarbon chain-water contact. Interpreta-

tion of the results thus obtained is, however, not always unequivocal mainly because of the difficulty in identifying the loci of solubilization of probes in the micellar pseudophase [1, 3, 4–6]. A more straightforward approach is to use a non-invasive, e.g., spectroscopic, technique to investigate the micellar system, then to interpret the results in terms of microscopic properties of interfacial water; NMR spectroscopy has been fruitfully used in this regard [7–9].

In the present work, we examine effects of the structure of surfactant on the degree of structure of water of hydration of aqueous micelles, relative to that of bulk water. This is achieved by measuring the deuterium isotope effect on the ^{1}H NMR chemical shift of the solvent (H_2O–D_2O mixtures). Isotopic substitution in the solvent involves minimum perturbation of the micellar system, and the resulting secondary isotope effect can be accurately measured by NMR.

It is convenient to describe isotope effects in terms of fractionation factors, φ, defined for a micelle, M, as

$$\varphi_M = (D/H)_{interface}/(D/H)_{bulk\ solvent} . \quad (1)$$

That is, for a micelle in a H_2O–D_2O mixture, the fractionation factor for interfacial water, φ_M, describes the isotopic composition at the aggregate/water interface relative to that in bulk solvent [9, 10].

We are interested in using φ_M to probe effects of the structure of surfactant on the degree of structure of interfacial water. In the present work we examine effects of the following surfactants: sodium tetradecyl sulfate ($C_{14}H_{29}$–SO_4^- Na^+, STS); sodium dodecyldioxyethylenesulfate (average structure, $C_{12}H_{25}$–$(OCH_2CH_2)_2$–SO_4^- Na^+, SDoEO$_2$S); 3-dodecylamido-1-(N,N-dimethyl)propane betaine (average structure, $C_{11}H_{23}CO$–$NH(CH_2)_3$–N^+ $(CH_3)_2$–CH_2–CO_2^-, CAPB); 3-N-cetyl-N,N-dimethylammonio-1-propane sulfonate ($C_{16}H_{33}N^+$–$(CH_2)_3$–SO_3^-, SB 3–16); dodecyldimethyl(n-butyl)ammonium bromide ($C_{12}H_{25}N^+(CH_3)_2C_4H_9Br^-$, Do(Me)$_2$BuABr); dodecyltri-$n$-propylammonium bromide ($C_{12}H_{25}N^+(C_3H_7)_3$–$C_4H_9Br^-$, DoPr$_3$ABr); and dodecyldimethyl(2-hydroxyethyl)-ammonium bromide ($C_{12}H_{25}N^+(CH_3)_2CH_2CH_2OH$ Br^-, Do(Me)$_2$HEtABr).

Previously, we have determined φ_M for sodium dodecylsulfate, SDoS, and for 3-N-dodecyl-N,N-dimethyl-ammonio-1-propanesulfonate (SB 3–12) [10a, b]. We examined micelles of STS, and of SB 3–16 in order to determine the effect, if any, of the increase of chain length of the surfactant on φ_M. In addition to carrying headgroups that were not examined before, SDoEO$_2$S and CAPB are extensively employed in consumer products, e.g., shampoos and shower gels because of their excellent detergency and foaming properties, and mildness to the skin [11]. Finally, the cationic surfactants studied either have a bulky headgroup, Do(Me)$_2$BuABr, and DoPr$_3$ABr, or carry the 2-hydroxyethyl moiety, Do(Me)$_2$HEtABr. Fractionation factors of the first two surfactants should indicate the effect of increasing hydrophobic character of the headgroup on the structure of interfacial water. A study of Do(Me)$_2$HEtABr was deemed important because our previous results have indicated that cetyltris(2-hydroxyethyl)ammonium chloride does not perturb the structure of interfacial water, because of strong H-bonding between the tris(2-hydroxyethyl) moiety and the counterion [10c].

Experimental

Materials

All chemicals were obtained from Aldrich or Merck, and were purified by standard procedures [12]. STS was synthesized by reacting tetradecyl alcohol with purified chlorosulfonic acid in ethyl ether. After stirring the mixture for two hours at 5–10 °C, then for additional 2 h at room temperature, the solvent was removed under reduced pressure, and the residue was neutralized with an aqueous solution of sodium carbonate, followed by removal of water by lyophilization. The waxy solid was extracted with absolute ethyl alcohol, the solvent removed, and the residue recrystallized from ethyl alcohol-acetone.

Do(Me)$_2$BuABr, DoPr$_3$ABr, and Do(Me)$_2$HEtABr were obtained by heating the appropriate trialkylamine with 20% excess 1-bromododecane in refluxing dry acetonitrile for 24 h, and were purified as given elsewhere [13]. SB 3–16 was prepared by reacting N,N-dimethylcetylamine (Hoechst, >99.5% pure) with 1,3-propane sultone in THF, and was purified as given elsewhere [10b]. Aqueous solutions of SDoEO$_2$S and CAPB were obtained from Hoechst do Brasil. These were evaporated to dryness under reduced pressure. The waxy solids were purified by extraction with absolute ethyl alcohol, as given for STS. The deuterium content of D_2O (Goss Scientific Instruments, used as received) was determined by ^{1}H NMR spectroscopy by the recommended procedure [14]. Distilled-deionized-H_2O was used throughout.

Purified surfactants were dried under reduced pressure, over P_2O_5 to a constant weight. Measurement of the pH of 0.05 M aqueous solutions of these surfactants indicated the absence of acidic and/or basic impurities. There were no minima in the surface tension versus surfactant concentration plots (Lauda T1C digital tensiometer) and the critical micelle concentrations agreed with literature values [13, 15–19].

Methods

In the NMR experiments, observed chemical shift of the solvent (H_2O or H_2O–D_2O) was measured as a function of [surfactant]. For solutions in H_2O, the spectrometer "lock" was D_2O contained in a Wilmad WGS-5BL coaxial insert. A Bruker AMX-500 spectrometer, operating at 500.13 MHz for proton was used.

All solutions were prepared by weight and then transferred to Wilmad 535-PP precision NMR tubes. These were left in the spectrometer sample compartment for 10 min for thermal equilibration. Thermal stability of the spectrometer probe was insured as given elsewhere [10], and from the observation that successive measurements of the same sample were always within the digital resolution limit. Chemical shifts were measured at 25 °C (35 °C for STS) from 1% internal dioxane. The spectra acquisition parameters were adjusted to obtain a digital resolution of 0.04 Hz/data point.

Results

For a micelle in H_2O–D_2O, e.g., that of an anionic surfactant with sodium counterion, the equation that relates the observed chemical shift of H_2O, δ_{obs}, to the mole fraction of surfactant, N_{Surf}, is given by [9, 10]:

$$\delta_{obs} = \delta_o + N_{Surf}\left\{\left[\frac{h_{Na_f}\delta_{Na_f}}{1 - \chi_D + \chi_D\varphi_{Na_f}}\right] + \left[\frac{h_M\delta_M}{1 - \chi_D + \chi_D\varphi_M}\right]\right\}, \quad (2)$$

where δ_o is the chemical shift of bulk solvent, h is the hydration number of the relevant species, χ_D is the atom fraction of deuterium in the solvent, and the subscripts (f) and (M) refer to free counterion, and micelle, respectively. The treatment is simpler for zwitterionic (and nonionic) micelles, and because of absence of counterions the corresponding equation is:

$$\delta_{obs} = \delta_o + N_{Surf}\left[\frac{h_M\delta_M}{1 - \chi_D + \chi_D\varphi_M}\right]. \quad (3)$$

Equations 2 and 3 predict a linear relationship between δ_{obs} and N_{Surf}, as shown in Fig. 1, and calculated φ_M are given in Table 1. Except for STS whose Krafft temperature is ca. 30 °C, all NMR experiments were carried out at 25 °C. The Krafft temperature of SB 3–16 is higher than 40 °C, at 50 °C the difference in δ_{obs} as a function of N_{Surf} was, however, too small to be useful, and we were unable to calculate φ_M for this surfactant. For CAPB, values of δ_{obs} are surprisingly insensitive to an increase in N_{Surf} or a change in χ_D (< 0.4 Hz). Negligible dependence of the slopes on the deuterium content of the solvent indicates that the fractionation factor for interfacial water is practically unity.

Discussion

For species (i) in H_2O–D_2O (e.g., free counterions or micelles) the fractionation factor is the equilibrium constant for the isotopic exchange reaction between bulk water molecules and those present in the hydration shell of (i). For example, if (i) carries no exchangeable hydrogens, and is solvated by one solvent molecule, the equilibrium constant is [14, 20]:

$$DOL + i(HOL) \rightleftharpoons HOL + i(DOL), \quad (4)$$

where L is an unspecified isotope of hydrogen. The deuterium/protium fractionation factor at (i), φ_i, expresses, therefore, the deuterium preference for the site in question (i.e., the hydration shell of i) relative to an average solvent site.

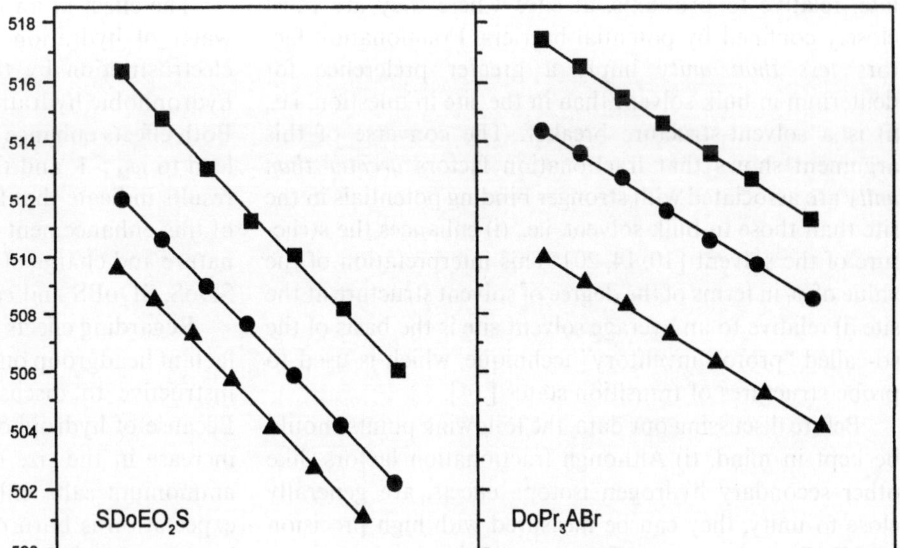

Fig. 1 Representative plots of Eq. (2), showing the dependence of the chemical shift of H_2O, δ_{obs}, on the mole fraction of surfactant, N_{Surf}, for $SDoEO_2S$ and $DoPr_3ABR$ in H_2O (■), at $\chi_D \approx 0.5$ (●), and at $\chi_D \approx 0.95$ (▶)

Table 1 Results of application of Eq. (2) and calculated micellar fractionation factors[a),b)]

Surfactant	χ_D	Intercept δ_o [Hz]	Slope N_{Surf}^{-1}/Hz	φ_M[c)]
STS	0	468.19	−1430.37	
	0.491	466.67	−1378.86	1.06
	0.945	462.54	−1348.43	1.06
SDoEO$_2$S	0	517.82	−2309.04	
	0.505	513.55	−2252.84	1.07
	0.945	511.07	−2230.22	1.05
Do(Me)$_2$-BuABr	0	517.64	−1342.07	
	0.502	516.64	−1315.30	1.08
	0.945	514.70	−1294.16	1.08
DoPr$_3$ABr	0	518.31	−1385.10	
	0.497	515.27	−1339.81	1.11
	0.946	510.90	−1310.28	1.11
Do(Me)$_2$-HEtABr	0	521.64	−1374.41	
	0.947	512.82	−1323.49	1.06

[a)] Abbreviations: STS, sodium tetradecylsulfate; SDoEO$_2$S, sodium dodecyl-dioxyethylene sulfate; Do(Me)$_2$BuABr, dodecyldimethyl(*n*-butyl)ammonium bromide; DoPr$_3$ABr, dodecyltri-*n*-propylammonium bromide; Do(Me)$_2$HEtABr, dodecyldimethyl(2-hydroxyethyl)ammonium bromide.
[b)] Chemical shift data were measured from internal dioxane, at 500.13 MHz, at 25 °C (35 °C for STS).
[c)] The uncertainty in these values is ±0.02, see calculations.

The next point is to show how the value of φ_i can be interpreted in terms of the degree of structure of water of hydration of (i), relative to bulk water. Note that for bulk solvent $\varphi = 1$, by definition. In a system where D and H equilibrate among a number of different sites, e.g., hydration shell of (i) and bulk solvent, deuterions accumulate, relative to protiums, at sites where they are most closely confined by potential barriers. Fractionation factors *less than unity* imply a greater preference for deuterium in bulk solvent than in the site in question, i.e., (i) is a solvent-structure breaker. The converse of this argument shows that fractionation factors *greater than unity* are associated with stronger binding potentials in the site than those in bulk solvent, i.e., (i) enhances the structure of the solvent [10, 14, 20]. This interpretation of the value of φ_i in terms of the degree of solvent structure at the site (i) relative to an average solvent site is the basis of the so-called "proton inventory" technique, which is used to probe structures of transition states [14].

Before discussing our data, the following points should be kept in mind: (i) Although fractionation factors, like other secondary hydrogen isotope effects, are generally close to unity, they can be measured with high precision by NMR spectroscopy. For example, in simple systems differences as small as 1% are physically significant [10, 14, 21]; (ii) Because the fractionation factor is an equilibrium constant, fast diffusion of water molecules between different sites in the system (i.e., bulk water and water of hydration of relevant species) has no bearing on the calculation of φ; (iii) Fractionation factors calculated by Eq. (2) refer to interfacial water of the *whole micelle*, e.g., tetradecylsulfate anions plus the associated sodium ions; (iv) We usually use *two* H$_2$O–D$_2$O solvent mixtures ($\chi_D \approx 0.5$, $\chi_D \approx 0.95$) to get *two* independent values of the fractionation factor in question. Both values agree, Table 1.

It is convenient to discuss our results together with other fractionation factors which we determined previously [10], see Table 2. Fractionation factors for SDoS and STS are the same, and a similar result was obtained for cationic CMe$_3$ACl and DoMe$_3$ABr [10d]. We conclude that modest changes in lengths of the hydrophobic tails of anionic and cationic surfactants have a negligible effect on φ_M, i.e., the observed differences in fractionation factors are due to differences in the structure of the surfactant headgroup. Although we were unable to determine φ_M of SB 3–16 because of its high Krafft temperature, we believe that the preceding conclusion is a general one. The high Krafft temperature of sodium cetylsulfate is the reason for using STS instead.

We compare results of the cationic surfactants Do(Me)$_2$BuABr, DoPr$_3$ABr, and Do(Me)$_2$HEtABr with those of DoMe$_3$ABr because they are structurally related. Substitution of a butyl group for one of the interfacial methyl groups of DoMe$_3$ABr resulted in a 2% increase in φ_M, whereas substitution of three *n*-propyl groups for the three interfacial methyl groups of DoMe$_3$ABr resulted in a 5% increase in the fractionation factor. Note that DoPr$_3$ABr has the second largest φ_M in Table 2, after CBu$_3$ABr.

The effect of an ionic interface on the structure of its water of hydration can be analyzed in terms of water electrostriction by the charged interface, and effects of hydrophobic hydration [22] of the surfactant headgroup. Both effects enhance the structure of interfacial water, i.e., lead to $\varphi_M > 1$, and the former will be analyzed first. Our results indicate that for simple surfactants, the magnitude of this enhancement does not seem to be sensitive to the nature and charge of the headgroup. Thus, φ_M for anionic SDoS, SDoBS and cationic CMe$_3$ACl is 1.06 [10a].

Regarding effects of hydrophobic hydration of the surfactant headgroup on the structure of interfacial water, it is instructive to discuss simple, i.e., non-aggregated ions. Because of hydrophobic hydration, it is expected that an increase in the size of the alkyl moiety of simple alkylammonium salts enhances the structure of water. This expectation is born out by IR and NMR data [23], and by our fractionation factors for $(CH_3)_3N^+$, $\varphi = 1.0$, and $(C_4H_9)_3N^+$, $\varphi = 1.06$ [10b]. Thus, formation of a micelle of either Do(Me)$_2$BuABr or DoPr$_3$ABr increases the water structure around the central ion, relative to the

Progr Colloid Polym Sci (1998) 111:151–157
© Steinkopff Verlag 1998

Table 2 Fractionation factors for aqueous micelles, φ_M, determined by ^{1}H NMR spectroscopy[a]

Name and structure		$T, °C$[c]	φ_M[d]
Anionic			
SDoS	Sodium dodecylsulfate, $C_{12}H_{25}SO_4^-\,Na^+$	27.0	1.06
SDoBS	Sodium dodecylbenzenesulfonate, $C_{12}H_{25}$-◯-$SO_3^-\,Na^+$	27.0	1.06
SPFO	Sodium perfluorooctanoate, $C_7F_{15}CO_2^-\,Na^+$	27.0	1.08
STS[b]	Sodium tetradecylsulfate, $C_{14}H_{29}SO_4^-\,Na^+$	35.0	1.06
SDoEO$_2$S[b]	Sodium dodecyldioxyethylenesulfate, $C_{12}H_{25}(OCH_2CH_2)_2OSO_3^-\,Na^+$	25.0	1.06
Cationic			
C(Me)$_2$PhACl	Cetyldimethylphenylammonium chloride, $C_{16}H_{33}N^+(CH_3)_2$-◯-Cl^-	35.0	1.08
C(Me)$_2$ PhPrACl	Cetyldimethyl-3-phenylpropylammonium chloride, $C_{16}H_{33}N^+(CH_3)_2$–CH_2–CH_2–CH_2–◯Cl^-	20.0	1.08
CPyCl	Cetylpyridinium chloride, $C_{16}H_{33}$–N^+◯ Cl^-	25.0	1.06
CQCl	Cetylquinuclidium chloride, $C_{16}H_{33}$–N^+⬡ Cl^-	20.0	1.07
CMe$_3$ACl	Cetyltrimethylammonium chloride, $C_{16}H_{33}N^+(CH_3)_3Cl^-$	27.0	1.06
CBu$_3$ABr	Cetyltri-n-butylammonium bromide, $C_{16}H_{33}N^+(C_4H_9)_3Br^-$	20.0	1.14
CHEtA$_3$Cl	Cetyltris(2-hydroxyethyl)ammonium chloride, $C_{16}H_{33}N^+(CH_2CH_2OH)_3Cl^-$	20.0	≈1
DoMe$_3$ABr	Dodecyltrimethylammonium bromide, $C_{12}H_{25}N^+(CH_3)_3Br^-$	25.0	1.06
Do(Me)$_2$BuABr[b]	Dodecyldimethyl(n-butyl)ammonium bromide, $C_{12}H_{25}N^+(CH_3)_2C_4H_9Br^-$	25.0	1.08
DoPr$_3$ABr[b]	dodecyltri-n-propylammonium bromide $C_{12}H_{25}N^+(C_3H_7)_3Br^-$	25.0	1.11
Do(Me)$_2$HEtABr[b]	dodecyldimethyl(2-hydroxyethyl) ammonium bromide, $C_{12}H_{25}N^+(CH_3)_2\,CH_2CH_2OHBr^-$	25.0	1.06
Nonionic and Zwitterionic			
TX-100	Polyoxyethylene (9.5) octylphenyl ether, C_8H_{17}-◯-$O(CH_2CH_2O)_{8.5}$–CH_2CH_2OH	27.0	0.95
SBMe 3–12	3-N-Dodecyl-N,N-dimethylammonio-1-propane sulfonate, $C_{12}H_{25}$–$N^+(CH_3)_2$–CH_2–CH_2–CH_2–SO_3^-	27.0	1.02
SBEt 3–14	3-N-Tetradecyl-N,N-diethylammonio-1-propane sulfonate, $C_{14}H_{29}$–$N^+(C_2H_5)_2$–CH_2–CH_2–CH_2–SO_3^-	30.0	1.03
SBBu 3–14	3-N-Tetradecyl-N,N-di-n-butylammonio-1-propane sulfonate, $C_{14}H_{29}$–$N^+(C_4H_9)_2$–CH_2–CH_2–CH_2–SO_3^-	30.0	1.07
CAPB[b]	3-dodecylamido-1-(N,N-dimethyl)propane betaine, $C_{11}H_{23}CO$–$NH(CH_2)_3$–$N^+(CH_3)_2$–CH_2–CO_2^-	25.0	≈1

[a] Data taken from Ref. [10].
[b] Data from the present work.
[c] The experiment temperature depends on the Krafft point of the surfactant.
[d] Uncertainties in reported fractionation factors are in the range of ±0.02, and are mainly due to uncertainties in hydration numbers and degrees of dissociation of aqueous micelles, see text for details.

corresponding case of non-aggregating model ions. As expected, the water structure enhancement is larger for DoPr$_3$ABr than for Do(Me)$_2$BuABr. We conclude that hydrophobic hydration of N-alkyl groups at the micellar interface is the main reason for the observed dependence of φ_M on the number, and chain length of alkyl headgroups of cationic surfactants. The fractionation factor of DoPr$_3$ABr indicates that the interfacial n-propyl groups do not "fold" back into the micellar core and away from water, by analogy with results in water-in-oil microemulsions [24]. *Complete* folding back should produce interfacial water whose structure is similar to that hydrating CMe$_3$ACl micelles, i.e., φ_M of DoPr$_3$ABr should be ca. 1.06, instead of 1.11, see Table 2.

The fractionation factor of SDoEO$_2$S is similar to that of SDoS, likewise, φ_M of Do(Me)$_2$HEtABr is similar to that of DoMe$_3$ABr. Barry and Wilson studied ethoxylated anionic surfactants, $C_{12}H_{25}(EO)_nSO_4Na$, and concluded that for surfactants with $n = 1$ and 2, the properties of the micelles (e.g., degree of micellar dissociation, α; critical micelle concentration, cmc; and area/interfacial headgroup) are dictated by the sulfate group [16]. This also agrees with the conclusion that water structure perturbation by short-chain alkyl sulfates, and sulfonates is dominated by anion–water interactions [10a, 25]. Provided that (n) is small, therefore, the structure of interfacial water of ethoxylated anionic surfactants seems to be determined by the sulfate headgroup.

Measurements of cmc, maximum surface excess concentration, minimum area per molecule at the aqueous solution, and ^{1}H NMR chemical shift of monomeric and aggregated surfactant have indicated that substitution of a 2-hydroxyethyl group for a methyl, or an ethyl headgroup has little effect on the surface and micellar properties of cationic surfactants [13, 19, 25]. For example, the Gibbs free energy of transfer of the 2-hydroxyethyl group from monomeric to micellized state is only -0.18 kJ mol^{-1}, which is very different from 3.2 kJ (moles of CH_2)$^{-1}$ calculated for the transfer of a CH_2 group in the hydrophobic tail of cationic surfactants [13, 26]. Thus, other data support our conclusion that φ_M for Do(Me)$_2$HEtABr and DoMe$_3$ABr are similar because substitution of a 2-hydroxyethyl group for one of the methyl groups of DoMe$_3$ABr has a negligible effect on the properties of the interface, hence on the structure of its water of hydration. Results of Table 2 show that φ_M for the structurally similar CHEt$_3$ACl is ≈ 1. For this surfactant, the ^{1}H NMR chemical shift of water was found to be neither sensitive to a variation of N_{Surf} nor of χ_D! That this is a micelle-related phenomenon is evident from the result on a short-chain analogue, methyltris(2-hydroxyethyl)ammonium chloride, which showed a linear dependence of δ_{obs} on (salt). The result for CHEt$_3$ACl was interpreted in terms of strong, specific interaction of the counterion with the *three* interfacial 2-hydroxyethyl groups [10c]. The difference between φ_M of Do(Me)$_2$HEtABr and CHEt$_3$ACl may be attributed to the fact that more than one interfacial hydroxy group is required for an efficient H-bonding of the counterion.

For CAPB, the ^{1}H NMR chemical shift of water is also neither sensitive to a variation of N_{Surf} nor of χ_D. Previous work indicated negligible perturbation of the structure of water by zwitterionic surfactants when they carry headgroups which form strong hydrogen bonds (e.g., carboxylate anion [27]). For CAPB, these intermolecular interactions between the headions (i.e., $-N^+(CH_3)_2-CH_2-CO_2^-$) of neighboring molecules in the micelle are significant [27], this charge attenuation results in weak overall hydration and electrostatic field effects, leading to the observed unity fractionation factor.

Calculations

Details of calculation of fractionation factors from ^{1}H NMR data have been given elsewhere [10], so that only the salient features will be discussed here, using STS as an example. In pure water, $\chi_D = 0$, so that Eq. (2) is reduced to Eq. (5), where allowance is made for fractional micellar ionization, α:

$$\delta_{obs} = \delta_o + N_{Surf}(\alpha h_{Naf}\,\delta_{Naf} + (1-\alpha)h_M\,\delta_M)\,. \tag{5}$$

The slope of a plot of δ_{obs} versus N_{Surf} is given by Eq. (6):

$$slope = \alpha h_{Naf}\delta_{Naf} + (1-\alpha)h_M\delta_M\,. \tag{6}$$

Value of α depend, to some extent, upon the method of estimation, and for SDoS this property has been determined by several authors, using a variety of experimental techniques, $\alpha = 0.25 \pm 0.05$ [28]. We used the same uncertainty limit in α for all surfactants studied, as follows: STS, 0.23 ± 0.05 [15]; SDoEO$_2$S, 0.35 ± 0.05 [16]; Do(Me)$_2$BuABr, 0.4 ± 0.05 [17, 26]; DoPr$_3$ABr, 0.35 ± 0.05 [13, 26]; and Do(Me)$_2$HEtABr, 0.25 ± 0.05 [13, 19].

Hydration number ranges were also used in our calculations, because of possible partial dehydration of ions at the micellar interface. For anionic surfactants, we used $h_M = 3$, and $h_{Na} = 6$ or 3; and for cationic micelles we used $h_M = 3$, and $h_{Br} = 6$ or 3 [7, 10].

Using: $\delta_{sodium}/\delta_{sulfate} = 0.8$, and $\delta_{bromide}/\delta_{cation} = 2.7$; $\varphi_{Naf} = 0.98$, and $\varphi_{Brf} = 0.96$; the above-mentioned ranges for α; and the above-mentioned hydration numbers, we obtained the following ranges for $h_{counterion}\delta_{counterion}/h_M\delta_M$: STS, 0.09 to 0.21; SDoEO$_2$S, 0.15 to 0.33; Do(Me)$_2$BuABr, 0.34 to 0.61; DoPr$_3$ABr, 0.28 to 0.51; and Do(Me)$_2$HEtABr, 0.20 to 0.36. The fractionation factors given in Table 1 were then calculated from Eq. (2).

Acknowledgements O.A. El Seoud thanks the FAPESP and FINEP research foundations for financial support, and the CNPq for research productivity fellowship. O.A. El Seoud, and R.C. Bazito thank the British Council for subsistence expenses in York. This work was done as a part of the cooperation Link between the Universities of São Paulo and York.

References

1. Fendler JH (1982) Membrane Mimetic Chemistry. Wiley, New York
2. Attwood D, Florence AT (1984) Surfactant Systems: Their Chemistry, Pharmacy, and Biology. Chapman & Hall, London
3. El Seoud OA (1994) In: Hinze WL (ed) Surfactants in Analytical Chemistry, Vol 1. JAI press, Greenwich, p 1
4. Bunton CA, Savelli G (1986) Adv Phys Org Chem 22:213
5. Menger FM (1984) In: Mittal K, Lindman B (eds) Surfactants in Solution, Vol 1. Plenum, New York, p 347
6. (a) Menger FM (1979) Acc Chem Res 12:111; (b) Wennerström H, Lindman B (1979) J Phys Chem 83:2931; (c) Dill KA, Flory PJ (1981) Proc Natl Acad Sci USA 78:676; (d) Menger FM, Mounier CE (1993) J Am Chem Soc 115:12222
7. (a) Lindman B, Wennerstrom H, Gustavsson H, Kamenka N, Brun B (1980) Pure Appl Chem 52:1307; (b) Lindman B (1984) In: Tadros Th F (ed) Surfactants. Academic Press, New York, p 83
8. Chachaty C (1987) Prog NMR Spectrosc 19:183

9. El Seoud OA (1977) J Mol Liq 72:85
10. (a) El Seoud OA, Farah JPS, Vieira PC, El Seoud MI (1987) J Phys Chem 91:1950; (b) El Seoud MI, Farah JPS, El Seoud OA (1989) Ber Bunsenges Phys Chem 93:180; (c) El Seoud OA, Blásko A, Bunton CA (1994) Langmuir 10:653; (d) El Seoud OA, Blásko A, Bunton CA (1995) Ber Bunsenges Phys Chem 99:1214
11. Balzer D, Varwing S, Weihrauch M (1995) Colloids Surfaces A 99:233
12. Perrin DD, Armarego LF (1988) Purification of Laboratory Chemicals, 3rd ed. Pergamon Press, London
13. Bazito RC, El Seoud OA, Barlow GK, Halstead TK, Ber Bunsenges Phys Chem, in press
14. Schowen KB (1978) In: Gandour RD, Schowen RL (eds) Transition States for Biochemical Processes. Plenum Press, New York, p 225
15. Elvingson C (1987) J Phys Chem 91:1455
16. (a) Barry BW, Wilson R (1978) Colloid Polymer Sci 256:44; (b) Barry BW, Wilson R (1978) Ibid 256:251
17. Zana R (1980) J Colloid Interafce Sci 78:330
18. Guoxi Z, Jinxin X (1995) Wuli Huaxue Xuebao 11:785; CA 123:350924p
19. Malliaris A, Paleos CM (1984) J Colloid Interface Sci 101:364
20. Albery JA (1975) In: Caldin E, Gold V (eds) Proton-Transfer Reactions. Chapman & Hall, London, p 275
21. Jarret RM, Saunders MJ (1985) J Am Chem Soc 107:2648
22. (a) Engberts JBFN (1982) Pure Appl Chem 54:1797; (b) Haak JR, Engberts JBFN (1986) J Am Chem Soc 108:1705
23. Symons MCR (1983) Chem Soc Rev 12:1, and references cited therein
24. (a) Clifford J, Pethica BA (1964) Trans Faraday Soc 60:1483; (b) Clifford J, Pethica BA (1966) Ibid 61:182
25. (a) Blackmore ES, Tiddy GTJ (1988) J Chem Soc Faraday Trans 2 84:1115; (b) Burczyk B, Wilk KA (1990) Progr Colloid Polym Sci 82:249; (c)Nowak JR, Pomianowski A, Szczena EN (1990) J Surface Sci Technol 6:287
26. Buckingham SA, Garvey CJ, Warr GG (1993) J Phys Chem 97:10236
27. (a) Beckett A, Woodward R (1963) J Pharm Pharmacol 15: 422; (b) Pottel R, Kaatze U, Müller SC (1978) Ber Bunsenges Phys Chem 82:1086; (c) Müller SC, Pottel R (1982) In: Mittal K, Fendler EJ (eds) Solution Behavior of Surfactants. Plenum Press, New York, p 485
28. Alonso EO, Quina FH (1995) Langmuir 11:2459

Progr Colloid Polym Sci (1998) 111:158–161
© Steinkopff Verlag 1998

ASSOCIATION COLLOIDS

H. Kuhn
H. Rehage

Self-diffusion processes of water molecules near the surface of sodium octanoate micelles studied by molecular dynamics computer experiments

H. Kuhn (✉) · H. Rehage
Institut für Umweltanalytik
Universität-GH-Essen
Universitätsstraße 3-5
D-45141 Essen
Germany

Abstract In this publication we have studied the diffusion properties of water molecules near the surface of sodium octanoate micelles. We performed computer experiments using molecular dynamics simulations. From these investigations we succeeded in calculating the self-diffusion coefficient. It turns out that water molecules which are directly connected to the surfactant head groups by hydrogen bonds show a significant smaller mobility than those solved in the bulk phase. The diffusion constants obtained from these investigations are in fairly good agreement with experimental data.

Key words Molecular dynamics simulation – micelles – self-diffusion coefficient of water

Introduction

In order to measure structural properties of micellar solutions, extensive experimental methods like static and dynamic light scattering, small-angle-neutron-scattering, Raman scattering or nuclear magnetic resonance spectroscopy are often used. In contrast to these experimental techniques it is also possible to obtain such informations from computer simulation experiments. These methods are especially useful to get a more detailed insight into the dynamic properties of aggregated surfactants.

Up to now, only a few molecular dynamics simulations on an atomistic scale were performed to investigate micelles in aqueous solutions [1–16]. In these types of computer experiments, structural quantities which depend on the atomic positions could be evaluated. It was also possible to calculate thermodynamic properties by using time averages. A recently published molecular dynamics simulation deals with the structure and electrostatics of the sodium octanoate–water interface [16]. In contrast to micellar structures close to the critical micellization concentration (CMC) which is the topic of this work, the authors directed their attention to the liquid crystalline mesophase of sodium octanoate dissolved in water.

In a preceding paper [17] we focused our attention on the molecular structure of the hydrophobic core of the micelle. We calculated typical features like the conformation distribution of alkyl chains, the shape of the micelle and its size. In a second publication we investigated the interactions of the micelle with solvent molecules [18]. In this paper we discussed the atomic forces between the hydrophilic head groups and the surrounding water molecules. By calculating the radial distribution functions between the polar head group atoms and the surrounding water molecules the average distance between the solvent and the polar head groups could be determined. Furthermore, the average hydrogen bond length was calculated and the distribution of solvent molecules could be evaluated. It was also possible to determine the average coordination number of solvent molecules surrounding the surfactant head groups. On detailed inspection it became clear that the interactions between water and negatively charged oxygen atoms of the polar head groups lead to rather short hydrogen bonds. This phenomenon can be explained by the polarization effect of the hydrogen bond

Progr Colloid Polym Sci (1998) 111:158–161
© Steinkopff Verlag 1998

through displacement of charge from the surfactant oxygen atom to the bound water molecule [18]. In other words, the hydrogen bond strengths between the head groups and the surrounding water molecules are larger than those between water molecules. This effect must have a significant influence of the mobility of water near the micellar surface.

In a recently published work of Canet et al. [19], the rate of the self-diffusion coefficient of water bound to a sodium octanoate micelle, D_m was compared with the corresponding value of bulk water, D_w. These experimental results revealed that the quotient D_m/D_w is in the range of 0.45–0.64 depending on the surfactant concentration. This implies that the self-diffusion coefficient of free bulk water is about twice as large as the one of the first hydration layer.

Method and theory

In order to get more informations of the diffusion properties of water molecules at the surface of the micelle, we have carried out molecular dynamics simulations. In agreement with the aggregation number, measured by experiments [20], the micellar aggregate consisted of 15 surfactant molecules which were surrounded by 813 water molecules. We used a cubic simulation box with periodic boundary conditions. The potential functions for calculating the atomic interactions, the simulation model and the simulation conditions are summarized in all details in ref. [17]. In contrast to former simulation we applied the SPC water parameters [21, 22] in the present model. Due to recent progress in computer technology we could extend the simulation time up to 400 ps. From the molecular dynamics trajectory we investigated the self-diffusion process of water molecules at a time interval between 350 and 400 ps. Additionally with the demand of saving computation time, we reduced the energy cut-off for the nonbonded van der Waals' interactions from 1.5 to 1.0 nm.

It is well known that the self-diffusion coefficient, D, of an atom can be calculated from the slope of the mean square displacement correlation function defined in Eq. (1).

$$D = \frac{1}{6}\frac{\Delta\text{MSD}}{\Delta t} \quad \text{with MSD} = \frac{1}{N}\sum_{i=1}^{N}\langle[\mathbf{r}_i(t) - \mathbf{r}_i(0)]^2\rangle \quad (1)$$

$\mathbf{r}_i(t)$ denotes the position vector of particle i and N is the number of particles. The expression describes the average of the mean square displacement of particle i calculated from all particle positions of the trajectory obtained from the molecular dynamics simulation. A particle position

can be either a position of an atom or a center of mass of a defined set of atoms. If the MSD(t) function is plotted, the self-diffusion coefficient can be obtained from the slope SL of the curve (SL $= \Delta$MSD)$/\Delta t$). It turns out that the exact value is just given by $D = \text{SL}/6$.

Results and discussion

In order to get more information on the dynamic properties of water molecules we compared the self-diffusion constants of pure water with those of bulk water in a micellar solution and with solvent molecules belonging to the first hydration shell of the aggregates.

The self-diffusion coefficient of pure water is determined from a 160 ps NPT- simulation of a box containing 266 water molecules. The SPC interatomic interaction potentials and parameters were assigned to each atom. The pressure P and the temperature T were set at 1 bar and 300 K. The atomic positions of the oxygen atoms could be extracted from the trajectory in the range of 115–160 ps. The mean square displacement of the water molecules was evaluated from these data and subsequently the self-diffusion coefficient could be obtained.

From the simulation of the sodium octanoate micelle in aequeous solution we calculated the self-diffusion coefficient from the oxygen positions of the bulk water. This result was compared with the self-diffusion coefficient from the oxygen positions of the water molecules which belong to the first solvation shell.

The radial distribution function between the carbon head group atoms and the water oxygen atoms, $g_{CO}(r)$, provides informations on the average coordination number of water molecules around the polar head groups and the radii of the solvation shells. From our simulation data we calculated the $g_{CO}(r)$ function which is summarized in Fig. 1.

From the coordination number function $n(r)$ in Fig. 1 it can be concluded that, on the average, the number of nearest water molecules is in the range of 7–8. From the first minimum of the $g(r)$ function, the radius of the first hydration shell can be determined (408 pm). From this result we calculated the self-diffusion constant of water molecules which are located at an average distance of 408 pm from the carbon head group atoms. A structure of the micelle also including the first hydration layer is shown schematically in Fig. 2. This equilibrium molecular structure corresponds to the atomic positions of the surfactant monomers and water molecules, extracted from the trajectory at the simulation time of 375 ps. In order to get more precise insight into the molecular structure, the hydrogen atoms are not shown in this picture.

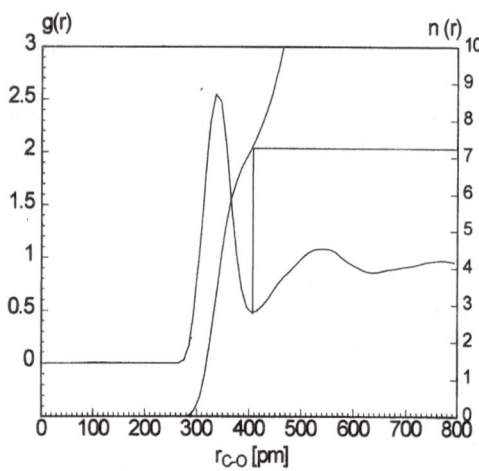

Fig. 1 Radial distribution function between the carboxylic carbon atoms and the oxygen atoms of water. The $n(r)$ function shows that the coordination number in the first hydration shell is in the range 7–8. The extent of the first hydration shell is 408 pm

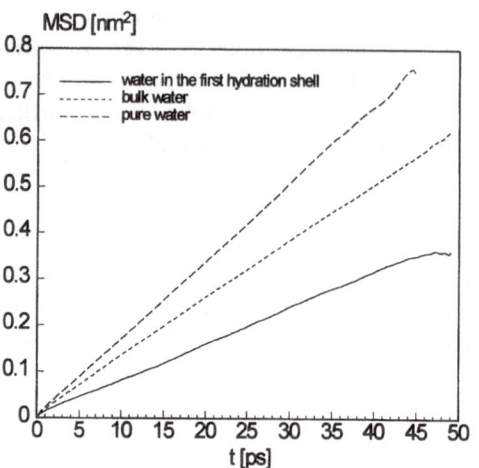

Fig. 3 Mean square displacement of the three systems: pure water, bulk water of the micellar solution and water in the first hydration shell

Fig. 2 Molecular structure of the micelle and the first hydration layer at a simulation time of 375 ps. The blue colored atoms correspond to the water oxgens which are located in the distance of 408 pm from the carbon head-group atoms

In Fig. 3 the corresponding MSD(t) functions of the three systems are summarized.

The calculated self-diffusion cofficients D, obtained from the slopes of the curves are presented in Table 1. The average displacement of an atom $\langle L \rangle$ is also given in Table 1.

From the simulation of pure water molecules it is obvious that the calculated self-diffusion coefficient agrees very well with experimental data. At 283 K, D was measured to $1.55 \times 10^{-5} \, \text{cm}^2 \, \text{s}^{-1}$ and at 314 K, D is equal to $3.32 \times 10^{-5} \, \text{cm}^2 \, \text{s}^{-1}$ [23].

Furthermore from Fig. 3 and Table 1 it becomes clear that the mobility of water in the first hydration shell is rather small. This phenomenon is correlated to the formation of hydrogen bonds which are observed at the negatively charged oxygen atoms of the surfactant molecules. Some solvent molecules are firmly attached to these polar head-groups. This effect could have the consequence of increasing average life-time of hydrogen bonds between water molcules and the surfactant head-groups.

Additionally we have taken into account the diffusion properties of water molecules in the second hydration layer. From Fig. 1 the radius of this hydration shell can be obtained to a value of 640 pm. The calculation of the self-diffusion coefficient of all water molecules which are contained in the first and second hydration shell resulted in $D = 2.195 \times 10^{-5} \pm 0.003 \, \text{cm}^2 \, \text{s}^{-1}$, equal to the self-diffusion coefficient of the bulk water molecules. It becomes clear that the diffusion of water molecules in the second hydration layer corresponds to the diffusion of the bulk water molecules. Therefore the hindered motion on the micellar surface is restricted to the water molecules in the first hydration shell. Only these molecules are able to form hydrogen bonds to the surfactant head-groups.

If the self-diffusion coefficient of water in the first hydration shell is divided by the corresponding value of pure water we get 0.43 which is in excellent agreement with experimental data [19].

Progr Colloid Polym Sci (1998) 111:158–161
© Steinkopff Verlag 1998

Table 1 Calculated self-diffusion coefficients and the averaged travelled distances of water molecules

System	$D \, [\text{cm}^2 \, \text{s}^{-1}]$ $\times 10^{-5}$	$\langle L \rangle$ [nm]
Pure water	2.780 ± 0.003	20.9
Bulk water in the micellar solution	2.113 ± 0.003	17.0
Water in shell with the distance of 408 pm from oxygen head-group atoms	1.190 ± 0.003	15.7

D was obtained from 1/6 of the slopes from the respective curves in Fig. 1.

Recently a molecular dynamics study of a fully hydrated phospholipid bilayer was reported by Damodaran et al. [24]. They calculated the mean square displacements for "bound" (i.e., closely associated with the bilayer surface) and bulk water molecules. In accordance to our molecular dynamics simulation results, the bound water molecules were found to have a diffusion constant which is smaller than the bulk water value by a factor of approximately 2. Also the absolute values for the diffusion constants (bound water: $D = 1.34 \times 10^{-5} \, \text{cm}^2 \, \text{s}^{-1}$, bulk water: $D = 2.4 \times 10^{-5} \, \text{cm}^2 \, \text{s}^{-1}$) are in the same range as our values.

Conclusions

From molecular dynamics simulation we succeeded in calculating self-diffusion coefficients of water molecules in micellar solution. It turns out that the diffusion of water molecules belonging to the first hydration layer of the sodium octanoate micelle is decreased by the factor of approximately 2. Taking into account the formation of hydrogen bonds, we could show that the hindered diffusion process is a direct consequence of these special forces [17, 18]. It turns out that the calculated values are in fairly good agreement with experimental results and molecular dynamics investigations of hydrated phospholipid bilayers.

Besides the above-mentioned property of water diffusion near the micellar surface, it is possible to determine the mobility of water molecules in the micellar core. We expect that the diffusion constants of these molecules should even increase. It is also interesting to determine the average life-time of different types of hydrogen bonds. Research programs of this type are still in progress.

Acknowledgment Financial support of this work by grants of the Deutsche Forschungsgemeinschaft (DFG: Re 681/4-3; Graduate Research Project: "The Improvement of the Water Cycle in Urban Areas for the Protection of Soil and Groundwater") and the *Forschungspool* of the University of Essen are gratefully acknowledged. We would also like to thank the Computercenter of the University of Essen for interesting discussions and technical assistance.

References

1. Haile M, O'Connell JP (1984) J Phys Chem 88:6363
2. Jönsson B, Edholm O, Teleman O (1986) J Chem Phys 85:2259
3. Woods MC, Haile JM, O'Connell JP (1986) J Phys Chem 90:1875
4. Watanabe K, Ferrario M, Klein ML (1988) J Phys Chem 92:819
5. Watanabe K, Klein ML (1989) J Phys Chem 93:6897
6. Wendoloski J, Kimatian SJ, Schutt CE, Salemme FR (1989) Science 243:636
7. Shelley J, Watanabe K, Klein ML (1990) Int J Quantum Chem Quantum Biol Symp 17:103
8. Karaborni S, O'Connell JP (1990) J Phys Chem 94:2624
9. Shelley J, Watanabe K, Klein ML (1991) Electrochim Acta 36:1729
10. Karaborni S, O'Connell JP (1993) Tenside Surf Det 30:235
11. Shelley JC, Sprik M, Klein ML (1993) Langmuir 9:916
12. Sprik M, Shelley JC (1993) Tenside Surf Det 30:243
13. Böcker J, Brickmann J, Bopp PJ (1994) J Phys Chem 98:712
14. Laaksonen L, Rosenholm JB (1993) Chem Phys Lett 216:429
15. MacKerell AD (1995) J Phys Chem 99:1846
16. Shelley JC, Sprik M, Klein ML (1997) Progr Colloid Polym Sci 103:146
17. Kuhn H, Rehage H (1997) Ber Bunsenges Phys Chem 101:1485
18. Kuhn H, Rehage H (1997) Ber Bunsenges Phys Chem 101:1493
19. Canet D, Mahieu N, Tekely P (1992) J Am Chem Soc 114:6190
20. Hayter JB, Zemb T (1982) Chem Phys Lett 93:91
21. Berendsen HJC, Postma JPM, van Gunsteren WF, Hermans J (1981) In: Pullman B (ed) Intermolecular Forces. Reidel, Dordrecht, Holland, p 331
22. Jorgensen WL, Chandrasekhar J, Madura JD, Impey RW, Klein ML (1983) J Chem Phys 79:926
23. Mills R (1973) J Phys Chem 77:685
24. Damodaran KV, Merz KM Jr, Gaber BP (1992) Biochemistry 31:7656

Progr Colloid Polym Sci (1998) 111:162–167
© Steinkopff Verlag 1998

COLLOIDAL SYSTEMS IN ENVIRONMENTAL SCIENCE

R. Szafranski
J.B. Lawson
G.J. Hirasaki
C.A. Miller
N. Akiya
S. King
R.E. Jackson
H. Meinardus
J. Londergan

Surfactant/foam process for improved efficiency of aquifer remediation

R. Szafranski · J.B. Lawson · G.J. Hirasaki
Prof. C.A. Miller (✉) · N. Akiya · S. King
Department of Chemical Engineering
Rice University
6100 Main Street
Houston, Texas 77005-1892
USA

R.E. Jackson · H. Meinardus
J. Londergan
Duke Engineering and Services
9111 Research Boulevard
Austin, Texas 78758
USA

Abstract Results of laboratory studies and preliminary results of a field demonstration are presented for a surfactant/foam process for remediation of ground water aquifers contaminated with organic liquid contaminants, especially chlorinated solvents. Injection of air with the surfactant solution generates foam in zones of high permeability or hydraulic conductivity. As the foam has a high resistance to flow of liquid, surfactant solution is diverted to zones of low permeability, thereby improving sweep efficiency of the process. This behavior was confirmed by laboratory experiments using a two-dimensional model packed with two layers of sands having different permeabilities. These experiments utilized sodium dihexyl sulfosuccinate as the surfactant at its optimal salinity with trichloroethylene (TCE) as a model contaminant. The same surfactant was used in the field at a site at Hill Air Force Base in Utah which had a TCE-rich contaminant. The surfactant/foam process reduced the amount of this contaminant in a line-drive pattern about 6.1 m long to a very low average saturation of 0.03%.

Key words Foam – remediation – surfactant

Introduction

Current methods for cleanup of ground water aquifers contaminated by chlorinated solvents are inadequate in most cases. Being denser than water, these solvents can penetrate the water table and continue downward until they reach layers of clays or other materials of low permeability. If these layers are inclined, drainage continues along their upper surfaces until stopped by a depression or other trap. Along the entire migration path trapped liquid is left behind which dissolves slowly into the ground water it contacts and constitutes a source of long-term contamination of the aquifer [1].

Surfactants can be useful in removing liquid chlorinated solvent contaminants. Research in this area has emphasized processes which solubilize them in micellar solutions [2] and those where interfacial tension is reduced to values low enough to mobilize them [3]. The latter approach is that taken during the 1960s through the 1980s in research on surfactant processes for enhanced oil recovery [4]. Whichever approach is used, the process is inefficient in a heterogeneous aquifer because flow occurs preferentially in zones of high hydraulic conductivity or permeability. Such zones are cleaned early in the process but continue to receive most of the injected surfactant solution which is therefore not effective in cleaning the still contaminated zones of low permeability. Even for an aquifer which has homogeneous properties but which is nonuniformly contaminated, the clean zones have a higher relative permeability to the injected surfactant solution, i.e., a lower resistance to flow. Hence, they receive more flow than the contaminated zones.

The approach taken in the surfactant/foam process described below is to inject air along with the surfactant solution. The air also flows preferentially in the zones of high permeability and forms a foam there. Because the foam has a high gas content, its relative permeability to the liquid phase is low. Accordingly, surfactant solution is diverted from foam-filled high permeability zones to contaminated low permeability zones, thereby improving the process efficiency. Foam flow has been the subject of considerable research in the petroleum industry as a means of reducing gas mobility during enhanced recovery processes involving injection of steam and supercritical carbon dioxide [5–7].

Laboratory results

Surfactant phase behavior

The surfactant used in this project was the anionic surfactant sodium dihexyl sulfosuccinate, which had been shown previously to form microemulsion phases with chlorinated solvents [8]. It was obtained from Cytec as the product Aerosol MA-80I, which is 80% active surfactant in water with some added isopropanol (IPA). Its critical micelle concentration (CMC) decreases from about 2 wt% in the absence of salt to about 0.8 wt% at 1.1 wt% NaCl [9], the salinity range of interest here.

Figure 1 shows the equilibrium phase behavior obtained at 22 °C for a 4 wt% (active) solution of MA-80I at various salinities with various amounts of added trichloroethylene (TCE), which was chosen as a representative contaminant. This partial phase diagram is based on observation of about 120 individual samples. When the volume fraction of TCE exceeds about 20%, the pattern of phase behavior seen with increasing salinity is the well-known Winsor I/III/II sequence [4, 10] with optimal salinity about 11 500 ppm (1.15 wt%) NaCl.

In the absence of TCE the surfactant–brine mixtures are micellar solutions. However, addition of relatively small amounts (about 0.5 vol%) of TCE causes the lamellar liquid crystalline phase and its dispersions to form, so that static and/or streaming optical birefringence is observed. When TCE content reaches some 2–4 vol%, depending on the salinity, a single isotropic phase is seen. This phase can solubilize up to 12 vol% TCE near the optimal salinity. At low salinities it is an oil-in-water microemulsion. At high salinities it exhibits streaming birefringence and presumably has a sponge-like microstructure. Also at high salinities a region where the sponge and lamellar phases exhibit macroscopic phase separation is seen.

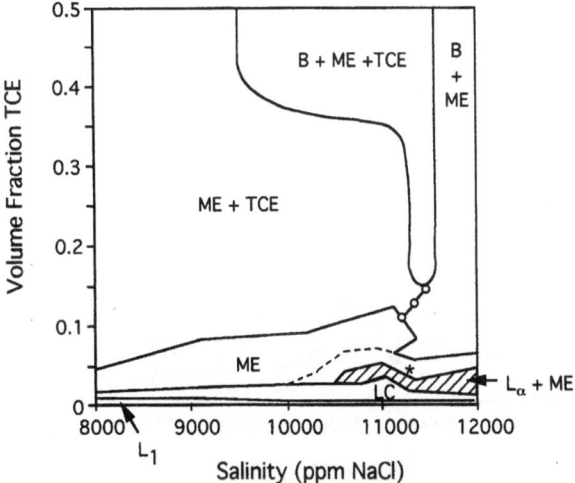

B = Brine L_1 = Micellar solution

ME = Microemulsion L_α = Lamellar liquid crystal

LC = Lamellar liquid crystal and its stable dispersions

* Streaming birefringence observed in ME phase to right of dashed line

–o–o– Probable small channel of ME phase (not confirmed)

Fig. 1 Partial phase diagram at 22 °C for 4 wt% (active) MA-80I at various salinities and TCE concentrations

This phase diagram is generally similar to those found in typical systems of interest in enhanced oil recovery (EOR) [11]. One difference is the absence of liquid crystal for the TCE-free solutions, a consequence of MA-80I being considerably more hydrophilic than surfactants of interest for EOR, which typically exhibited birefringence near optimal salinity in the absence of oil. A second difference is the existence of the continuous microemulsion region spanning all salinities in Fig. 1. For petroleum sulfonate/hydrocarbon/alcohol/brine systems, single-phase regions at low and high salinity were separated by various multiphase regions near optimal salinity including some where one phase was the lamellar liquid crystal [11].

Figure 2 shows the phase behavior when 4.5 wt% IPA was added to the surfactant formulation of Fig. 1. No regions containing liquid crystal are seen in Fig. 2, a consequence of the cosolvent effect of the alcohol. It is noteworthy that the IPA also lowers the optimal salinity (to about 9500 ppm or 0.95 wt% NaCl), indicating that the surfactant/alcohol films are less hydrophilic than those of the surfactant alone. IPA had the opposite effect on the surfactants investigated for EOR. The difference is a result of using a more hydrophilic surfactant here.

The phase behavior experiments were repeated at 15 °C because a temperature in this range was expected for the field test. The reduction in temperature caused a slight decrease in the optimal salinity [9], the expected result for

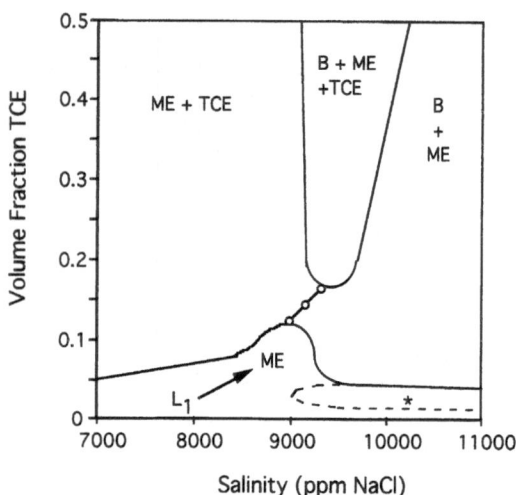

B = Brine   ME = Microemulsion

L_1 ➝ ME = Micellar solution changing to microemulsion

*   Streaming birefringence observed in ME phase
to right of dashed line

–o–o–   Probable small channel of ME phase (not confirmed)

Fig. 2 Partial phase diagram at 22 °C for 4 wt% (active) MA-80I with 4.5 wt% IPA at various salinities and TCE concentrations

an anionic surfactant. Also the range of TCE concentrations over which the lamellar liquid crystalline phase was present increased. In particular, 4.5 wt% IPA was no longer sufficient to prevent liquid crystal formation at salinities near and above the optimal value of about 8700 ppm (0.87 wt%) NaCl [9].

The presence of liquid crystal was of concern because its dispersions can have relatively high apparent viscosities, as found in previous studies of EOR systems [12]. One sample from the LC region of Fig. 1 exhibited an apparent viscosity of about 60 cp (0.060 kg m^{-1} s^{-1}) at a shear rate of 1.5 s^{-1}, which is representative of flows encountered in aquifers. From this perspective it would have been preferable to use a formulation containing enough IPA to prevent liquid crystal formation even at low temperatures. However, it was found in experiments discussed below that IPA destabilized the foam formed with this surfactant and adversely affected performance of the surfactant/foam process. Some samples of fluids produced during laboratory sand pack tests of the surfactant/foam process described below did exhibit birefringence, indicating that liquid crystal was present. However, TCE production did not seem to be inhibited, and it was concluded that the effect of the liquid crystal on process performance was minimal.

Finally, the strong effect of divalent cations on phase behavior should be mentioned. At 22 °C the optimal

conditions at 4 wt% MA-80I occurs at 2000 ppm (0.2 wt%) CaCl$_2$, well below the 11500 ppm for NaCl given above. Moreover, replacement of a relatively small fraction of sodium ions by calcium shifts the system to more lipophilic conditions. There is convincing evidence that such replacement actually occurred at some locations during the field demonstration described briefly below as a result of ion exchange between the surfactant solution and clay minerals in the aquifer material just above an impermeable clay layer. This effect was recognized during research on EOR processes using surfactants [13] and must be considered in the design of surfactant processes for aquifer remediation.

Experiments with two-dimensional sand pack

Laboratory studies were conducted with a two-dimensional model approximately 500 mm long, 89 mm high and 19 mm thick. The front surface was made of glass to permit observation of the flow. It was packed with two layers of sand of equal height, one layer having larger particles and thus a greater permeability than the other. It was equipped with transducers to measure the pressure at various positions along the pack.

In one set of experiments, for example, the upper layer was a coarse sand with a permeability of about 200 darcies (hydraulic conductivity 2×10^{-3} m s^{-1}), and the lower layer was a finer sand with a permeability of about 10 darcies (hydraulic conductivity about 10^{-4} m s^{-1}). TCE dyed red to make it clearly visible was introduced into the sand pack and reduced to residual saturation by waterflooding. This was the initial condition for each of the three experiments described next which were conducted at 22 °C.

In the first experiment a solution of 4 wt% (active) of MA-80I at its optimal salinity of 1.15 wt% was injected continuously. TCE in the upper layer of coarse sand was readily mobilized, and nearly all of it appeared in the effluent as a separate phase during the first 0.5–0.6 pore volume (PV) of fluid injection. However, the TCE in the lower layer of finer sand was removed very slowly, as shown in Fig. 3. Indeed, only 65% of the TCE initially present had been recovered after injection of 26 PV of surfactant solution. The reason is that most of the surfactant solution flowed through the higher permeability layer which had already been cleaned.

The second experiment used the same surfactant solution. However, after about 0.3 PV of surfactant solution had been injected, air was injected. Its inlet pressure was controlled so as not to exceed the outlet pressure (which was atmospheric) by more than about 5000 Pa. This pressure was selected because the resulting pressure gradient of

Progr Colloid Polym Sci (1998) 111:162–167
© Steinkopff Verlag 1998

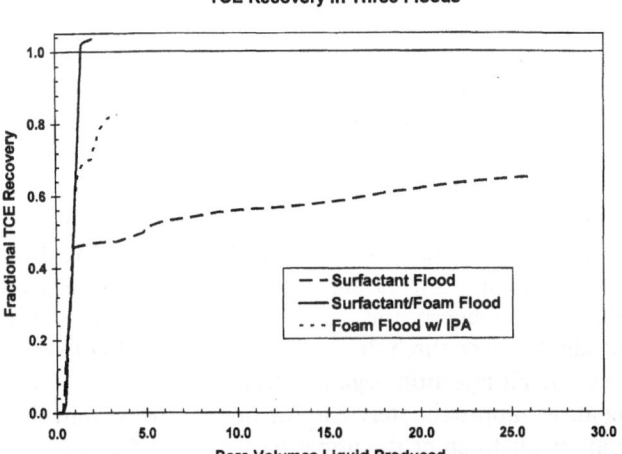

Fig. 3 Fractional TCE recovery during surfactant flood and pressure-regulated foam floods for formulations with and without IPA. All formulations were at optimal salinity

about 10^4 Pa m^{-1} was about that desired for the subsequent field demonstration. About 0.5 PV of liquid was displaced by the air, mostly in the upper coarse sand, as it generated foam there. When injection of surfactant solution was resumed, about half of it flowed through the finer sand, which contained little foam, because the foam in the coarse sand greatly increased resistance to liquid flow there. As a result, TCE continued to be produced. Alternate injection of air and surfactant solution continued. After four cycles of air injection and about 1 PV total of injected surfactant solution, all TCE had been removed from the model to within the accuracy of a mass balance. No TCE was visible in the model at this time although some solubilized TCE was still being produced. As Fig. 3 shows, the rate of TCE production was nearly constant during the process and had a value close to that during initial displacement of TCE from the coarse sand by the surfactant solution. About 96% of the TCE was produced as a separate phase. A series of photographs giving a graphic view of the displacement process for both the surfactant and surfactant/foam experiments may be found elsewhere [9, 14].

For the third experiment the surfactant solution contained 4 wt% (active) MA-80I and 4.5 wt% IPA at the optimal salinity of 0.95 wt% NaCl (Fig. 2). In this case the foam was visibly weaker than in the absence of IPA. Although air was injected with the same control on inlet pressure as before, the pressure drop across the model decreased from about 5000 to about 2000 Pa as soon as air injection ceased and liquid injection resumed. After 17 cycles of liquid and air injection and a total of 3 PV of surfactant solution injected only 84% of the TCE initially present had been recovered (see Fig. 3). Based on this

result it was decided not to include IPA in the formulation for the field test.

The injection of air at a constant inlet pressure rather than at a constant rate is a key factor in practical application of the surfactant/foam process. Experiments in which air was injected into horizontal columns 25 mm in diameter which had been packed with coarse sand and saturated with surfactant solution showed that an air velocity exceeding about 60 m s^{-1} was required for immediate generation of a strong foam to occur. However, once generated the foam could be propagated at much lower velocities [9]. It would not be feasible in practice to inject air at a constant velocity of 60 m s^{-1} because the foam generated would soon raise the pressure gradient to unacceptably high values. This problem can be solved by using the pressure-regulated scheme described above in which inlet pressure is controlled. Photographs taken during initial injection of air in the second experiment described above showed that initial air velocity in the coarse sand layer was at least 75 m s^{-1}. However, velocity decreased rapidly because the foam generated increased the resistance to flow of air. Nevertheless, foam was generated throughout the coarse sand layer, and the desired pressure gradient was not exceeded.

Field demonstration

The first field demonstration of the surfactant/foam process was carried out in the spring of 1997 at Hill Air Force Base near Ogden, Utah. It was conducted by Duke Engineering & Services based on a process design provided by Rice. Only some preliminary results are given here. A more detailed account is given elsewhere [9, 15].

The site was used from 1967–1975 to dispose of spent degreasing solvents, primarily TCE. The contaminant is located in the lowest part of the saturated zone of an alluvial sand aquifer having the form of a channel confined on its sides and below by thick clay deposits. The bottom of the channel is about 15 m below the surface and 5.5 m below the water table. Analysis of samples during initial borings and subsequently during drilling of the wells for the test showed that the contaminant was present in only about the bottom 1.3 m of the channel and that permeabilities ranged from about 10 to more than 100 darcies (hydraulic conductivities from 10^{-4} to more than 10^{-3} m s^{-1}). Heterogeneity was thus suitable for use of the surfactant/foam process.

Three injection and three extraction wells were completed in a line drive pattern 6.1 m long and spanning the channel (see Fig. 4). The outer wells in each row were about 4 m apart. Two monitoring wells were located within the pattern at positions about $\frac{1}{3}$ and $\frac{2}{3}$ of the distance

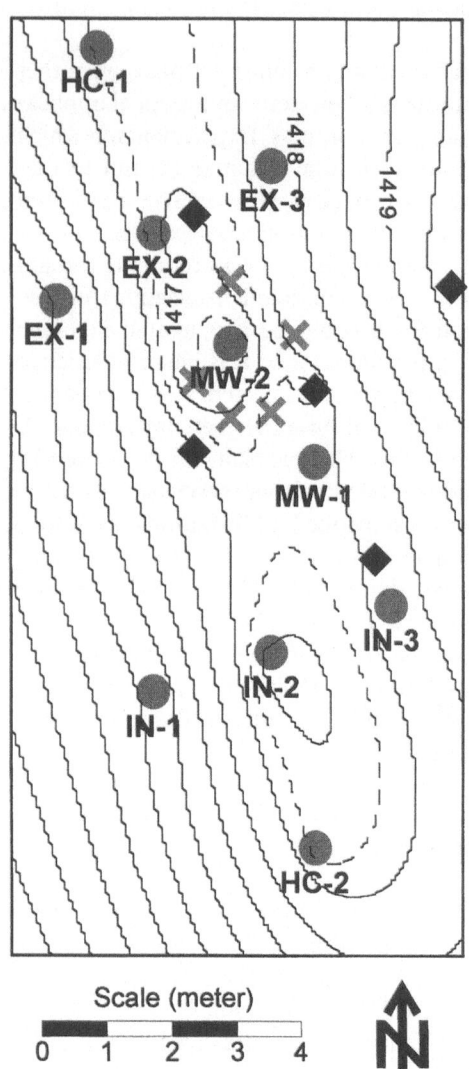

Fig. 4 Contour map of clay aquitard showing well pattern for field demonstration. The contours are labelled with their elevations in meters above sea level, solid lines representing whole and half-meters, dashed lines representing quarter-meters. Injection, extraction, monitoring, and hydraulic control wells are designated by IN, EX, MW, and HC respectively. Locations of the initial and final borings are designated by diamonds and X's, respectively

between injection and extraction wells as shown in Fig. 4. Each monitoring well could be sampled at three elevations. Finally, two hydraulic control wells were located along the channel but outside the pattern at a distance of 3 m from the rows of injectors and extractors respectively. Water was injected into these wells during the test to minimize flow of surfactant solution outside the pattern.

An initial partitioning interwell tracer test (PITT) [16] with isopropanol and *n*-heptanol as the nonpartitioning and partitioning tracers yielded an estimate of 0.079 m³ for

the amount of liquid contaminant initially present, a value reasonably consistent with the contaminant distribution found in the wells and initial borings whose locations are shown in Fig. 4. Following the initial PITT a surfactant-free solution containing about 1 wt% NaCl, the optimal salinity for mobilizing the contaminant with MA-80I at 12 °C, was injected for one day. The total volume of this solution was approximately equal to the swept volume of the pattern (31.4 m³) as determined by the PITT. Then injection of a 3.5 wt% surfactant solution at the same salinity commenced and was continued at the same rate for slightly over three days. After some 8 h of surfactant injection, air injection began with each well in turn receiving air for approximately 2 h. Air pressure was controlled to allow air to enter the upper part of the 1.5 m screened interval while surfactant solution continued to flow into the lower part. The total amount of surfactant solution injected was about 3.2 times the swept volume of the pattern. Afterwards, a more dilute NaCl solution (0.8 wt%) was injected for 12 h, followed by a waterflood to break the foam and remove most of the surfactant, and finally a second PITT to determine the amount of contaminant remaining.

No significant problems were encountered with air injection, and pressure rose at the injection wells, indicating an increase in the resistance to flow. Moreover, foam was observed in samples from the two upper screened intervals of both monitoring wells throughout much of the test. Even in the bottom screened intervals of the monitoring wells, which were located just above the clay aquitard, foam was seen at various times during the test.

The contaminant produced during the test (in addition to that which would have been produced in dissolved form if no surfactant had been used) was about 0.14 m³ based on analysis of many effluent samples. As this value significantly exceeds the 0.079 m³ thought to be present initially (see above), it appears that contaminant from outside was able to enter the pattern during the test, probably from the region beyond the injection wells which was known to be contaminated. Data from both monitoring wells seem consistent with this tentative conclusion. However, various data are still being studied and possible scenarios are being simulated numerically to provide more information on this matter.

In spite of the apparent influx of contaminant, both the final PITT and five borings taken at the end of the test (Fig. 4) showed that very little contaminant (0.006–0.010 m³) remained at the end of the test. The average final contaminant saturation in the pattern was about 0.03%, the same as the lowest value achieved previously with surfactant remediation, which was during a test conducted in a nearby portion of the same aquifer in the summer of 1996 by INTERA and the University of Texas [17]. This test involved injection of about 3 PV of a solution containing MA-80I at a higher concentration with some added IPA. While both tests were conservatively designed, it is worth noting that the surfactant/foam test used only about 60% as much surfactant per unit of swept volume.

Summary

Both laboratory studies and a recent field demonstration show that the surfactant/foam process can reduce the amount of liquid chlorinated solvent present in a ground water aquifer to very low values. Foam provides a method of improving efficiency of surfactant remediation processes by substantially reducing the proportion of injected surfactant solution which flows through zones of high permeability or hydraulic conductivity and increasing the proportion which flows through zones of low permeability or hydraulic conductivity.

Acknowledgements This work was supported by the advanced applied technology demonstration facility, sponsored by the U.S. Department of Defense.

References

1. Pankow J, Cherry J (1996) Dense Chlorinated Solvents, Waterloo Press, Portland, OR
2. Fountain J, Starr R, Middleton T, Beikirch M, Taylor C, Hodge D (1996) Ground Water 34:910–916
3. Pennell K, Pope G, Abriola L (1996) Env Sci Technol 30:1328–1335
4. Miller C, Qutubuddin S (1987) In: Eicke H, Parfitt G (eds) Interfacial Phenomena in Apolar Systems. Marcel Dekker, New York, pp 117–185
5. Chambers T, Radke C (1991) In: Morrow N (ed) Interfacial Phenomena in Petroleum Recovery. Marcel Dekker, New York, pp 191–255
6. Hirasaki G (1989) J Petrol Technol 41:449–456
7. Rossen W (1996) In: Prud'homme R, Khan S (eds) Foams. Marcel Dekker, New York, pp 413–464
8. Baran J, Pope G, Wade W, Weerasooriya V, Yapa A (1994) Env Sci Technol 28:1361–1366
9. Szafranski R (1997) Laboratory Development of the Surfactant/Foam Process for Aquifer Remediation. PhD Thesis, Rice University
10. Bourrel M, Schechter R (1988) Microemulsions and Related Systems. Marcel Dekker, New York
11. Miller C, Ghosh O, Benton W (1986) Colloids and Surfaces 19:197–223
12. Benton W, Baijal S, Ghosh O, Qutubuddin S, Miller C (1987) SPE Res Eng 2:664–670
13. Hirasaki G (1982) SPEJ 22:181–192
14. Hirasaki G, Miller C, Szafranski R, Lawson J, Akiya N (1997) SPE Preprint 37257 presented at Intl Symp Oilfield Chem, Houston
15. Hirasaki G, Miller C, Szafranski R, Tanzil D, Lawson J, Meinardus H, Jin M, Londergan J, Jackson R (1997) SPE Preprint 39292 presented at Ann Tech Conf, San Antonio
16. Jin M, Delshad M, Dwarakanath V, McKinney D, Pope G, Sephernoori K, Tilburg C, Jackson R (1995) Water Resour Res 31:1201–1211
17. Brown C, Delshad M, Dwarakanath V, Jackson R, Londergan J, Meinardus H, McKinney D, Oolman T, Pope G (1998) Demonstration of surfactant flooding of an alluvial aquifer contaminated with DNAPL, to appear in ACS Symp Ser

Progr Colloid Polym Sci (1998) 111:168–173
© Steinkopff Verlag 1998

D.A. Sabatini
J.H. Harwell
R.C. Knox

Surfactant selection criteria for enhanced subsurface remediation: Laboratory and field observations

Dr. D.A. Sabatini* (✉) · R.C. Knox*
Geologisches Institut
Lehrstuhl für Angewandte Geologie
Sigwarstr. 10
D-72076 Tübingen
Germany

J.H. Harwell
The School of Chemical Engineering and
Materials Science, and
The Institute for Applied Surfactant
Research
The University of Oklahoma
Norman, OK 73072
USA

*The School of Civil Engineering and
Environmental Science
The University of Oklahoma
Norman, OK 73072
USA

Abstract Traditional approaches to ground water remediation have proven ineffective, especially when trapped oil phases exist. Surfactants are being widely evaluated to enhance remediation of such ground water contamination episodes. Successful implementation of surfactant enhanced subsurface remediation requires careful consideration of fundamental surfactant properties. The economic viability of this technology requires targeting the residual contamination, minimizing surfactant losses, and recovery and reuse of the surfactant. The relative efficiencies and advantages of surfactant solubilization and mobilization systems are described, as well as means to optimize these systems. Unit processes for contaminant-surfactant separation and surfactant reuse are summarized, along with unique surfactant impacts on these separation processes. Factors affecting the environmental acceptability of surfactants are discussed. Finally, results of field demonstrations are presented that reinforce concepts presented. This study thus provides an introduction to surfactant chemistry and highlights key factors critical to the successful design and implementation of surfactant enhanced subsurface remediation systems.

Key words Remediation – surfactants – economics – losses – reuse – field demonstrations

Introduction

Pump and treat ground water remediation is often inefficient when a trapped oil phase (residual saturation) is present. The ground water simply flows past the trapped oil, being unable to overcome the capillary forces and physically displaces the oil. The water solubility of trapped oil constituents is low enough to render pump and treat remediation inefficient (potentially requiring hundreds to thousands of flushings even under equilibrium conditions), yet high enough to contaminate the ground water above the allowable levels. Surfactant enhanced subsurface remediation is a leading technique for enhancing pump and treat remediation of such sites. This enhancement can be due to micellar-enhanced aqueous concentrations (i.e.,
contaminant partitioning into the hydrophobic core of surfactant aggregates known as micelles), or due to ultra-low interfacial tensions realized along with middle phase microemulsion systems. This study begins with a brief overview of these enhancement mechanisms, followed by a discussion of the economic and environmental factors critical to their successful implementation. Finally, field demonstration results are briefly presented to reinforce the concepts presented.

Surfactant fundamentals

Surfactants (<u>surf</u>ace-<u>act</u>ive-<u>agents</u>) are classified by their charge (cationic, anionic, nonionic, zwitterionic – having

Progr Colloid Polym Sci (1998) 111:168–173
© Steinkopff Verlag 1998

both cationic and anionic groups), their origin (biosurfactants from plant or microbial production versus synthetic surfactants), their regulatory status (direct or indirect food additive status, acceptable for discharge to wastewater treatment systems or for use in pesticide formulations), etc [1]. Surfactants are also characterized by their hydrophilic-lipophilic balance (HLB) as water soluble (high HLB) or oil soluble (low HLB).

Above a critical concentration surfactant monomers self-aggregate into micelles (above the critical micelle concentration (CMC)). Organic compounds partition into the hydrophobic interior of micelles, thereby increasing the apparent water solubility of the contaminant. This partitioning is often considered to be linear in nature (Henry's Law), although deviations from linearity have been reported [1, 2]. The linear distribution coefficient (K_m) is the ratio of micellar to aqueous activity, and increases as contaminant hydrophobicity increases. For example, K_m values for PCE are roughly an order of magnitude greater than TCE, and likewise for TCE and DCE (log K_m values of T MAZ 60 and DCE, TCE and PCE are 2.9, 3.8 and 4.9, respectively [3]). Solubility enhancement also increases as surfactant concentrations increase above the CMC. Thus, to dramatically improve contaminant extraction we will operate significantly above the CMC (an order of magnitude or more).

Significant reductions in the oil–water interfacial tension virtually eliminate the capillary forces which cause the oil to be trapped, thereby allowing the oil to readily flush out with the water. The minimum interfacial tension occurs in middle phase microemulsion systems. By adjusting the surfactant system it is possible to initiate a transition from normal micelles (aqueous phase, Winsor Type I), to middle phase microemulsions (Winsor Type III), to reverse micelles (oil phase inverted micelles, Winsor Type II), as shown in Fig. 1. For a high HLB surfactant system (right side of figure) the surfactant resides in the water phase as normal micelles, and a portion of the oil phase partitions into the micellar phase. For low HLB surfactant systems (the left side of the figure) the surfactant resides in the oil phase as reverse micelles. In between we observe three phases; the water and oil phases and a new "middle" phase (so designated because of its intermediate density). The interfacial tension reaches a minimum within the middle phase region while solubilization reaches a maximum (they are inversely related [14]).

Middle phase microemulsions can be formed in several ways, as denoted at the top of Fig. 1. For ionic surfactants, increasing salinity or hardness can produce the middle phase system, a strategy commonly utilized in surfactant enhanced oil recovery [5]. However, introduction of high salt concentrations may not be desirable in aquifer restoration. Middle phase systems can also be achieved by alter-

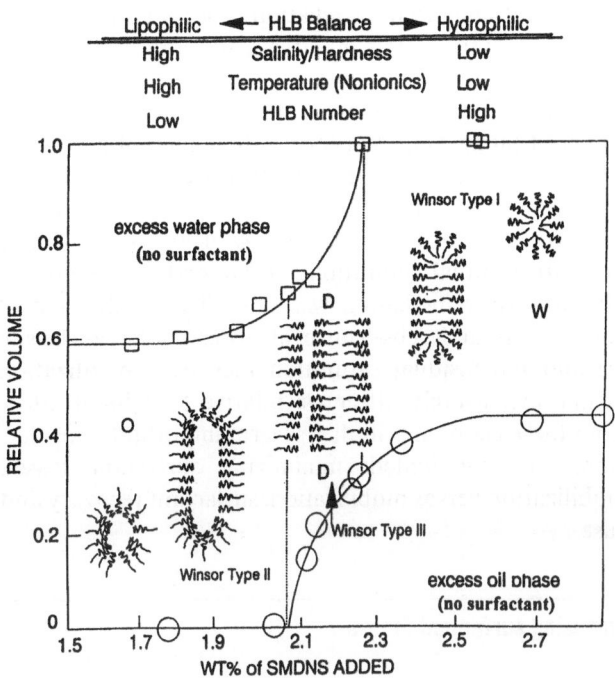

Fig. 1 Surfactant-oil phase behavior diagram (1,2-DCE, 0.5 wt% AOT and scanning with sodium mono- and di-methyl naphthalene sulfonate); W: surfactant-water phase; O: surfactant-oil phase; D: middle phase; (Adapted from [3])

ing the hydrophilic–lipophilic balance (HLB) of a binary surfactant system. By varying the mass ratio of two surfactants, one with an HLB above and one below the desired level, it is possible to achieve the desired HLB value – this is the scheme used to achieve the middle phase system in Fig. 1 for 1,2-DCE. At low SMDNS concentrations AOT partitions into the oil phase while at high SMDNS concentrations the system is over-optimized and the surfactants reside in the water phase (Type I system). In between a middle phase microemulsion exists. This is the inverse of a salinity scan, where increasing salt concentrations causes a high-HLB, water-soluble surfactant to partition into the oil phase. Additional discussion of solubilization and mobilization mechanisms can be found in the literature [6–10].

Economic viability

Obviously, the implementation of surfactant enhanced subsurface remediation is highly dependent on the economics of this approach. In looking at two cases, Krebs-Yuill et al. [11] determined that surfactant solubilization can be more economical than conventional pump-and-treat. The two most important conclusions were: (1) surfactant capital costs constitute the single largest cost in a surfactant-enhanced remediation process, and (2) decontamination of the surfactant-stream and surfactant reuse is critical.

A subsequent economic study evaluated surfactant-enhanced pump-and-treat for residual zone of 2 acres or less [12]. It was shown that the target cost of $5 M was met for all sites of less than 1 acre in size, and for 1 acre sites with residual saturation of 5% or less. For 1/4 acre sites with residual saturations of 5% or less, the cost of solubilization remediation was only 40% of the cost of pump-and-treat. Krebs-Yuill et al. observed that as the site size and the residual saturation increased mobilization became increasingly attractive relative to solubilization. From these economic studies several important considerations are highlighted: minimizing surfactant losses, solubilization versus mobilization, surfactant recovery and reuse.

Minimizing surfactant losses

Minimizing physicochemical surfactant losses is a major factor in surfactant selection (e.g., precipitation, sorption, phase separation). These losses can decrease the efficiency of the system, thereby negatively impacting the economics of the system, and can potentially render the process ineffective.

Nonionic surfactants typically have lower CMC values, and thus require less surfactant addition to form micelles. While nonionic surfactants do not precipitate with counterions, they are generally susceptible to higher levels of sorption and may be more suscept to phase separation than anionic surfactants [13–15]. Precipitation-resistant anionic surfactants may thus be ideal candidates for environmental remediation. Dowfax surfactants, with two sulfonate groups, are less susceptible to precipitation than monosulfonated surfactants (e.g., SDBS, [13]). Steol surfactants have demonstrated similar precipitation resistance based on the addition of ethylene oxide groups (EO's) to the basic SDS structure [14]. Rouse et al. [14] also demonstrated that sorption of the Dowfax surfactants was less than that for SDBS (seven-fold lower), while the Steol surfactants were an order of magnitude lower in sorption than SDS [14]. The sorption decreases were equally if not more dramatic when comparing the Dowfax and Steol surfactants with nonionic surfactants.

Having demonstrated that the Dowfax and Steol surfactants are less susceptible to precipitation and sorption, their contaminant solubilization potential is obviously important. The Dowfax surfactants were shown to be equally if not more efficient in enhancing the solubility of contaminants relative to their monosulfonated equivalents [13]. Likewise, the Steol surfactants showed a comparat-

ively slightly higher solubilization relative to their non-ethoxylated equivalents [14]. Thus, the importance of considering surfactant losses during surfactant selection is evident.

Solubilization versus mobilization

In solubilization, the more hydrophobic compounds experience a greater relative solubility enhancement, as discussed above. Since micellar partitioning varies more between contaminants for any surfactant than between surfactants for a given contaminant [3], other factors may significantly impact surfactant selection for solubilization (e.g., cost. susceptibility to losses, toxicity, etc.). Micellar solubilization (and thus K_m) does however vary as a function of contaminant type (nonpolar versus polar/ionic) and aqueous contaminant concentrations [16, 17].

In contrast to solubilization, surfactant(s) selection is much more dependent on contaminant composition when considering mobilization. Our initial efforts to achieve middle phase microemulsions with chlorinated solvents were unsuccessful. As the HLB of the surfactant system was varied from 2.1 to 40, a transition from a Type II to a Type I system was observed. However, a clear middle phase was not achieved in the transition region, instead a liquid crystal phase occurred. When a branched surfactant, Aerosol OT (AOT), was used along with sodium mono and di methyl naphthalene sulfonate (SMDNS), a liquid crystal phase did not form and a middle phase microemulsion occurred [3].

The same approach that was used for forming a middle phase system in Fig. 1 was used with PCE, TCE and 1,2-DCE individually [3] and in binary and ternary mixtures of these chlorocarbons [15]. For mixtures, regular solution theory proved to be able to predict phase behavior in the ternary system better than ideal solution theory [15]. It is especially encouraging that, given the analyses from binary systems, we can predict results for ternary systems of variable DNAPL composition.

While deviations from ideal middle phase conditions will decrese the system efficiency (much more than for solubilization), it does not necessarily negate the improved efficiency relative to solubilization. Shiau et al. [18] demonstrated that deviating just outside the ideal middle phase composition did not totally negate the enhanced removal efficiency of the system, due to the fact that interfacial tensions are still very low even if not at their minimum. These results also demonstrate that solubilization and mobilization are not discrete systems, but rather are "regions" in a continuum of surfactant systems.

Figure 2 shows results of one-dimensional column studies for solubilization and mobilization, and demonstrates

Progr Colloid Polym Sci (1998) 111:168–173
© Steinkopff Verlag 1998

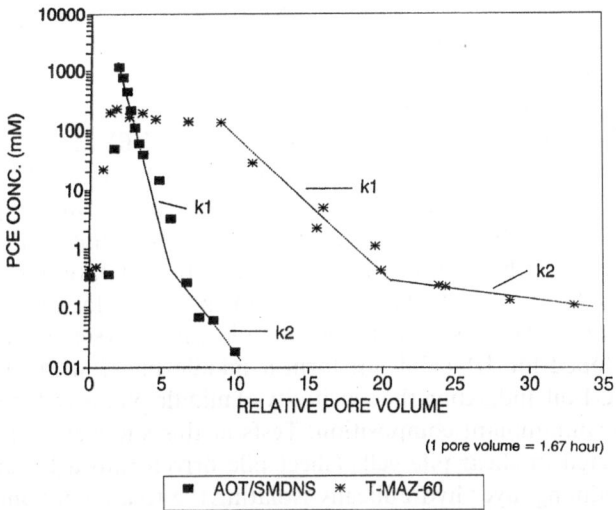

Fig. 2 PCE elution in column studies for solubilization (T-MAZ 60) and mobilization (AOT/SMDNS) systems

that mobilization achieved higher concentrations and eluted the PCE more quickly than solubilization (> 97% extracted via mobilization in ca. 3 pore volumes [18]. The lower maximum concentration, plateau and tail on the solubilization curve indicates the reduced extraction rate and thus a slow approach to complete PCE elution for solubilization (likely due to interfacial area constraints); while ca. 85% of the PCE is eluted within 10 pore volumes, less than 90% has been eluted by 30 pore volumes. This illustrates the advantages of mobilization over solubilization. By contrast water-based dissolution would require over a thousand pore volumes for this system; this illustrates the relative advantages of surfactant-enhanced remediation using solubilization or mobilization.

Surfactant-contaminant separation and surfactant reuse

A key to selecting the unit process for surfactant-contaminant separation is the contaminant volatility. Numerous systems can be used for volatile contaminants; however, fewer alternatives exist for nonvolatiles. Air stripping with antifoams, membrane air stripping, vacuum stripping and vacuum distillation, pervaporation, membrane liquid extraction, and precipitation are alternatives for treating volatile contaminants. The strong affinity of surfactants for activated carbon and ion exchange resins eliminates them as viable candidates for this separation. Contaminants will partition into surfactant adsorbed onto the media (admicelles), thereby preventing the desired separation (surface adsolubilization is similar to aqueous solubilization [16]).

For high volatility contaminants conventional packed-column air stripping can be used. Lipe et al. [19] showed that air stripper efficiency declined with increasing surfactant concentration, and that the impact was more dramatic for more hydrophobic contaminants and for surfactants with increased solubilization potential. The reduced contaminant activity, which results from micellar solubilization, is responsible for these observations. Predictions using air stripper design equations modified to account for micellar solubilization agreed well with experimental results. Using the modified design equations will allow proper design of air strippers to achieve the desired effluent concentrations, taking into account surfactant impacts on the system. Foaming in packed-tower air strippers can be addressed by adjusting the air to water ratio [19], by using an antifoam, or by using a hollow fiber system. The use of hollow fiber columns can mitigate the foaming problem [19] without sacrificing efficiency and without requiring the use of antifoams. In much the same way liquid–liquid extraction in hollow fiber membranes can be used for non-volatile contaminants [20]. For cases where surfactant reconcentration is necessary, ultrafiltration (UF) can be effective [19].

Given that surfactant reuse is crucial to the economic viability of the system, contaminant-surfactant separation and surfactant reuse needs to be a key consideration in the surfactant selection process. The ultimate goal is a surfactant system that is highly efficient in extracting the contaminant, has favorable phase behavior in the subsurface, and can efficiently be decontaminated and reused. Failure in any of these areas can render the system uneconomical and/or infeasible.

Environmental acceptability

Surfactants are not typically regulated as environmental contaminants and as such health based standards do not exist. This concern might suggest the use of surfactants with US Food and Drug Administration direct food additive status. [13, 15]. Not only have these surfactants been approved for direct human consumption, they are commonly combinations of fatty acids and sugars, thereby alleviating concerns as to their degradation products. However, caution must be exercised when considering the value of this designation. For example, high-performance surfactants (e.g., improved resistance to sorption and precipitation) may not have obtained, or even pursued, direct food additive status. Manufacturers are hesitant to invest the financial resources necessary to obtain direct food additive status; as such, numerous surfactants may qualify for this status, but have not undergone the necessary testing. Surfactants with indirect food additive status have been approved for contacting food products (i.e. for washing the product or use in packaging); a number of these have better operating characteristics versus surfactants with direct food additive status [19]. Obviously, as this technology progresses, improved methods for assessing environmental friendliness are needed.

Surfactant degradation raises several issues; what are the rates under various redox conditions, what are the metabolities, and what are the concerns associated with the answers to these questions. While a significant amount of data exists on the degradation of surfactants in aqueous aerobic systems, much less data exists on the degradation of surfactants in soil environments under varying redox conditions. A number of ongoing efforts are evaluating these issues; however, additional research is needed. The ease of surfactant degradation requires a balance; the surfactant should not be so degradable that it is prematurely lost or that it causes bioplugging, yet it should be sufficiently biodegradable so as not to persist after the remediation is completed.

Field demonstrations

A few highlights from field demonstrations will be presented, with an emphasis on factors discussed above. The first field demonstration was conducted at a US Coast Guard facility in Traverse City, Michigan. A significant amount of aviation fuel had been accidentally released at the site, along with a much smaller amount of PCE. While a decade of bioremediation activities had reduced the petroleum hydrocarbon levels, detectable concentrations of the aviation fuel and PCE were still evidenced in monitoring wells. The history of work at this site, and the existing infrastructure, made it a good site for a pilot scale demonstration. However, the decade of remediation activities raised concerns as to the potential for surfactant enhanced contaminant elution due to the reduced contaminant levels.

Because of the highly conductive nature of the system, and the fact that surfactant recovery was a primary goal of the research, a vertical circulation well system was used (injection in an upper portion and extraction in the lower portion of a common borehole, with packers separating the respective screen areas). During the surfactant test 540 gallons of 60 mM Dowfax 8390 ($\sim 10 \times$ CMC) was injected. The surfactant was only slightly retarded, and the surfactant recovery was in excess of 95%. Both these factors demonstrate the value of using a "high performance" surfactant (i.e., prior research had shown much greater sorption and loss mechanisms for other surfactants). Elevated levels of both PCE and petroleum hydrocarbons were realized after surfactant breakthrough

(40-fold and 90-fold, respectively). These results were especially encouraging given the decade of bioremediation activities that had occured at the site [21].

Two sets of experiments were conducted at Operable Unit 1 (OU1) at Hill Air Force Base (HAFB). HAFB is located about 35 miles north of Salt Lake City, Utah. A jet fuel (JP-4) leak near chemical disposal pits containing PCBs, PAHs, pesticides, etc. caused the contamination at the research site. Over the decades the contamination had significantly weathered to the point that it no longer resembles JP-4. For example, middle phase systems developed for JP-4 did not form a middle phase with the OU1 oil, indicating the sensitivity of middle phase systems to contaminant composition. Tests at this site were conducted in sheat pile cells (sheet pile driven into a lower confining layer hydraulically isolating the treatment zone from the surrounding media), 3 m × 5 m in cross section with roughly 3 m of saturated zone.

We conducted both a solubilization and a mobilization study in two different cells at HAFB OU1. In the solubilization cell ten pore volumes of 4.3 wt% Dowfax 8390 was flushed through the cell. Greater than 95% recovery of the surfactant was achieved, along with the removal of roughly 50% of the contaminant (obviously additional flushing would have removed more contaminant, but this was the pre-established baseline). The Dowfax 8390 proved to be a very robust system, although not as efficient as the middle phase system (as expected). In the mobilization cell a surfactant system of 2.2 wt% AOT, 2.1 wt% Tween 80 and 0.43 wt% calcium chloride was used. In this case 6.6 pore volumes of surfactant removed roughly 85% to 90% of the contaminant. The improved elution was obvious, both visually and by measured concentrations. While this system was more efficient, it required much higher operator skill to both design and implement.

At another location at HAFB PCE contamination exists (OU2). OU2 was the site of a surfactant foam flooding test by Rice University, the University of Texas and Interra, Inc. We collected 500 gallon aliquots from their in situ test and conducted surfactant decontamination studies on these aliquots. Both packed tower and hollow fiber air stripping was evaluated on these aliquots. The packed tower air stripper was 2.3 m tall by 0.2 m diameter, while the hollow fiber system was 0.9 m tall by 0.1 m diameter (the relative sizes represent the higher interfacial mass transfer area per unit volume for the hollow fiber system). TCE removal efficiencies ranged from 90% to 95% in the packed tower and 98–100% in the hollow fiber, while PCE removals were 80–90% in the packed tower and 90–100% in the hollow fiber system. These results are as expected for the less volatile and more hydrophobic PCE and are in keeping with design predictions based on laboratory studies. Thus, these systems can be designed to regenerate the

Progr Colloid Polym Sci (1998) 111:168–173
© Steinkopff Verlag 1998

surfactant for reinjection (likely 90–95% decontamination will be required). Foaming was mitigated by lower liquid loading rates in the packed tower system, which obviously results in higher capital costs; foaming was not a problem in the hollow fiber system.

Summary

This study has provided an overview of surfactant fundamentals, and highlighted issues important to the selection and design of surfactant systems. Careful consideration of surfactant, contaminant, media and fluid properties are critical to the design of a successful, and optimal, surfactant system. These properties should be evaluated not only individually, but collectively. By integrating the design approach in this manner, using laboratory batch, column and modeling studies, surfactant systems can be developed that will significantly improve the remediation system. Results from field demonstrations helped to reinforce the importance of these concepts.

References

1. Rosen MJ (1989) Surfactants and interfacial phenomena, 2nd ed. Wiley, New York, NY
2. Edwards D, Luthy R, Liu Z (1991) Envir Sci Tech 25(1):127–133
3. Shiau B, Sabatini D, Harwell J (1994) Ground Water 32(4):561–569
4. Pope G, Wade W (1995) in: Surfactant-enhanced subsurface remediation: Emerging technologies. Sabatini D, Knox R, Harwell J (eds) ACS Symp Ser 594. American Chemical Society, Washington DC, 142–160
5. Bourrel M, Schechter R (1988) Microemulsions and related systems. Surfactant science series, Vol 30. Marcel Dekker, New York
6. West C, Harwell J (1992) Envir Sci Tech 26:2324–2330
7. Pennel K, Abriola L, Weber W (1993) Envir Sci Tech 27(12):2341–2351
8. Baran J, Pope G, Wade W, Weerasooriyaa V, Yapa A (1994) Envir Sci Tech 28(7):1361–1366
9. Fountain J, Waddell-Sheets C, Lagowski A, Taylor C, Frazier Byrne M (1995) in: Surfactant-enhanced subsurface remediation: Emerging technologies. Sabatini D, Knox R, Harwell J (eds) ACS Symp Ser 594. American Chemical Society, Washington DC, 177–190
10. Sabatini D, Knox R, Harwell J (eds) (1995) Surfactant enhanced subsurface remediation: Emerging technologies. ACS Symp Ser 594. American Chemical Society, Washington DC
11. Krebs-Yuill B, Harwell J, Sabatini D, Knox R (1995) in: Surfactant-enhanced subsurface remediation: Emerging technologies. Sabatini D, Knox R, Harwell J (eds) American Chemical Society, Washington DC
12. Krebs-Yuill B, Harwell J, Sabatini D, Quinton G, Shoemaker S (1996) Economic study of surfactant-enhanced pump-and-treat remediation. 69th Annual Water Envir Federation Conf, Dallas, Texas, 5–9 October
13. Rouse J, Sabatini D, Harwell J (1993) Envir Sci Tech 27(10):2072–2078
14. Rouse J, Sabatini D, Brown R, Harwell J (1996) Water Envir Res 68(2):162–168
15. Shiau B, Sabatini D, Harwell J, Vu (1996) Envir Sci Tech 30(1):97–103
16. Nayyar S, Sabatini D, Harwell J (1994) Envir Sci Tech 28:1874–1881
17. Rouse J, Sabatini D, Deeds N, Brown E, Harwell J (1995) Envir Sci Tech 29(10):2484–2489
18. Shiau B, Sabatini D, Harwell J (1997) Removal of chlorinated solvents in subsurface media using edible surfactants: column studies. J Envir Engr Div ASCE (in review)
19. Lipe M, Sabatini D, Hasegawa M, Harwell J (1996) Ground Water Monit Remed 16(1):85–92
20. Hasegawa M, Sabatini D, Harwell J (1997) J Envir Engr Div ASCE 123(7):691–697
21. Knox R, Sabatini D, Harwell J, Brown E, West C, Blaha F, Griffin C (1997) Surfactant remediation field demonstration using a vertical circulation well. Ground Water 35(6):948–953

Progr Colloid Polym Sci (1998) 111:174–178
© Steinkopff Verlag 1998

A. Hild
J.-M. Séquaris
H.D. Narres
M.J. Schwuger

Adsorption of anionic surfactant Na-dodecylsulphate (SDS) at the polyvinylpyrrolidone (PVP)-modified Na-montmorillonite surface

A. Hild · Dr. J.-M. Séquaris (✉)
H.D. Narres · M.J. Schwuger
Institute of Applied Physical Chemistry
Research Centre Jülich
D-52425 Jülich
Germany
E-mail: J.M. Sequaris@fz-juelich.de

Abstract An enhanced adsorption of Na-dodecylsulphate (SDS) at Na-montmorillonite is observed through a specific binding to an adsorbed layer of nonionic polyvinylpyrrolidone (PVP) polymer. These results are supported by electrokinetic measurements and the fluorescence pyrene probe method. At high PVP loadings, a negative electrophoretic mobility increase and the fluorescence detection of PVP/SDS aggregates confirm that a PVP coiled structure promotes the interactions with anionic SDS molecules outside the range of electrostatic repulsion forces arising from the negatively charged montmorillonite surface.

Key words Na-dodecylsulphate – polyvinylpyrrolidone – montmorillonite – surface modification – adsorption

Introduction

A growing use of synthetic polymer–surfactant mixtures is found in different technological applications such as for example the solubilization of poorly soluble organic derivatives [1, 2]. Furthermore, the transport and immobilization mechanisms of pollutants by natural amphiphilic substances have been investigated with mimetic substances. In this context, the role played by the inorganic soil colloids as sorbent surfaces has been also considered [3, 4]. Especially, an understanding of the electrical charge effects on the formation of adsorbed amphiphilic macromolecular structures is essential. Thus, it is generally admitted that no adsorption or a minute adsorption of negatively charged Na-dodecylsulphate (SDS) surfactant molecules can take place at the similar negative Na-montmorillonite surface, a typical swelling clay mineral [5]. On the other hand, specific interactions between the nonionic polymer polyvinylpyrrolidone (PVP) and SDS surfactant in solution [6a, b, 7] are generally reported. In the same way, a high PVP loading at the Na-montmorillonite surface [8–11] is observed. It is thus of interest to investigate the role played by the adsorption of a nonionic polymer polyvinylpyrrolidone (PVP) to avoid the unfavourable electrostatic interactions between the amphiphilic SDS surfactant molecule and the clay mineral.

In this work, direct evidence is shown for an enhanced SDS adsorption at PVP-modified montmorillonite surfaces. Its dependence on the surface PVP-loading is discussed briefly [10–11]. Results obtained with the microelectrophoretic and the fluorescence pyrene probe methods are also reported giving insights into the binding mechanisms of SDS at PVP-modified montmorillonite.

Materials and methods

Materials

Na-montmorillonite is characterized by procedures as described elsewhere [10, 11]. BET surface area and cation exchange capacity (CEC) are 96 m^2/g and 87.5 meq/100 g respectively. Na-montmorillonite has a total surface area of 760 m^2/g. The unlabelled and ^{3}H-labelled Na-dodecylsulphate (SDS) were purchased from Fluka and Hartmann

Analytik (Braunschweig), respectively. Polyvinylpyr-rolidone K30 of Mw 44000 g/mol was obtained from BASF. Other chemicals are Merck products. Water purified through Millipore filters was used in all experiments.

Methods

PVP surface modification of Na-montmorillonite: Na-montmorillonite modified with increasing PVP loadings up to saturation (0.55 g PVP/g Na-montmorillonite) was prepared in 0.01 M NaCl. After a first adsorption step, the excess free PVP was discarded by centrifugation according to a repeated washing procedure described elsewhere [10, 11].

Adsorption isotherms of SDS: SDS concentrations of up to 20 mM were added in 10 g/l PVP-modified Na-montmorillonite. After an equilibration of 64 h on a horizontal shaker at room temperature and centrifugation (20000 rpm, 30 min), the amount of adsorbed ^{3}H labelled SDS is calculated by the difference method. A liquid scintillation counter Beckman LS 5000TA was used.

Microelectrophoresis: A Lazer Zee Meter (Pen Kem Model 501) was used. The amount of PVP-modified Na-montmorillonite in the electrophoretic measurements was 0.1 g/l. The solid phase in adsorption experiments was diluted with SDS solutions of the same composition as in the supernatant.

Fluorescence spectroscopy: Fluorescence measurements were performed with a Perkin Elmer luminescence spectrometer LS 50. The pyrene fluorescent marker concentration was 1.5 μM.

Results and discussion

PVP polymer-modified montmorillonite

The surface of Na-montmorillonite, a clay mineral with a high surface area (760 m^2/g), is first modified by the adsorption of the nonionic polymer polyvinylpyrrolidone (PVP). Different montmorillonite samples with increasing polymer loadings from 0 → 0.55 g PVP/g montmorillonite were thus prepared and characterized by X-ray diffraction, microelectrophoresis, light absorbance, sedimentation volumes and viscosity [10, 11]. The main results of this characterization can be briefly summarized as follows. At low PVP loadings ($\leq$ 0.3 g PVP/g montmorillonite), a flocculation between particles covered by PVP with a flattened conformation was observed by light absorption measurements. This can be shown by the low extent of the PVP adsorbed layer thickness and an increase of the

interlayer distance d_{001} [8–11] between montmorillonite plates in dried flocculates measured with microelectrophoresis and X-ray diffraction. At high PVP loadings (> 0.3 g PVP/g montmorillonite), these flocculates formed by montmorillonite plates with intercalated PVP polymer at the inner surface are sterically stabilized by the adsorption of coiled PVP at the outer surface. This can be shown in particular by the increase of the PVP adsorbed layer thickness measured with microelectrophoresis. An increase of the PVP-modified montmorillonite particle symmetry due to the adsorbed PVP can also be deduced from the large decrease of the intrinsic viscosity $[\eta]$ of the sample solutions.

Keeping in mind that adsorbed structures of PVP polymer with flattened or more coiled conformations respectively occur at low PVP or high PVP loadings, SDS binding at the polymer-modified montmorillonite surface was investigated.

Adsorption of SDS

In Fig. 1, the adsorption isotherms expressed in mM SDS/g montmorillonite were reported in the case of unmodified montmorillonite and PVP-modified montmorillonite with different PVP loadings (0.1, 0.3, 0.45 and 0.55 g PVP/g montmorillonite). Results of SDS binding at unmodified montmorillonite and PVP-modified montmorillonite with a loading of 0.1 g PVP/g montmorillonite do not indicate a sensitive adsorption of SDS. The lowest amount of adsorbed SDS can be detected after a modification of 0.3 g PVP/g montmorillonite while a sensitive

Fig. 1 Adsorption isotherms of SDS at unmodified and PVP-modified montmorillonite. ▲-▲, 0 g PVP/g montmorillonite and 0.10 g PVP/g montmorillonite; ♦-♦, 0.30 g PVP/g montmorillonite; ×-×, 0.45 g PVP/g montmorillonite; ●-●, 0.55 g PVP/g montmorillonite; montmorillonite 10 g/l; 0.01 M NaCl, pH 5.5

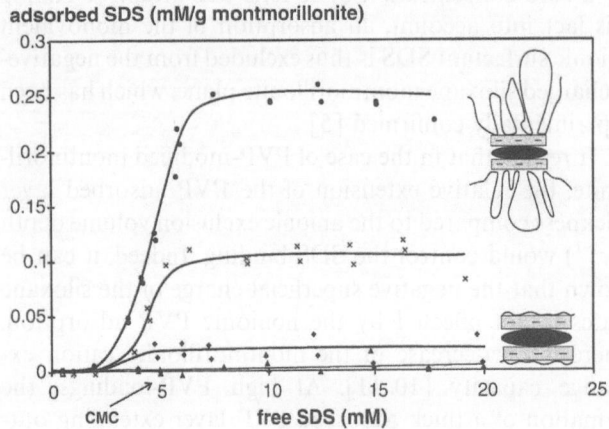

adsorbed SDS (mM/g montmorillonite)

increase of adsorbed SDS amounts is measured for the two highest PVP loadings of 0.45 g PVP and 0.55 g PVP/g montmorillonite. It must also be noted that a sensitive binding starts at a free SDS concentration of 2 mM which is lower than the critical micelle concentration (CMC) of SDS, 5 mM, in 0.01 M NaCl.

The cooperative character of the binding isotherms can be modelled by a monotonous increase of the aggregation number of surfactants along the binding isotherm as is generally found in a one-dimensional nearest neighbor lattice model [12–14]. In this model, the binding constant K of an isolated site and μ, the cooperative parameter, are considered where $K\mu$ is the binding constant of a surfactant with a site adjacent to an occupied site. For the highest PVP loading of 0.55 g PVP/g montmorillonite, K and μ values of 5.3 ± 0.8 M^{-1} and 39 ± 5 can be calculated, respectively.

The SDS binding, exclusively observed with high PVP loadings, indicates that the conformation of the adsorbed PVP plays a decisive role for the interaction. Thus, as previously characterized by the microelectrophoretic method, a thick adsorbed layer at the outer surface of PVP-montmorillonite particles seems to be required for the SDS binding. On the other hand, at low PVP loadings a flattened conformation for PVP at the outer and inner (intercalation) surfaces does not promote any SDS binding (see also schematic representations of PVP adsorption in Fig. 1).

The dependence of the SDS binding on the conformation of the adsorbed PVP can be related to the existence of an anionic exclusion volume at the montmorillonite surface. Indeed, an electrostatic anion exclusion is the general rule in the interfacial region between the high negatively charged montmorillonite siloxane plate and the solution. The depth of this exclusion volume for the co-ions was treated theoretically and is given by twice the Debye length ($2\kappa^{-1}$) for 1:1 electrolyte [15]. In 0.01 M NaCl, a maximum distance of 6 nm from the surface is thus calculated which delimits a volume where the conditions for a zero concentration of co-ions was attained. Taking this fact into account, an adsorption of the monovalent anionic surfactant SDS is thus excluded from the negatively charged siloxane montmorillonite plane, which has been experimentally confirmed [5].

It results that in the case of PVP-modified montmorillonite, the relative extension of the PVP adsorbed layer thickness compared to the anionic exclusion volume depth ($2\kappa^{-1}$) would control the SDS binding. Indeed, it can be shown that the negative superficial charge of the siloxane plates is not affected by the nonionic PVP adsorption. There is no decrease of the montmorillonite cation exchange capacity [10, 11]. At high PVP-loadings, the formation of a thick adsorbed PVP layer extending out-

side the anionic exclusion volume can thus explain a specific binding of SDS. A PVP adsorption hydrodynamic layer up to 30 nm with a PVP molecular weight of 44 000 g/mol was estimated in the case of an adsorption at the non-swelling clay mineral kaolinite [16]. On the other hand, at low PVP loadings, the relatively stronger electrostatic repulsion forces would impede a specific binding in a more flattened PVP conformation.

Microelectrophoresis

In Fig. 2A, the electrophoretic mobilities of unmodified montmorillonite and PVP-modified montmorillonite with different PVP loadings (0.1, 0.3, 0.45 and 0.55 g PVP/g montmorillonite) are reported as a function of the total concentration of SDS in solution. An increase of the negative electrophoretic mobility is found in the presence of SDS which confirms its binding at the particle surface. Thus, in the presence of PVP, a relative decrease of the electrophoretic mobility due to the shear plane shift towards the solution [16] is nearly compensated by the SDS adsorption. In Fig. 2B, these relative electrophoretic mobility variations ($\Delta\mu$) of unmodified montmorillonite and PVP-modified montmorillonite with different PVP loadings (0.1, 0.3, 0.45 and 0.55 g PVP/g montmorillonite) are reported as a function of the adsorbed concentration of SDS expressed in mM/g montmorillonite. In a general way, an increase of the negative electrophoretic mobilities

Fig. 2 (A) Effect of the total added SDS concentration on the electrophoretic mobility of unmodified and PVP-modified montmorillonite. (B) Effect of the adsorbed SDS concentration on the relative variation of the electrophoretic mobility ($\Delta\mu$) of unmodified and PVP-modified montmorillonite. ▲-▲, 0 g PVP/g montmorilllonite; ■-■, 0.10 g PVP/g montmorillonite; ◆-◆, 0.30 g PVP/g montmorillonite; ×-×, 0.45 g PVP/g montmorillonite; ●-●, 0.55 g PVP/g montmorillonite; 0.01 M NaCl, pH 5.5

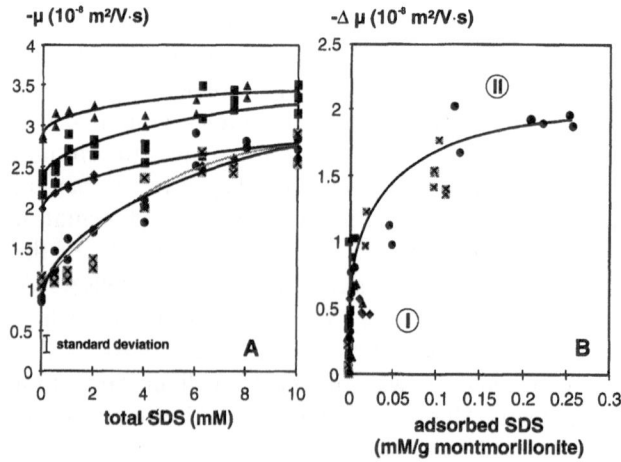

Progr Colloid Polym Sci (1998) 111:174–178
© Steinkopff Verlag 1998

is observed by increasing the adsorbed amount of SDS. Two effects due to SDS adsorption can however be distinguished, which are related to the nature of the binding sites. Thus, for a minute adsorption of SDS (region I in Fig. 2B), a stepped increase of the negative electrophoretic mobility for unmodified and PVP-modified montmorillonite particles can be assigned, in 0.0 1M NaCl, to an electrostatic interaction of SDS with positively charged aluminol binding sites. They are located at the edge surface which is estimated about 1–2% of the total montmorillonite surface area [4]. This sensitive microelectrophoretic detection can be explained by a neutralization and a possible overcompensation of the positively charged edge surface sites by negatively charged surfactants small aggregates [17]. On the other hand, the SDS binding detection in the PVP adsorbed layer at higher PVP loadings (0.45 and 0.55 g PVP/g montmorillonite, region II in Fig. 2B) is marked by a smooth increase of the negative electrophoretic mobility. The formation of negatively charged SDS/PVP aggregates gives a polyelectrolyte character to the adsorbed layer which complicates the interpretation of the electrophoretic mobility, i.e. the definition of the associated zeta potential [18]. An increase of the charge density due to the specific binding of the organic anions SDS at the PVP surface as well as a shear plane shift associated with the polymer layer conformational changes must now be taking into account.

Fluorescence pyrene probe method

The fluorescence pyrene probe technique [1, 19, 20] for detecting the onset of polymer-surfactant interactions is based on micropolarity changes of the medium surrounding the pyrene molecules. These are expressed in variations of the fluorescence emission band intensities at λ_{em} 372 nm (I_1) and 381 nm (I_3), respectively while the excitation wavelength was 310 nm. Thus, the so-called I_1/I_3 intensity ratio has typical values of about 1.8–1.9 in polar water solvent and drops to about 0.6 in nonpolar hydrocarbon. In the case of pyrene solubilization in the hydrophobic environment of SDS micelles, I_1/I_3 ratio values of about 1.2 can be found. This method allowed polarity variations to be detected at SDS concentrations much lower than the CMC when SDS interacts with PVP [10, 11, 19]. Thus, the presence of SDS–PVP aggregates which solubilize pyrene molecules can be assumed by a rapid transition of the I_1/I_3 ratio value from 1.8 to 1.6 in the presence of 2 mM SDS. A decrease to 1.45 can be further observed in the presence of 10 mM SDS. In comparison with the obtained I_1/I_3 ratio value of about 1.2 in micellar solution, it was concluded that the pyrene environment in the PVP/SDS aggregates is more polar or hydrophilic.

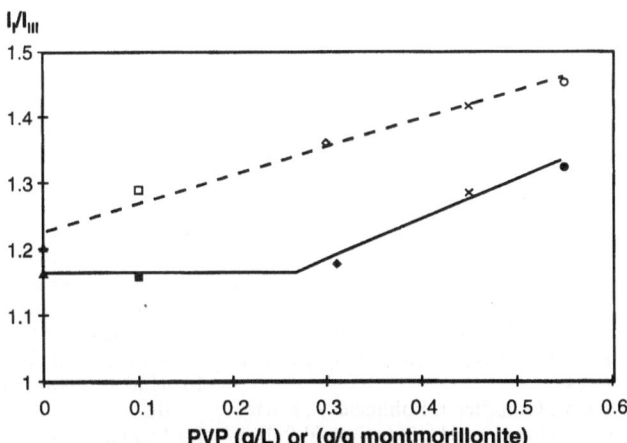

Fig. 3 Effect of the PVP concentration, in free solution (---) or adsorbed at montmorillonite (——), on the ratio value of the fluorescence intensities I_1/I_3 from pyrene contained in SDS micelles. 10 mM SDS; 1 g/l montmorillonite; 0.01 M NaCl, pH 5.5

In order to detect adsorbed PVP–SDS aggregates, the polarity environment variations of pyrene molecules in a micellar solution of SDS were observed, in Fig. 3, after the addition of PVP-modified montmorillonite. For comparison, the I_1/I_3 ratio values obtained with the corresponding PVP concentrations in solution are also reported. In this case, a smooth transition of the I_1/I_3 ratio values along the PVP concentration from 1.2 to 1.45 characterizes the formation of PVP–SDS aggregates at the expense of SDS micelles. In the case of adsorbed PVP, this transition only occurs for PVP loadings higher than 0.3 g PVP/g montmorillonite when a thick adsorbed PVP layer is "effective" for the SDS binding, which confirms the binding isotherms and microelectrophoretic results.

Conclusion

Results show that the montmorillonite surface modification with a nonionic polymer PVP enhances the adsorption of organic anions such as SDS. It results in a sorbent surface with amphiphilic surfactant/polymer aggregates which immobilizes poorly soluble compounds such as pyrene molecules. Thus, by choosing the appropriate polymer loading, unfavourable electrostatic adsorption forces can be avoided, which enlarges the application of organoclays from nonionic and cationic surfactants to anionic ones. By analogy, under soil conditions, the modification of clay mineral surfaces by natural macromolecules and widely distributed anionic amphiphilic biomolecules could explain some trapping mechanisms for organic pollutants.

178

A. Hild et al.
Adsorption of SDS at the PVP-modified Na-montmorillonite surface

References

1. Goddard ED (1986) Colloids Surfaces 19:255–300
2. Myers D (1988) In: Surfactant Science and Technology. VCH, New York, pp 153–182
3. Klumpp E, Schwuger MJ (1997) In: Schwuger MJ (ed) Detergents in the Environment. Surfactant Science Series, Vol 65. Marcel Dekker, New York, pp 39–64
4. Klumpp E, Lewandowski H, Séquaris JM (1998) Prog Colloid Polym Sci, in press
5. Ilic M, Gonzàlez J, Pohlmeier A, Narres HD, Schwuger MJ (1996) Colloid Polym Sci 274:966–973
6. (a) Schwuger MJ, Lange H (1968) Proc 5th Internatl Congr Surface Active Substances. Barcelona, Vol 2, pp 955–964; (b) Nüßlein H, Schumann K, Schwuger MJ (1979) Ber Bumsenges Phys Chem 83:1229–1238
7. Arai H, Murata M, Shinoda K (1971) J Colloid Interface Sci 37:223–227
8. Francis CW (1973) Soil Sci 115:40–54
9. Hild A, Séquaris J-M, Narres HD, Schwuger MJ (1997) Mitteilgn Dtsch Bodenkundl Gesellsch 83:25–28
10. Hild A (1998) PhD Thesis. Univ Düsseldorf
11. Sequaris J-M, Hild A, in preparation
12. Satake I, Yang JT (1976) Biopolymers 15:2263–2275
13. Applequist J (1977) J Chem Educ 7:417–419
14. Séquaris JM (1997) Langmuir 13:653–658 and references cited therein
15. Pashley RM, Quirk JP (1997) Soil Sci Soc Am J 61:58–63
16. Hild A, Séquaris J-M, Narres HD, Schwuger MJ (1997) Colloids Surfaces A 123–124:515–522
17. Sastry NV, Séquaris J-M, Schwuger MJ (1995) J Colloid Interface Sci 171:224–233
18. Donath E, Walther D, Shilov VN, Knippel E, Budde A, Lowack K, Helm CA, Möhwald H (1997) Langmuir 13:5294–5305
19. Turro NJ, Baretz BH, Kuo P-L (1984) Macromolecules 17:1321–1324
20. Winnik FM, Regismond STA (1996) Colloids Surfaces A 118:1–39

Progr Colloid Polym Sci (1998) 111:179–183
© Steinkopff Verlag 1998

A. Pohlmeier
S. Haber-Pohlmeier
H. Rützel
H.D. Narres

Sorption of Cu²⁺ and Zn²⁺ at Na-montmorillonite: Analysis by means of affinity spectra

A. Pohlmeier (✉)
MOD, Research Centre Jülich
D-52425 Jülich
Germany

S. Haber-Pohlmeier
Faculty of Chemistry
Physical Chemistry II
University of Bielefeld
D-33501 Bielefeld
Germany

H. Rützel · H.D. Narres
Research Center Jülich
D-52425 Jülich
Germany

Abstract The sorption isotherms of Cu^{2+} and Zn^{2+} at Na-mont-morillonite are determined electro-chemically by the combination of DPASV and DPP and analyzed for the first time by affinity spectra. To obtain these distribution functions an inverse integral transformation is performed numerically by the constrain-ed regularization technique CONTIN with the integral adsorption equation, employing a smoothing regularizer.

Affinity spectra are obtained with a main process ($>94\%$) characterized by a mean pseud-Langmuir sorption coefficient of $\log K = 4.23$ and 4.08 for the binding of Zn^{2+} and Cu^{2+}, respectively. The exchange coefficient for Zn^{2+} is calculated as $\log K_{ex} = -0.14$. The affinity spectrum for Cu^{2+} is broader than for Zn^{2+}, and expresses a considerable tailing to the low affinity range. This indicates a greater heterogeneity, probably caused by precipitation of copper hydroxides at higher concentrations of Cu^{2+}. Additionally, pH-dependent investigations are presented, indicating no pH-dependent binding sites for Cu^{2+} and Zn^{2+} at montmoril-lonite.

Key words CONTIN – copper – heterogeneity – isotherm – montmorillonite – zinc

Introduction

The sorption at the interface soil colloids/water is one of the most important processes controlling the mobility of toxic metal ions in the environment. It is described physicochemically by equilibrium coefficients, obtained by analysis of sorption isotherms [1]. However, it is known that this classical description by discrete coefficients does not describe the processes satisfactorily, since the surface of many soil colloids is structurally and chemically hetero-geneous, e.g. due to inhomogeneous charge distributions [2, 3]. In these cases the discrete coefficients must be replaced by distribution functions of the equilibrium coef-ficient, leading to affinity spectra [2]. One way to obtain such spectra is the fitting of appropriate isotherm equa-tions like the generalized Langmuir–Freundlich law to the

experimental data yielding a mean affinity coefficient and a heterogeneity parameter. However, the corresponding monomodal, symmetric Sips distribution [4] implicitly confines the variety of solutions.

The second way is the use of numerical inverse trans-formation techniques. These possess the advantage that they make no a priori assumptions about number and shape of peaks in the distribution function. Different prin-ciples are employed like singular value decomposition, expectation maximization or regularization and they have been applied successfully to, e.g. the binding of Cu^{2+} to humic acid [5] and the binding of Ca^{2+} at iron hydroxide [6]. In previous studies we have applied the regularization technique CONTIN [7–9] to the analysis of the equilib-rium and kinetics of Cd^{2+} sorption at montmorillonites [10–12]. This program has the great advantage of offering an objective criterion for the determination of the relative regularization strength, i.e. the choice of the regularization parameter α, as it will be described in the Theory section. Therefore, CONTIN is used in this work with the modification described in a previous paper [12] for the analysis of the sorption of the heavy metals Cu^{2+} and Zn^{2+} with Na^+ at the important colloidal soil mineral montmorillonite. The aim is to obtain model independent affinity spectra, which will allow statements about the mean affinity of montmorillonite for the metal ions, the heterogeneity, and different types of binding sites. The Na-form of montmorillonite is chosen instead of e.g. Ca^{2+}, or Mg^{2+}, since in the latter case three different metal ions would compete about the binding sites due to the obligatory buffering with Na-containing buffers (e.g. PIPES/NaOH). This study focuses on the equilibrium properties of the exchange process, the kinetics will be the subject of separate paper.

Material and methods

Material

The metal ions were used as nitrate salts from Fluka, and montmorillonite was kindly provided by Südchemie AG, Germany. Its Na^+-form was prepared by threefold ion exchange with 1 M $NaNO_3$ and subsequent dialysis for 6 weeks. Finally the $<2\,\mu m$ fraction was separated by sedimentation (60 cm, 48 h) and characterized by the determination of the cec ($9.8 \times 10^{-4}\,mol\,g^{-1}$) and BET ($89\,m^2\,g^{-1}$). As buffers were used Acetate, PIPES, HEPES and TRIS.

Isotherm

For sorption experiments at constant pH the solutions were buffered at pH = 6.3 with 5 mM PIPES/6 mM NaOH. The binding isotherm was recorded electrochemically as follows: To suspensions containing $0.1\,g\,l^{-1}$ Na-montmorillonite increasing amounts of $M(NO_3)_2$ were added from a stock solution. An equilibration time of 20 min was checked to be sufficient for complete equilibration. After this time the concentrations of free Cd^{2+}-ions in the suspensions were determined in situ by differential pulse anodic stripping voltammetry (DPASV) in the concentration range below $2 \times 10^{-6}\,mol\,l^{-1}$ and by differential pulse polarography (DPP) above $1 \times 10^{-6}\,mol\,l^{-1}$. Conditions employed for the DPASV were a deposition time of 120 s at $E = -1.25\,V$ (Zn) and $-0.80\,V$ (Cu), respectively, and a scan-rate of $5\,mV\,s^{-1}$. The DPP analysis was performed with a drop time of 1 s. The voltammograms were evaluated by comparing the peak heights with

calibration curves. pH-dependent measurements were performed using different buffers (Acetate, PIPES, HEPES, TRIS) adjusted with $NaOH/HNO_3$, and unbuffered, using $NaNO_3$ as background electrolyte. The amount of bound metal ions $n_M(c_M)$ was calculated from the total and the free concentration.

Theory

It is supposed that at the surface of montmorillonite a great amount of independently reacting local binding sites of the type "j" is located, where following reaction may take place:

$$M^{2+} + Na_2Z_j \underset{K_{j,ex}}{\leftrightarrows} MZ_j + 2Na^+, \qquad (1)$$

where M^{2+} represents the bivalent metal ion Zn^{2+} or Cu^{2+}. Z_j denotes one type of binding site with the local exchange coefficient $K_{j,ex}$, defined in Eq. (2):

$$K_{j,ex} = \frac{n_{Z_jM}\,c_{Na}^2\,f_I^2}{n_{Z_jNa_2}\,c_M\,f_{II}}, \qquad (2)$$

where n denotes the concentration at the surface of montmorillonite in $mol\,g^{-1}$, c is the concentration in solution in $mol\,dm^{-3}$, and f are the activity coefficients calculated by Davis' equation. The amount of bound M at one type of binding site is not determinable, since its absolute value is very small. The only measurable quantity is the total amount of bound metal at montmorillonite, which is the sum over all accessible binding sites:

$$n_M = \sum_{j=1}^{N} n_{Z_jM} = \sum_{j=1}^{N} n_{j,max} \frac{K_j c_M}{1 + K_j c_M}. \qquad (3)$$

In Eq. (3) $n_{j,max}$ is a weighting factor, i.e. the total number of binding sites of type "j", and the local pseudo-Langmuir equilibrium coefficient K_j is related to the exchange coefficient by Eq. (4). Equation (4) means that for sufficient high concentrations of Na^+, which are used in this study, c_{Na} is nearly constant and K_j is direct proportional to K_{ex}.

$$K_j = K_{j,ex} \frac{f_{II}}{f_I^2\,c_{Na}^2}, \qquad (4)$$

It is important to notice that Eq. (3) is generally valid, independently of the exact sorption mechanism, if the above-mentioned condition is fulfilled. The only necessary condition is that the local binding sites react independently, which is very plausible, regarding the small density of binding sites at montmorillonites.

As mentioned above, the single terms in the sums on the right-hand side of Eq. (3) are not measurable, only the total amount of bound metal ions, n_M is. If the number N of different types of binding sites is high enough, Eq. (3) may be substituted by the integral binding equation [2, 5, 6]:

$$n_M = \int_{K=0}^{\infty} S(K) \frac{Kc_M}{1 + Kc_M} dK + \varphi , \qquad (5)$$

where $S(K)$ is the distribution function of the binding sites with the affinity coefficient K. If n_M is measured as a function of c_M, $S(K)$ may be determined by solving Eq. (5) in a finite range between K_{min} and K_{max}. Unfortunately, this is an ill-posed problem leading to an infinite number of solutions within the noise φ. So the number of probable solutions must be reduced, and the most probable one is to be chosen. In this work the regularization technique CONTIN of S. Provencher [7–9] is employed, where the standard least-squares condition is extended by the regularizor, which is for the purpose of this study the second derivative of the distribution function, and a parameter α that determines the strength of regularization. CONTIN takes advantage of the statistical F-test by comparing the variance of a regularized solution with the unregularized one and deciding, whether its increase is only due to random. A parameter $prob1(\alpha)$ is calculated according to Eq. (6):

$$prob1(\alpha) \equiv P[F_1(\alpha), N_{DF}(\alpha_0), N_y - N_{DF}(\alpha_0)] , \qquad (6)$$

where $P(F)$ denotes the F-distribution with $F_1(\alpha)$ given by

$$F_1(\alpha) = \frac{V(\alpha) - V(\alpha_0)}{V(\alpha_0)} \frac{N_y - N_{DF}(\alpha_0)}{N_{DF}(\alpha_0)} \qquad (7)$$

and N_y is the number of data points and N_{DF} the degrees of freedom. Provencher recommends that this parameter should lie in the range of 0.1–0.9 with an optimum value of 0.5 [8]. This range prevents on the one side too much bias of α and on the other side too many artifacts to distort the solution. In an iterative way the program calculates a set of solutions with varying α and finally presents the so-called "chosen solution" as that closest to 0.5. The distribution function $S(K)$ of this solution is plotted versus $\log K$ and termed in the following as affinity spectrum.

Results and discussion

Figure 1 shows the sorption isotherms of (a) Zn^{2+} and (b) Cu^{2+} at Na-montmorillonite. Plotted is the total amount of bound heavy metal ions n_M as a function of the corresponding concentrations of free ions, c_M. The combination of the two electrochemical techniques DPASV and DPP allows the reliable determination of the isotherm

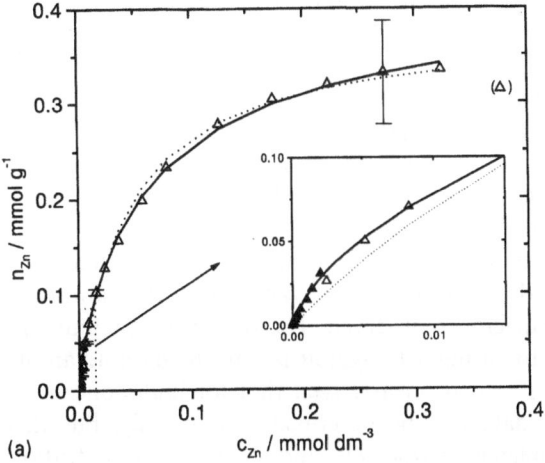

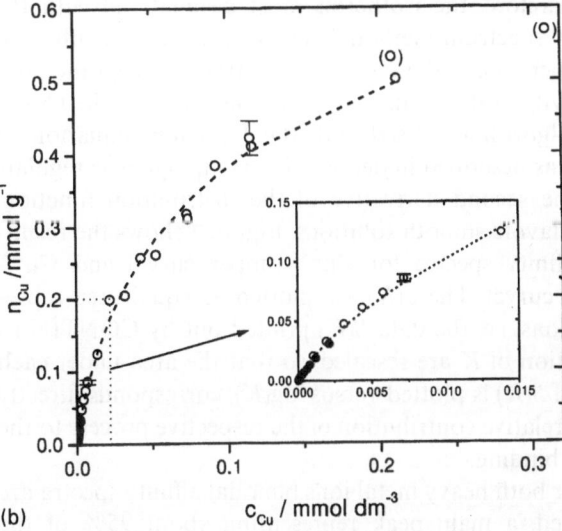

Fig. 1 Sorption isotherms of (a) Zn^{2+} and (b) Cu^{2+} at 100 mg dm⁻³ Na-montmorillonite. The pH = 6.3 was adjusted with 5 mM PIPES/6 mM NaOH at $T = 25 °C$. Insets: zoom of the low concentration range. Open and full symbols represent measurements analyzed with the DPP and DPASV, respectively. The lines (-----, ——) are calculated according to Eq. (5) using the affinity spectra shown in Fig. 2. The line (·······) in Fig. 2(a) is the Langmuir fit with $\log K = 4.3$

over a wide concentration range from 1×10^{-8} mol dm⁻³ to 4×10^{-4} mol dm⁻³. Both techniques yield similar results in the overlapping region. Two facts should be noted:

(i) The resulting error bars are calculated by error propagation assuming a mean uncertainty of $\Delta c_M = \pm 2\%$. The strong increase of the error with increasing adsorbed amount is caused by the indirect calculation of n_M, which is proportional to the difference between the total concentration of M^{2+} in the suspension and the amount of free metal ions according to: $n_M \propto (c_{total,M} - c_M)$.

(ii) For copper, no plateau is reached even at high values of c_M. The maximum amount of bound metal ions should not exceed the total cation exchange capacity of Na-montmorillonite, which is for divalent metal ions 0.48 mmol g^{-1}. A probable reason is the precipitation of copper-hydr(oxides) at the surface at high copper concentrations. Model calculations showed, that at pH = 6.3 and concentration $c_{Cu} = 1 \times 10^{-4}$ mol dm^{-3} a great fraction of copper exists as various hydroxides and CuO precipitates in solution. A pure ion-exchange mechanism is only probable at a copper concentration smaller than 3×10^{-5} mol dm^{-3}. For zinc no exceeding of the cation exchange capacity is observed. The data were preliminary analyzed by fitting a Langmuir law to the data. Example for Zinc, it can be clearly seen that for high concentrations the data can be described satisfactorily, but this fails completely in the low and medium concentration range.

Therefore, the isotherms must be analyzed with the affinity spectrum method. This means, Eq. (5) is fitted to the experimental data with no constraints except the non-negativity of the affinity distributions. For this the CONTIN algorithm is used with the Langmuir equation as kernel as described in detail in [12] employing as regularizor the second derivative of the distribution function, which favors smooth solutions. Figure 2 shows the resulting affinity spectra for Zn^{2+} (upper curve) and Cu^{2+} (lower curve). The $S(K)$ are plotted as equal area representations, i.e. the data $S(K)$ printed out by CONTIN as a function of K are rescaled, so that the area under each peak, if $S(K)$ is plotted versus $\log(K)$, corresponds directly to the relative contribution of the respective process to the whole binding.

For both heavy metal ions bimodal affinity spectra are obtained, a main peak representing about 95% of the whole ion exchange process and a small high affinity peak of about 6% (Zn) and 2.5% (Cu). The weighted mean value of $\log K$ for the main process, calculated as the ratio of the moments M_1/M_0 of the distribution function [8], amounts to $10^{4.23}$ dm^3 mol^{-1} for Zn^{2+} and $10^{4.08}$ dm^3 mol^{-1} for Cu^{2+}, and is indicated by the arrows in Fig. 2. That means that Zn^{2+} is bound slightly stronger than Cu. Assuming a pure ion exchange mechanism for the sorption of zinc, from these data the mean exchange coefficient, as defined by Eq. (2) is calculated according to Eq. (4) as -0.14 mol dm^3 for Zn. This value lies roughly in the range reported for some bivalent ions at various montmorillonites [1, 13–15] but is somewhat smaller than that obtained for the exchange of Cd^{2+} [12]. A reason could be the smaller ionic radius of this first-row transition metal ion compared to the bigger Cd^{2+} ion. This leads to a stronger hydration of Zn^{2+}, which could prevent inner sphere complexation.

The peaks are considerably broader than expected for a homogeneous surface. This proves that the exchange of

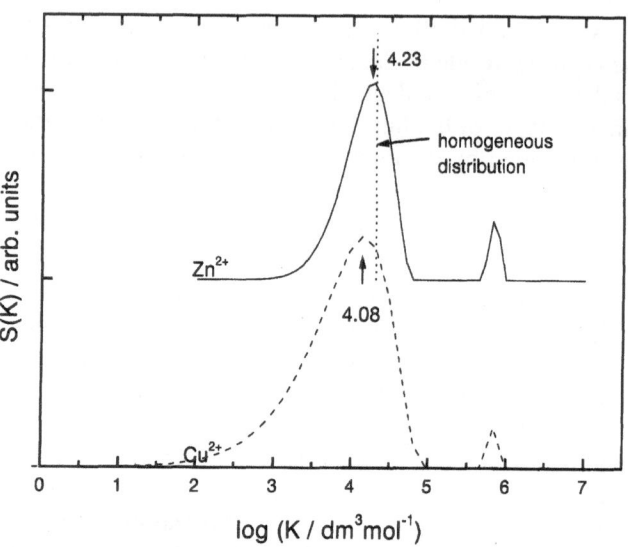

Fig. 2 Affinity spectra $S(K)$ for the ion exchange of Zn^{2+} and Cu^{2+} at Na-montmorillonite, calculated from the isotherm data from Fig. 1, according to Eq. (5). The arrows indicate the mean values of $\log K$, defined by Eq. (4). The line ($\cdots\cdots$) represents the "spectrum" of the Langmuir fit for the Zn^{2+}-montmorillonite system

heavy metal ions at montmorillonite is influenced by the heterogeneity of the binding sites located at the surface of the clay mineral. The symmetric shape and the width of the main peak is for Zn^{2+} comparable to that for Cd^{2+} [12], whereas for Cu^{2+} the peak is broader and shows a strong asymmetry to the low affinity side. The reason lies in the course of the isotherm, which exceeds the cation exchange capacity (see Fig. 1b), and probably reflects the further binding of Cu^{2+} by precipitation as hydr(oxides) at higher concentrations.

A distinction between binding sites located at the outer surface of montmorillonite and the interlayer space is not possible. The second calculated peak, possibly indicating high affinity processes, contributes only to about 6% (Zn) and 2.5% (Cu) to the whole ion exchange, whereas the whole area of the interlayer space should lie at least in the range of about 30%, if the mean aggregation number of the montmorillonite platelets amounts to about 2. The question arises, if these high affinity binding sites, which are nearly homogeneous with respect to the free sorption energy, could be located at the edges of the clay platelets, which carry pH-dependent charges [13]. Since the high affinity sites will be occupied at first during the increase of c_M, pH-dependent measurements are performed at low concentrations of Zn^{2+} and Cu^{2+}. Figure 3 shows the results for $c_{total,M} = 9 \times 10^{-7}$ mol dm^{-3}. Plotted is the mean exchange coefficient calculated by Eq. (2), since the concentration of Na$^+$ ions varies for different buffers and pH. No systematic dependence of $K_{ex}(pH)$ is observable. This

Progr Colloid Polym Sci (1998) 111:179–183
© Steinkopff Verlag 1998

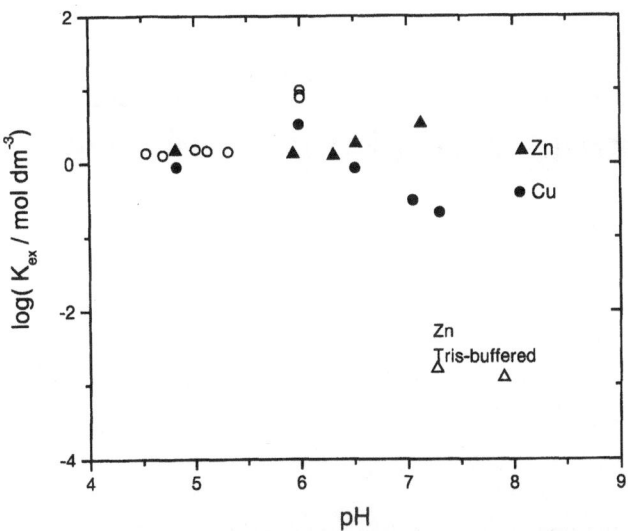

Fig. 3 Results of pH-dependent measurements for the sorption of Zn^{2+} (▲, △) and Cu^{2+} (●, ○) in 100 mg dm^{-3} Na-montmorillonite. Plotted is the exchange coefficient, calculated by Eq. (2) as a function of pH. Buffers: Acetate for pH < 5 (only copper), PIPES for pH < 6.8, and HEPES for pH > 6.8. For Cu^{2+}, also unbuffered solutions were investigated (○). (△) are K_{ex} (pH) for Zn^{2+}, obtained in 5 mM TRIS buffer (see text)

indicates that the binding at the edges plays no significant role for the exchange of Zn^{2+} and Cu^{2+} at Na-montmorillonite. The data obtained in the buffer TRIS should be regarded with caution, since TRIS is a cationic buffer that can bind at the surface of clay minerals and displace other cations. Keeping in mind these results, the obtained high affinity peaks at $\log K = 5.8$ should be regarded with caution. Further investigations in the very low concentration range including the direct determination of metal ions present at the mineral/electrolyte interface are necessary to ensure the existence of pH-independent high affinity binding sites.

Summary

The sorption reactions of Zn^{2+} and Cu^{2+} at Na-montmorillonite are investigated at constant and varying pH. The isotherms obtained at constant pH must not be described with the Langmuir law, i.e. a discrete exchange coefficient is insuffient to describe the binding of Zn^{2+} and Cu^{2+} over a wide concentration range due to the heterogeneity of the surface. Therefore, the concept of the affinity spectrum is applied for the analysis. For this purpose an inverse transformation of the integral sorption equation is performed by means of the modified program CONTIN, yielding the distribution functions $S(K)$ of a pseudo-Langmuir coefficient K. The program employs the regularization technique with a smoothing regularizer, where the relative regularization strength is adjusted by means of the F-test, to ensure that the solution is not oversmoothed. For both metal ions broad distributions are obtained, reflecting the heterogeneity of the montmorillonite surface. These results coincide with that obtained earlier for the exchange of Cd^{2+}. For Cu^{2+} a considerable asymmetry of the affinity spectrum due to the binding beyond the cation exchange capacity is observed at higher concentrations, possibly caused by the precipitation of Cu^{2+}-hydr(oxides). The pH-dependent measurements show no systematic change of the exchange coefficient with pH, indicating that pH-dependent binding sites at the edges of montmorillonite do not contribute to the whole binding process.

Acknowledgements The authors thank H. Ruf, Max Planck Institute for Biophysics, Frankfurt, Germany, and A. Topp, University of Cologne, Germany for the latest version of CONTIN.

References

1. Bolt GH (1979) Soil Chemistry: B Physico Chemical Models. Elsevier, Amsterdam
2. Buffle J (1988) Complexation Reactions in Aquatic Systems: An Analytical Approach. Ellis Horwood, Chichester
3. Malla PB, Robert M, Douglas LA, Tessier D, Komarneni S (1993) Clays Clay Min 41:412
4. Sips R (1948) J Chem Phys 16:490
5. Nederlof MM, van Riemsdijk WH, Koopal LK (1994) Environ Sci Technol 28:1048
6. Cernik M, Borkovec M, Westall JC (1995) Environ Sci Technol 29:413
7. Provencher SW (1982) Comput Phys Commun 27:213
8. Provencher SW (1982) Comput Phys Commun 27:229
9. Ruf H (1993) Adv Colloid Interface Sci 46:333
10. Pohlmeier A (1994) Progr Colloid Polym Sci 95:113
11. Pohlmeier A, Rützel H (1996) J Colloid Interface Sci 181:297
12. Haber-Pohlmeier S, Pohlmeier A (1997) J Colloid Interface Sci 188:377
13. Jasmund K, Lagaly G (1993) Tonminerale und Tone. Steinkopf Verlag, Darmstadt, Germany
14. Tang L, Sparks DL (1993) Soil Sci Soc Am J 57:42
15. Fletcher P, Sposito G (1989) Clays Clay Min 24:375

Progr Colloid Polym Sci (1998) 111:184–188
© Steinkopff Verlag 1998

Th. Hofmann
U. Schöttler

Behavior of suspended and colloidal particles during artificial groundwater recharge

Th. Hofmann (✉) · U. Schöttler
Institute for Water Research
Zum Kellerbach 46
D-58239 Schwerte
Germany
E-mail: ifw@poboxes.com

Abstract Transport phenomena in the subsurface environment are often treated on the basis of a two-phase system: an immobile solid phase and a mobile dissolved phase. Many organic and inorganic substances, e.g. PAH, biogen material and heavy metals, have a strong tendency to adsorb readily onto the solid phase, thus, considered to be virtually immobile. However, this solid phase, generally assumed to be immobile, could be mobile as colloidal or suspended particles in the aquifer. These particles can be assumed to be chemically similar to the surface of the aquifer material and hence may have a huge sorption capacity due to their large surface area. Thus, colloidal and suspended particle transport must be taken into consideration in describing and predicting the movement of substances. Particles are abundant in the subsurface environment. Particles are associated with geologic matrices or form due to geochemical reactions. The geochemistry and hydraulic boundary conditions of the aquifer determine the formation, mobilization and transport of particles. They can be mobilized by a hydraulic or geochemical gradient and reach a concentration up to 25 mg/l at the monitored test site "Insel Hengsen" located within the Ruhr valley, Germany.

Key words Colloids – particles – groundwater – artificial recharge – transport

Colloidal and suspended particles: significance during artificial groundwater recharge

Artificial groundwater recharge is a widespread technique to treat surface water for drinking water purposes and to increase the amount of the natural groundwater reserves. Infiltration can be done by wells, infiltration ponds, slow sand filters or the use of bankfiltrate. During the infiltration process colloidal and suspended particles and many contaminants are retained or reduced. This is due to several chemical, physical and biological processes along the infiltration and flow path of the water. Apart from real elimination many contaminants are held back by filtration. Filtration includes not only straining filtration but also physico-chemical collection processes.

In addition, hydrochemical changes induced by the artificial infiltration may result in a redox gradient in the aquifer or filter. This gradient can induce not only the formation of particles by precipitation but also the re-mobilization of particles and contaminants which were fixed in mineral coatings.

The elimination of organic and inorganic contaminants is hereby a function of the geochemical and biological aquatic conditions. Furthermore, the element specific characteristics and the hydraulic conditions of the infiltration plant must be taken into consideration.

Progr Colloid Polym Sci (1998) 111:184–188
© Steinkopff Verlag 1998

Many organic and biogene substances, e.g. polycyclic aromatic hydrocarbons, geosmine, toluene, m-xylene, have a strong tendency to adsorb to a solid phase (and also to particles, which are assumed to be chemically similar to the surface of the immobile aquifer material). Inorganic trace substances, like heavy metals and radionuclides can show the same tendency [1–5]. We observed at the "Insel Hengsen" test site significant amounts of iron, manganese, copper, lead, nickel, and cadmium, associated with particles or building up particles themselves [1]. The anoxic groundwater at our test site showed a level of humic substances up to 1.5 mg/l, which were 70% of the total DOC. Organic macromolecules like humic substances could enhance the transport of hydrophobic substances [6]. In the case of the PAH a mobilization was noticed by sorption onto bacteria [7]. A size-range effect was observed: very strong sorption occurred on small particles due to their large surface area and number [8, 9]. Thus,

particles may alter the transport of these contaminants during artificial groundwater recharge and must be taken into consideration.

Particles, which are retained due to filtration or have formed due to a geochemical gradient can build up depots in the subsurface environment. These depots, together with the adsorbed contaminants, can be mobilized by hydraulic changes which are frequent in an operating water work. During "shock pumping" tests in the observation wells at the "Insel Hengsen" test site – in order to simulate hydraulic changes in a water work we increased the pumping rate abruptly – we mobilized PAH to a concentrated level 10 times higher than the background value and chromium, copper and lead two to six times higher than the background value (i.e., the concentration of these substances after 24 h of careful pumping).

In this article macromolecules, microorganisms, and all mobile solids are referred to as particles. They cover

Fig. 1 Size range of suspended and colloidal particles, separation and detection techniques

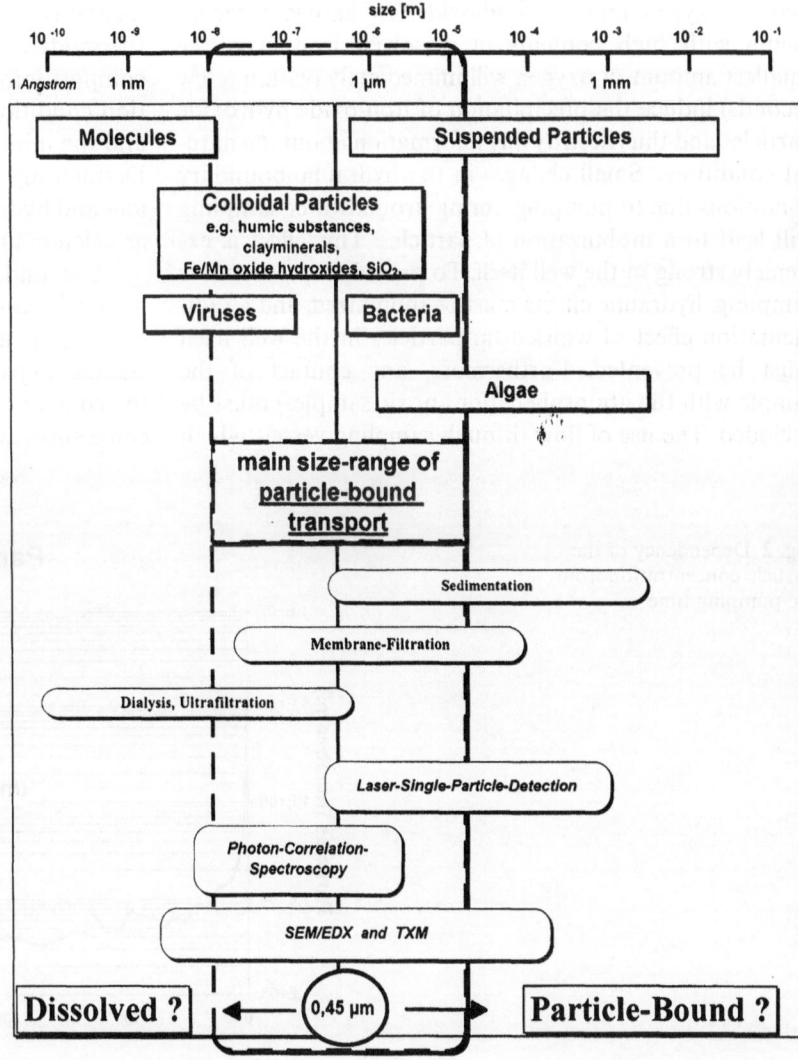

a size range from several nanometers up to 20 micrometers at the investigated test site "Insel Hengsen", located within the Ruhr valley. They could have colloidal or suspended characteristics (Fig. 1). Due to the wide range of sizes of particles, common filtration techniques in groundwater quality monitoring (i.e. filtration through 450 nm membrane filters) may exaggerate the content of dissolved elements and underestimate the total element content.

Sampling of particles

The understanding of processes in the aquifer is limited to our sampling techniques which are used to characterize them. Sampling is of special interest if one is interested in information about the natural particle content of the groundwater [3, 10]. In general, common sampling procedures are linked to changes in the hydrochemical boundary conditions, e.g. changes in the Eh, pH, temperature, exposure to daylight, and the content of dissolved gases, mainly oxygen and carbondioxide. In anoxic environments with high contents of dissolved iron even the smallest amount of oxygen will immediately (within a few seconds) induce the precipitation of iron oxide hydroxide particles and thus destroy any information about the natural conditions. Small changes in the hydraulic boundary conditions due to pumping during groundwater sampling will lead to a mobilization of particles. This effect is extremely strong in the well itself. To provide representative sampling, hydraulic effects must be minimized, and a sedimentation effect of whirled up particles in the well itself must be prevented. Furthermore, any contact of the sample with the atmosphere (for anoxic samples) must be excluded. The use of flow through sampling vessels which

can be preflushed with an inert gas is strongly recommended. Figure 2 shows the dependency of the particle number (larger 500 nm) from the pumping time in an observation well. A minimum of 5 h of purging was necessary to provide stable conditions in the monitored observation well.

Sample preparation and measurement needs the same precautions as groundwater sampling. Especially, the oxidation of anoxic samples must be prevented. Promising methods for size range detection lower than 1 μm are photon-correlation spectroscopy and the field-flow-fractionation technique [11].

Origin of particles in subsurface environments

In the groundwater environment many different particles could be expected. Besides the transport of particles from the vadose zone, the aquifer material itself can consist to a large part of small particles in the size range lower than 20 μm. These particles can be deposited during the sedimentation of the aquifer or during the alteration of thermodynamically unstable primary minerals [3]. The composition of these particles depends on the sedimentation conditions, the chemistry of the pore water solution and the mineral content of the original geologic material. Dominating inorganic particles are aluminosilicates, oxides and hydroxides of aluminium, iron, manganese as well as calcium and magnesium carbonates and silica [1, 7].

A second group of particles in the aquifer are formed by precipitation or mobilization as a result of a geochemical gradient along the flow path. Particles can form due to changes of parameters like pH, Eh, electric conductivity or the content of dissolved gases. Mineral phases, which become supersaturated along the flow path may precipitate

Fig. 2 Dependency of the particle concentration from the pumping time

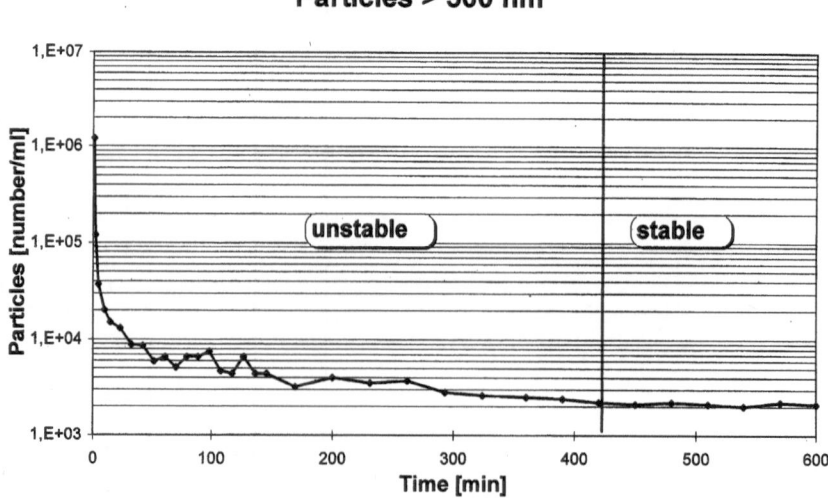

Particles > 500 nm

Progr Colloid Polym Sci (1998) 111:184–188
© Steinkopff Verlag 1998

and form particles. These particles can be initially very small – for example, we observed the formation of iron oxide hydroxide particles smaller than 50 nm during initial oxidation of previous anoxic water – but may grow and built up agglomerates. Changes in the hydrochemical environment are frequently observed using the technique of artificial groundwater recharge. Infiltration of river water, which is enriched in organic matter, could lead to anoxic conditions in the aquifer by the microbial reduction of O_2, NO_3^-, NO_2^-, MnO_2, Fe(III)-oxide, SO_4^{2-} and CO_2 [1]. The same effects are observed in contaminant plumes. Iron and manganese coatings of the aquifer material will be dissolved during this process and liberate formerly fixed particles and elements in these coatings. Nevertheless, the total content of iron and manganese, released by dissolution of these cements, is limited by the solubility of carbonates and sulfides. Thus, precipitation of iron and manganese carbonates and sulfides can also lead to the formation of new particles if supersaturation is reached. These particles can change their crystal structure during transport in the groundwater and modify their physical and chemical properties. For example, the transition of ferrihydrite to goethite leads to larger crystals and hence different sorption and transport properties [12].

Geochemical induced precipitation and mobilization can also occur during the mixing of two water types with a different carbon-dioxide content (mixing corrosion). Infiltration of water with a lower ionic strength – or when the ion balance of the water is shifted from dominating bivalent cations to monovalent cations – will lead to a swelling of the electric double layer and hereby induce a mobilization of particles.

Biogene substances belong to a third group of "particles". Organic macromolecules and humic substances can be released from the aquifer material or brought in by the infiltrated surface water. To the matrix attached bacteria may become unattached or can be infiltrated. They might release extracellular matter or cell fragments. Another rare but very important and conceivably dangerous group are spores and viruses. Most of these organic substances and micro-organisms have a strong tendency to adsorb readily to the solid phase and therefore also to inorganic particles or build up aggregates with them. This effect can give rise to serious problems during the disinfection of drinking water.

Transport, aggregation and filtration of particles

Suspended or colloidal particles can be mobile in the aquifer if they are not susceptible to filtration and if they are stable in respect to agglomeration and dissolution. Filtration effects can be either straining filtration or physico-chemical collection. While the maximum size of mobile particles is limited by straining filtration and pore velocity – for example, in the monitored test site "Insel Hengsen" 20 μm – concentration and size distribution of particles in the nanometer size range are more controlled by physico-chemical filtration. A minimum filtration rate can be observed at a size range close to 1 μm, where straining filtration is less effective in most pore aquifers and physico-chemical filtration does not yet reach its maximum [13].

The collection of particles on the aquifer material strongly depends on the surface charge. Opposite surface charge of the particle and the aquifer matrix will enhance collection. Heterogeneous charged particles favor coagulation and consequently sedimentation. This effect is even more complicated regarding clay minerals, because the edges and faces may show opposite charge. A charge reversal of particles can occur due to adsorption processes

Fig. 3 Charge reversal of ferrihydrite particles by adsorption of humic substances

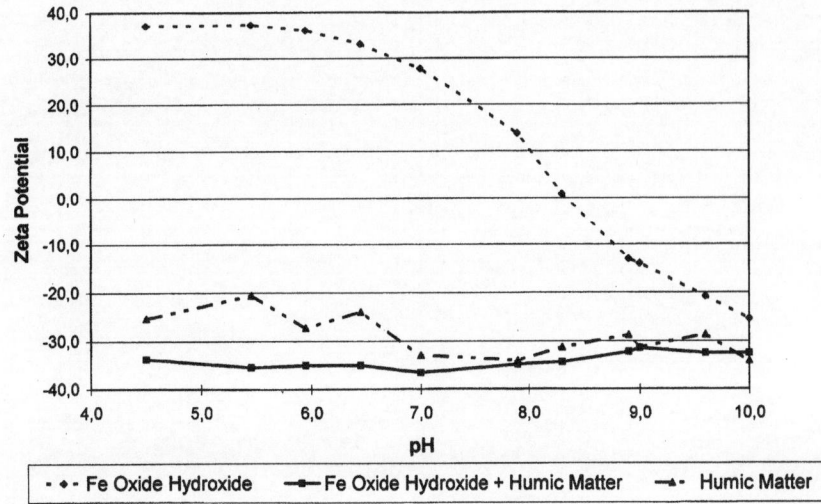

of polyelectrolytic substances. Therefore, the adsorption of humic substances to positive charged particles like aluminosilicates, calcium carbonates, and oxides can cause a charge reversal and facilitate the transport (Fig. 3).

The effects dominating these processes are a complex function of the surface charge, density, size, size distribution, surface and water chemistry and are difficult to predict [3, 4, 7].

Conclusions

In our current work we assume that many substances have a strong tendency to adsorb to solid constituents and generally are considered to be retarded or immobile. Suspended and colloidal particles are expected to be chemically cognate to the surface of the immobile aquifer material. Due to their large surface area they can have an enormous sorption potential. To understand transport phenomena and secure drinking water quality – especially operating artificial groundwater recharge plants where hydraulic and chemical variations can be frequent – particle bound transport has to be taken into consideration. Particle linked transport of substances cannot be investigated with common filtration techniques applied for drinking water analyses (i.e., filtration through 450 nm membranes). The numerous physico-chemical, biological and hydraulic mechanisms which influence suspended and colloidal particle transport in complex natural environments are as of now not fully understood.

Acknowledgements This project is supported by the EC Environmental and Climate Research Programme (ENV4-CT95-0071) and the DVGW (Deutscher Verein des Gas und Waserfaches e.V.).

References

1. Schöttler U, Schulte-Ebbert U (1995) Schadstoffe im Grundwasser – Band III. Verhalten von Schadstoffen bei der Infiltration von Oberflächenwasser am Beispiel des Untersuchungsgebietes "Insel Hengsen" im Ruhrtal bei Schwerte. Deutsche Forschungsgemeinschaft, Bonn
2. Kögel-Knabner I, Maxin C, Totsche K, Danzer J (1993) Mitteilungen der Deutschen Bodenkundlichen Gesellschaft 71:253–256
3. McCarthy JF, Zachara JM (1989) Environ Sci Technol 23:496–502
4. Stumm W (1987) Aquatic Surface Chemistry. Wiley, New York
5. Backhus DA, Gschwend PM (1990) Environ Sci Technol 24:1214–1223
6. Abdul AS, Gibson TL, Rai DN (1990) Environ Sci Technol 24:328–333
7. Jenkins MB, Lion LW (1993) Appl Environ Microbiol 59:3306–3313
8. Hellman H (1991) Deutsche Gewässerkundliche Mitteilungen 35:46–52
9. Ledin A (1993) PhD thesis, Linköping Studies in Arts and Science, 91, Linköping
10. US Department of Energy (1988) Role of Colloidal Particles in the Subsurface Transport of Contaminants. Subsurface Science Program, DOE/ER-0384
11. Barth HG, Flippen RB (1995) Anal Chem 67:257R–272R
12. Cornell RM, Schwertmann U (1996) The Iron Oxides. Structure, Properties, Reactions, Occurence and Uses. VCH, Weinheim
13. Bedbur E (1993) Laboruntersuchungen zum Einfluß sedimentologischer und hydraulischer Parameter auf die Filterwirkung gleichförmiger Sande. Dissertation, Christian-Albrechts-Universität, Kiel

Progr Colloid Polym Sci (1998) 111:189–192
© Steinkopff Verlag 1998

A. Eckelhoff
A.V. Hirner

On the influence of surfactants on the mobility of contaminants

A. Eckelhoff · Prof. A.V. Hirner (✉)
Institute of Environmental
Analytical Chemistry
University of Essen
Universitätsstraße 3-5
D-45141 Essen
Germany

Abstract The adsorption behavior of Orthic Luvisol and the solubilizing potential of three different products of sodium dodecyl sulfate (SDS) was examined in batch experiments. Two PCBs were used as representative pollutants for persistent, hydrophobic substances. For the soil surface layer, adsorption occurs at the hydrophobic tail of the surfactant, whereas for the other horizons adsorption is located at the hydrophilic head group. These two different adsorption mechanisms of dodecyl sulfate (DS) to the surface lead to the appearance of an adsorption maximum in the adsorption isotherms for the lower soil, which is caused by the formation of a double layer. Behind the maximum, the double layer is destroyed by the transition of the surfactant aggregates in solution. The surfactant employed in these experiments mobilizes the hydrophobic PCBs better than water. Below the cmc (critical micelle concentration) mobilization of the PCBs is caused by surfactant monomers due to detergent effect, and above the cmc by insertion in surfactant aggregates. The vesicle-like droplets of a dispersed lamellar phase are able to solubilize more PCB than spherical micelles, which are formed as soon as the double cmc in solution is achieved.

Key words Mobilization – anionic surfactants – PCB – solubilization – adsorption – SDS

Introduction

Today environmental research, national as well as international, focuses intensively on the problem of the global distribution of contaminants like PCB and PAH. The environmental relevance of this pollutants is justified by their toxic, mutageneous and persistent properties [1]. Determination of the complete contaminant amount is not important for the systematic assessment of their toxicity, but rather the one of the mobile part. In aqueous systems the mobilization of hydrophilic substances is appointed by their solubility in water, whereas the one of hydrophobic substances is influenced by the presence of surfactants as amphiphilic solubilizing molecules.

In many applications surfactants, e.g. detergents, are used so they are wide spread in the environment. They include many natural substances like fatty acids, fulvic acids, proteins, humic acids or phospholipids. In principle it is conceivable that natural surfactants, formed by decomposing straw, appear in concentrations above the cmc which then are able to transport the contaminants through soil to ground water [2].

This work is devoted to the systematic analysis of the influence of surfactants on the mobility of persistent contaminants, and is especially aimed at defining the underlying mechanisms. The experiments are carried out with the surfactant sodium dodecyl sulfate as a model substance for natural surface active agents since it has already been shown that this surfactant mobilizes hydrophobic organic pollutants to a degree which is comparable to natural solvents [3, 4].

Materials and methods

In batch experiments, according to the S4 – elution (DIN 38414 [5]), the adsorption behavior of each horizon of Orthic Luvisol and the solubilizing ability of the surfactant was analyzed. For this purpose the contaminated soil samples were eluted over a period of 24 h by different surfactant concentrations. The chosen soil/liquid ratio was 1:10. Subsequently, the adsorbed amount of dodecyl sulfate in soil and the content of the PCBs in the equilibrium solution were examined [6].

The most important parameters of the Orthic Luvisol are arranged in Table 1. As hydrophobic contaminants 3,3',4,4'-tetrachlorobiphenyl (PCB 77) and 2,3,3',4,5-penta-chlorobiphenyl (PCB 106) were chosen. Their water solubility is in the range of 29–52.5 μg/l [7]. Three different products of the surfactant sodium dodecyl sulfate were used. These were SDS99% which consists of 99% dodecyl sulfate and contains no other homologues; SDS95%, which is a mixture of 67% dodecyl sulfate, 27% myristyl sulfate and 5% cetyl sulfate as well as the technical product texapon, produced by Henkel. This contains 35% dodecyl sulfate and 30% dodecanol, the remainders are water and sodium sulfate. Table 2 shows the cmc of each product measured by surface tension experiments.

Results

Figures 1–3 illustrate the adsorption behavior of the three products for the different horizons. DS adsorption grows

Table 1 Soil parameters of each horizon

Horizon (HZ)	Soil type	C_{org} [%]	KAK [mmol(z)/kg]	A_s [m²/g]
HZ1/A_h	Ut3	1.8	79.2	10.9
HZ2/A_l	Ut4	0.5	95.6	19.7
HZ3/B_t	Ut4	0.4	90.0	25.5
HZ4/$C(B_v)$	Ut4	0.3	84.8	25.5

A_s = specific surface.

Table 2 cmc of the different SDS products

SDS-product	cmc [mg/l]
SDS99%	1772
SDS95%	698
Texapon	746

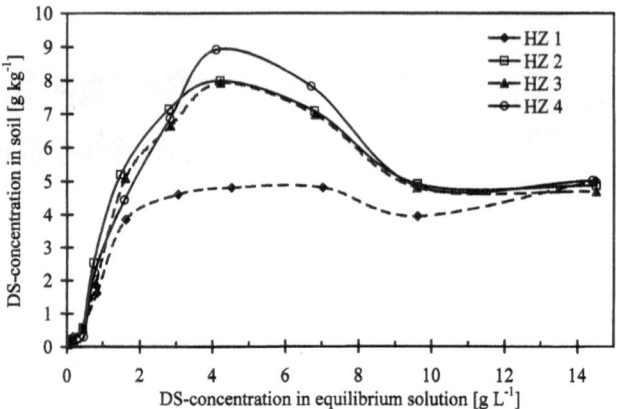

Fig. 1 Adsorption isotherms of DS for SDS99%

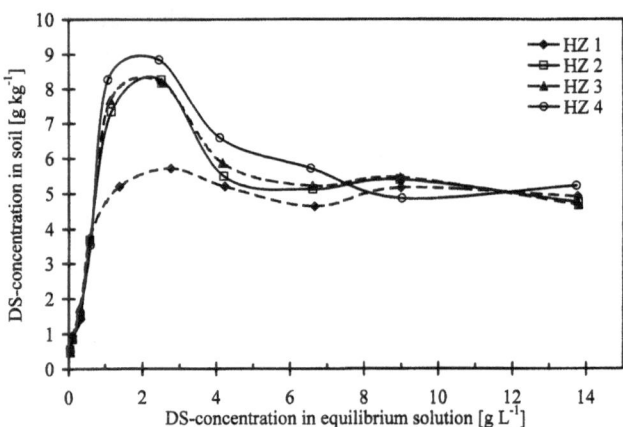

Fig. 2 Adsorption isotherms of DS for SDS95%

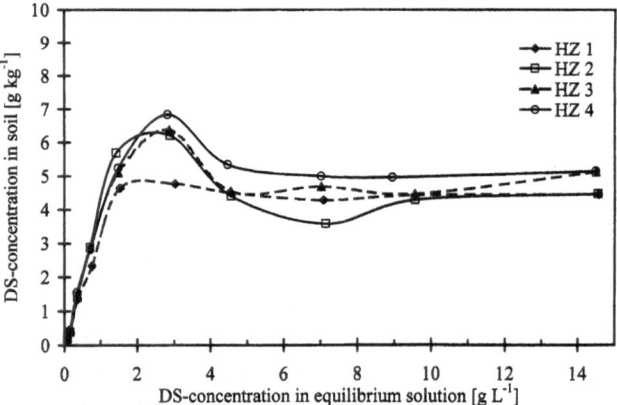

Fig. 3 Adsorption isotherms of DS for Texapon

as surfactant concentration increases in any of the soil layers. For the horizons 2–4 a maximum can be observed, which does not exist for horizon 1. Qualitatively, all three products show the same plots. Quantitatively, they are

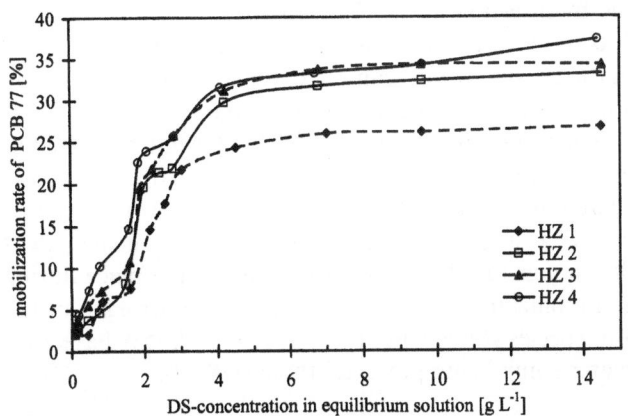

Fig. 4 Solubilization of PCB 77 by SDS99%

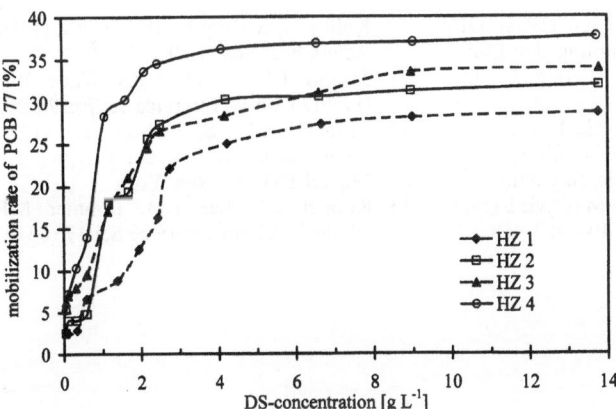

Fig. 5 Solubilization of PCB 77 by SDS95%

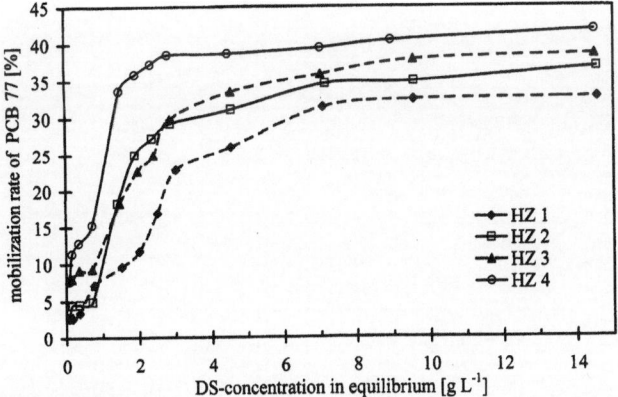

Fig. 6 Solubilization of PCB 77 by Texapon

different in the adsorbed amount of DS, whereas SDS99% and SDS95% show roughly identical results. For texapon, however, the adsorbed DS is evidently lower, because the available alcohol also adsorbs. In addition, maxima ap-

pear at different DS-concentrations in the equilibrium solution being due to the cmc of different products. A decreasing cmc lowers the equilibrium concentration of SDS in the maximum of the adsorption isotherm.

Figures 4–6 show the solubilizing power of each SDS-product for PCB 77. The mobilization rate of PCB 77 increases slowly with increasing surfactant concentration. When the cmc is achieved, an evident growth is noticeable. When the double cmc is reached in solution, the mobilization rate changes insignificantly. The trend of the solubilizing plots is identical for all three products. Analogous curves were obtained for PCB 106.

Discussion

The adsorption results show that the upper soil fraction close to the surface exhibit different behavior than do the lower soil fractions. The adsorption isotherm of the surface sample approximately follows Langmuir behavior, while lower samples show a maximum in the isotherm plot. In order to explain the underlying mechanisms, model experiments were carried out simultaneously with the dye Sudan Black, and the most important cations in the soil as well as the solubility of the pollutants in the surfactant were determined [6]. Based on our experiments, the following adsorption mechanism for DS were established: for the soil surface layer, the adsorption occurs at the hydrophobic tail of the surfactant, whereas for the other compounds adsorption is located at the hydrophilic head group. Due to lateral interactions between the adsorbed surfactant molecules, DS adsorption grows as surfactant concentration increases in all of the soil layers. Under these conditions of increasing DS adsorption, the isotherm gradient becomes higher due to the formation of hemimicelles. This adsorption mechanisms correspond to the different soil parameters of each horizon. The first horizon has got the highest organic carbon concentration, whereas the other horizons contain less organic carbon in similar amounts (see Table 1).

When the cmc of each SDS product is reached, the isotherm shows inflection of the curve at the point where the surfactant forms a monolayer on the soil. In solution the surfactant molecules compose vesicle-like droplets of a dispersed lamellar phase with dodecanol, which is always present by the hydrolysis of DS. Following this, the lower soil fractions (2–4) form a double layer resulting from the hydrophobic interactions between the alkyl chains of the surfactant. When the double cmc in solution is reached, this double layer is destroyed by the transition of the surfactant aggregates and this causes the formation of a maximum in the adsorption isotherms. These transitions are characteristic of SDS [8].

The maximum in isotherm plots has been explained in the literature by the precipitation of low soluble salts, the adsorption of homologous in technical-grade surfactant products, and artifacts resulting from the measurement method [9–11]. Based on our observations, these reasons cannot explain the isotherm maxima. However, the four region model, as described elsewhere [12, 13], explains the observed behavior best.

The surfactants employed in this study mobilize the hydrophobic PCBs better than water. Below the cmc, solubilization of the PCBs is caused by surfactant monomers due to the hydrophobic interactions, and above the cmc by insertion in micelles. The observed solubilization plots agree with the results of the adsorption isotherms. Characteristic points can be detected in the adsorption plots as well as in the mobilization plots.

The results also show, that PCB 77 is not mobilized as well as PCB 106. This can be explained with the slightly different molecular structure of the two PCBs in question. PCB 77 is adsorbed more strongly to the soil than PCB 106. In general, the mechanisms which take place during the mobilization process are comparable for each of the three species studied, but texapon and SDS95% have a far greater solubilizing potential than SDS99%.

References

1. Wania F, Mackay D (1996) Environ Sci Technol 30:390–396
2. Wirsig G (1992) Umweltmagazin 3:56–57
3. Busche U, Hirner AV (1997) Acta Hydrochim Hydrobiol 25:248–252
4. Pestke FM, Bergmann C, Rentrop B, Maaßen H, Hirner AV (1997) Acta Hydrochim Hydrobiol 25:242–247
4. Deutscher Normausschuß (1994) DIN 38414, Teil 4: Bestimmung der Eluierbarkeit mit Wasser. Verlag Chemie, Weinheim
6. Eckelhoff A (1997) Thesis, Essen
7. Maaß V (1987) Thesis, Hamburg
8. Struller B (1991) Thesis, Bayreuth
9. Kallay N, Pastuovic M, Matijevic E (1985) J Colloid Interface Sci 106: 452–458
10. Kallay N, Xi-Jing F, Matijevic E (1986) Acta Chem Scand 40:257–260
11. Trogus FJ, Schecher RS Wade WH (1979) J Colloid Interface Sci 70:293–305
12. Christian SD, Scamehorn JF (1995) Solubilization in surfactant aggregates. Marcel Dekker, New York
13. Koopel LK, Lee EM, Böhmer MR (1995) J Colloid Interface Sci 170:85–97

Progr Colloid Polym Sci (1998) 111:193–201
© Steinkopff Verlag 1998

J. Thieme
J. Niemeyer

Interaction of colloidal soil particles, humic substances and cationic detergents studied by X-ray microscopy

Dr. J. Thieme (✉)
Forschungseinrichtung Röntgenphysik
Geiststraße 11
D-37073 Göttingen
Germany

J. Niemeyer
Fachbereich VI Geowissenschaften
Universität Trier
Trier
Germany

Abstract The interaction of soil colloids with humic substances and cationic detergents has been studied by X-ray microscopy. The major advantages of X-ray microscopy for these studies are the much higher resolution than achievable with light microscopy and the ability to image colloidal structures directly in aqueous media. Of particular interest was, how the coagulation force of cationic detergents will change the structure formed by the soil colloids when humic substances are present. The humic substances form a network-like structure around the soil colloids, thus increasing tremendously the surface available for chemical and physical reactions.

Therefore, much more cationic detergent was necessary to introduce a coagulation of the structures formed by the soil colloids. In addition, humic substances and cationic detergents coagulate forming sphere shaped structures, where size, density and state of aggregation is a function of the concentration of the detergents. X-ray microscopic images give evidence for these results.

Key words X-ray microscopy – colloidal soil particles – humic substances – detergents

Introduction

X-ray microscopy is able to image particles and structures with colloidal dimensions directly in their aqueous environment with approximately a tenfold better resolution than with light microscopy. This can be done without preparational steps as drying or staining [1]. The capabilities of X-ray microscopy have been demonstrated with a great variety of colloidal dispersions [2–4]. Therefore, it is a unique technique for the investigation of the structure formed by soil colloids in aqueous media and the investigation of the interaction of these colloids with humic substances and detergents. As humic substances are colloidal particles themselves, the interaction of these colloids with surfactants can also be observed.

Methods

X-rays within the wavelength range between the K-absorption edges of oxygen at $\lambda = 2.34$ nm and carbon at $\lambda = 4.38$ nm are very well suited for X-ray microscopy studies of aqueous colloidal systems [5]. The photoelectric absorption and phase shift are the two dominating processes of interaction of X-rays with matter. The radiation is weakly absorbed by water but strongly absorbed by iron oxides, silicates, organic matter, etc. resulting in a good amplitude contrast of objects in aqueous environments. These differences are even larger when looking at the phase shift of X-rays penetrating water or other materials [6]. The graph in Fig. 1 shows the linear absorption coefficient of water, the phyllosilicate smectite, and the

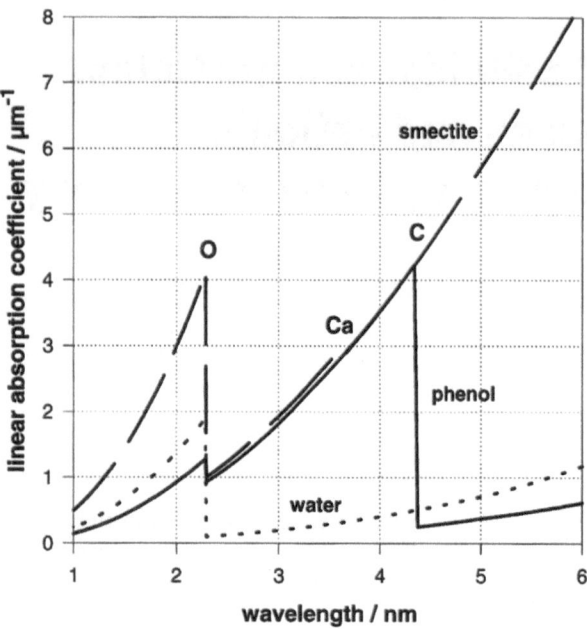

Fig. 1 Linear absorption coefficient of water, smectite and phenol as function of wavelength in the soft X-ray wavelength region

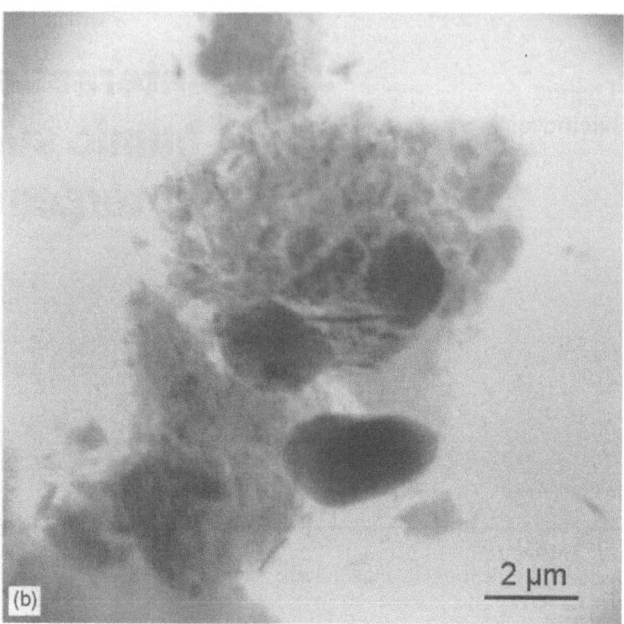

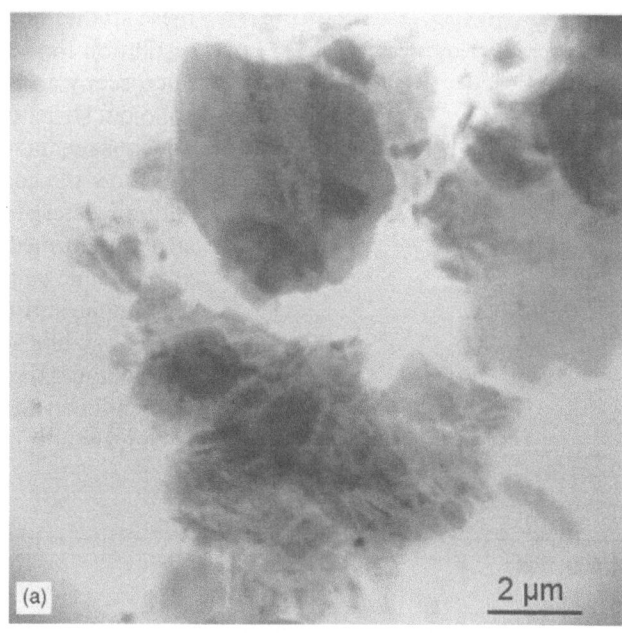

Fig. 2 X-ray micrographs of colloidal particles and aggregates from the A_H-horizon of a CAV in Rosdorf near Göttingen

organic molecule phenol. The difference in absorption between the particles consisting of mineral or organic material and the water gives rise to amplitude contrast in the X-ray microscopic images shown in this paper. Thus, it is possible with an X-ray microscope to image these objects directly in aqueous media. Details of the X-ray microscope used for the experiments described here are found elsewhere [7].

The soil samples used in the work presented here are colloidal particles from the A_H-horizon of a calcareous aquic vermudoll (CAV) in Rosdorf near Göttingen (FAO: Calcaric Phaeozem) [8]. In all the experiments a 1% dispersion (weight by weight) in deionized water has been used.

Humic substances are anionic polyeletrolytes. They were extracted from this A_H-horizon following the procedure given by Stevenson [9]. Prior to the extraction of the humic substances, proteins and carbohydrates were removed by hydrolysis of the soil sample (50 g) with 500 ml 2 M hydrochloric acid for 6 h. After five times washing with distilled water and centrifugation, this sample was extracted with 500 ml 0.01 M sodiumpyrophosphate solution, adjusted to pH 7, in an overhead shaker for 24 h. After centrifugation and removal of the soil sample, 6 M hydrochloric acid was added to the supernatant until the formation of precipitate occurred. The humic substances were obtained by another centrifugation step followed by six washing and centrifugation cycles. Thereafter the humic substances were freeze-dried and stored in an ex-

cicator. The humic substances have been used in a 1% dispersion in bidistilled water.

The detergents used for the described experiments are Dodecyltrimethylammoniumbromide (DTB) and Hexadecyltrimethylammoniumbromide (CTB). Both detergents have been used in a 1% solution in bidistilled water.

Progr Colloid Polym Sci (1998) 111:193–201
© Steinkopff Verlag 1998

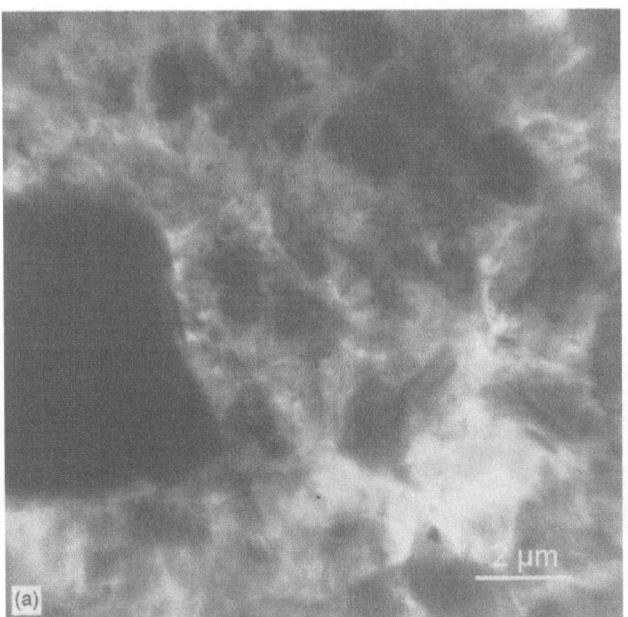

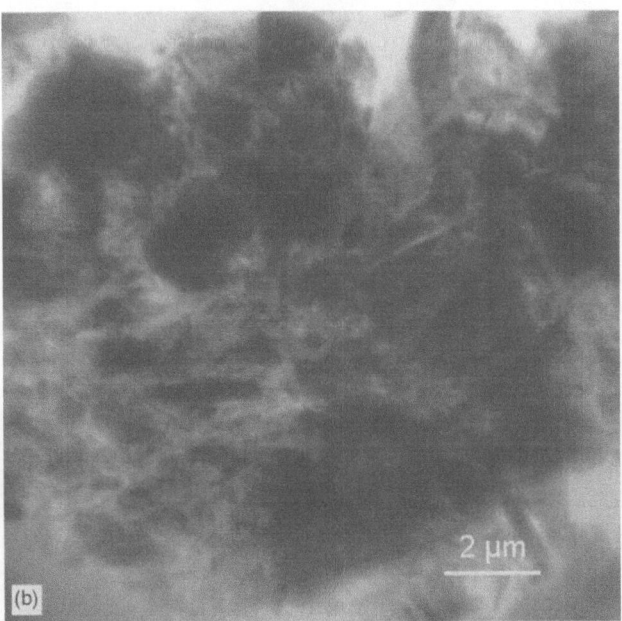

Fig. 3 Aggregates of colloidal particles from the A_H-horizon of a CAV in Rosdorf near Göttingen after the addition of the cationic detergent DTB, imaged with an X-ray microscope

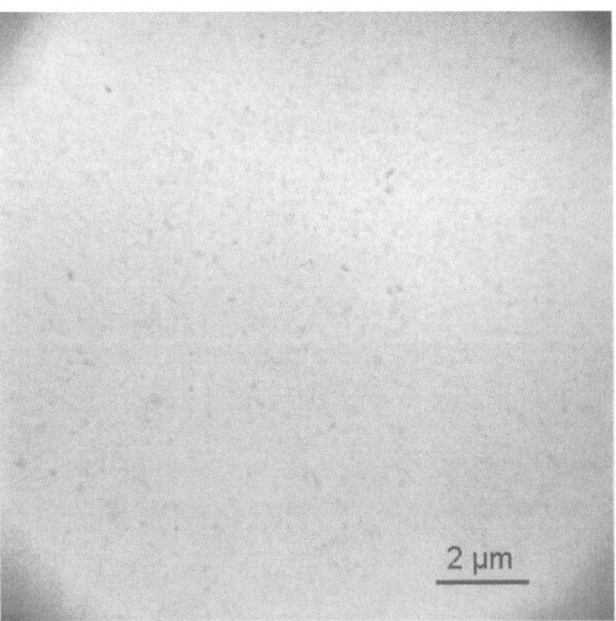

Fig. 4 X-ray micrograph of a 1% dispersion of humic substances in bidistilled water

Results

To study the interaction of colloidal soil particles with humic substances and with detergents, first of all the structure formed by soil colloids in aqueous media was investigated. The experiments presented in this paper have been performed with colloidal particles derived from CAV as

can be seen in Fig. 2a and b. The appearance of the particles and the small aggregates shown in these X-ray micrographs are typical. The particles are in loose contact, the aggregates show an open structure.

In the next step, 10 μl of a 1% solution of DTB has been added to a 1 ml aliquot of a 1% dispersion of CAV. Changes in the structure formed by the colloidal aggregates could already be seen with the naked eye. The aggregates appeared to be much larger. In the X-ray microscope the colloids revealed a much denser appearance, the formerly open structure was now much more compact, as shown in Fig. 3a and b. The cationic detergent DTB acts as a coagulating agent, which contracts the formerly open inner structure of the colloidal aggregates and, in addition, increases the size of the aggregates due to its coagulating force.

In a second experiment, 50 μl of this 1% solution of DTB was added to another 1 ml aliquot of the dispersion. Again the aggregates appeared much larger, even larger and thicker as in the first experiment, when observed with the naked eye. However, with the X-ray microscope no significant difference in the appearance of the inner structure could be seen. An increase in compactation of the particles may not be possible anymore, so that the excess of DTB only leads to larger and thicker aggregates.

The humic substances are colloidal particles themselves, so before investigation the interaction with other colloids, a 1% dispersion of these particles has been imaged in the X-ray microscope. Figure 4 shows an X-ray

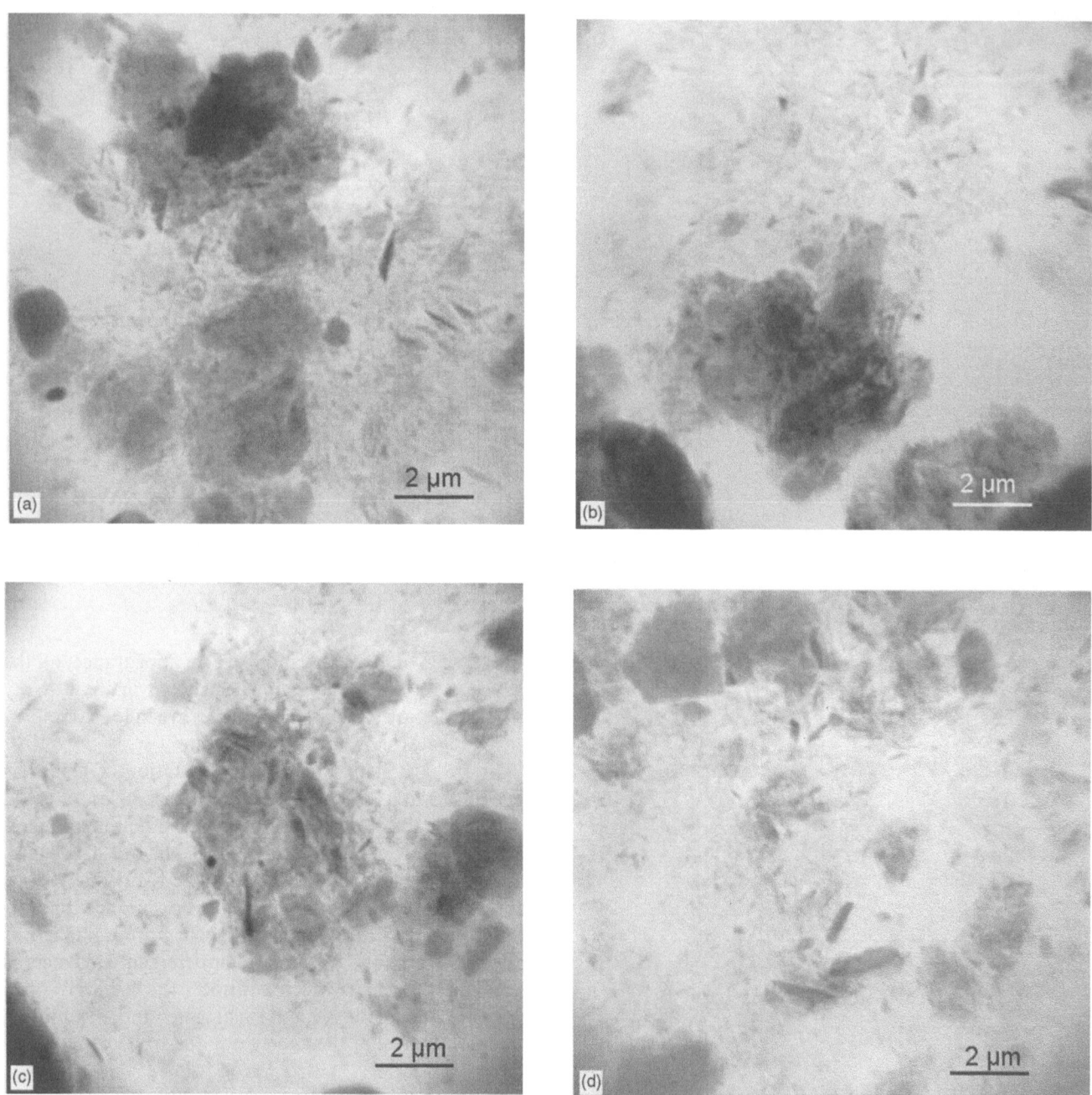

Fig. 5 X-ray micrographs visualizing the interaction of soil colloids with humic substances forming a network like structure

micrograph of this dispersion, showing the single colloidal humic particles.

To visualize the interaction of humic substances with the CAV, 50 μl of a 1% dispersion of humic substances has been added to a 1 ml aliquot of the 1% CAV dispersion. A network-like structure appeared in the X-ray micrographs, connecting single soil particles and extending from small aggregates. The differences in the original soil dispersion can be seen very well comparing Figs. 2b and 5a, where colloidal aggregates of similar shape were imaged. In Fig. 5c and d the entangling of small soil particles within the network structure of the humic substances is clearly visible. The addition of 100 μl of the 1% dispersion of humic substances to an 1 ml aliquot of CAV has led to an increase in the size of the network. It extended over a larger area, but did not show a change in shape, the

Progr Colloid Polym Sci (1998) 111:193–201
© Steinkopff Verlag 1998

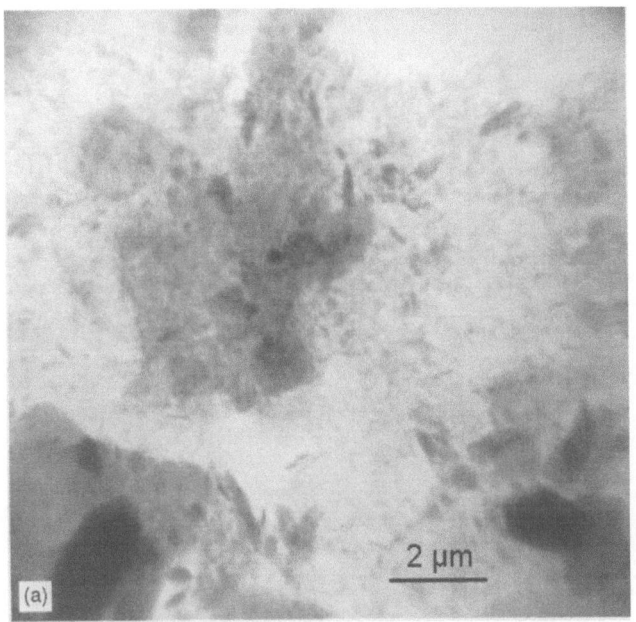

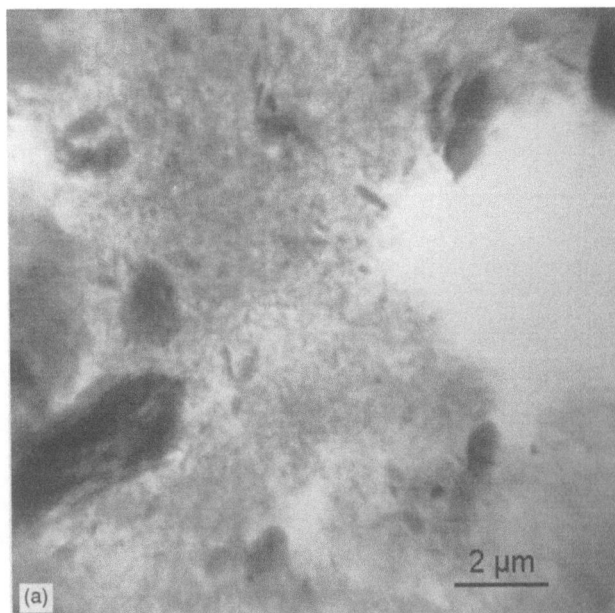

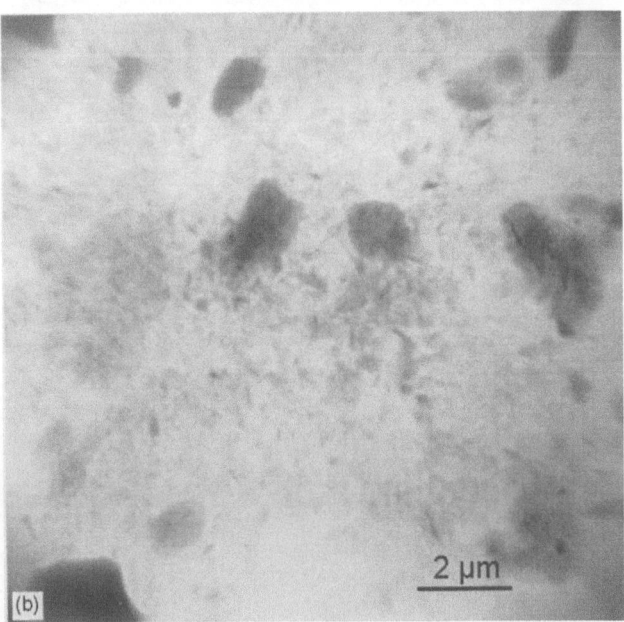

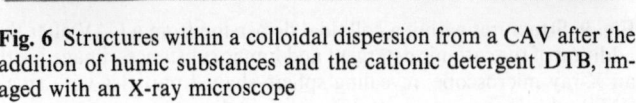

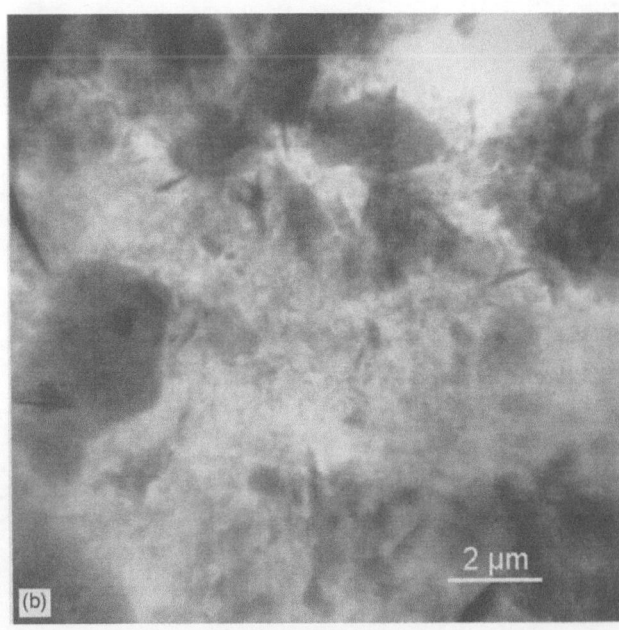

Fig. 6 Structures within a colloidal dispersion from a CAV after the addition of humic substances and the cationic detergent DTB, imaged with an X-ray microscope

Fig. 7 Structures within a colloidal dispersion from a CAV after the addition of humic substances and increased amounts of cationic detergent DTB, imaged with an X-ray microscope, revealing denser structures

density of the network was the same. In both experiments no influence was seen on the inner structure of the other colloidal aggregates.

The next step in this sequence of experiments was to introduce both, humic substances and cationic detergents, to the soil sample. For this, to a 1 ml aliquot of the 1%

dispersion of CAV 50 μl of the 1% dispersion of humic substances was added and after that 10 μl of the 1% DTB solution. The X-ray images showed no difference in appearance compared to Fig. 5. No compaction of the soil particles occurred. The reason might be, that the very large inner surface of the network of humic substances adsorbed

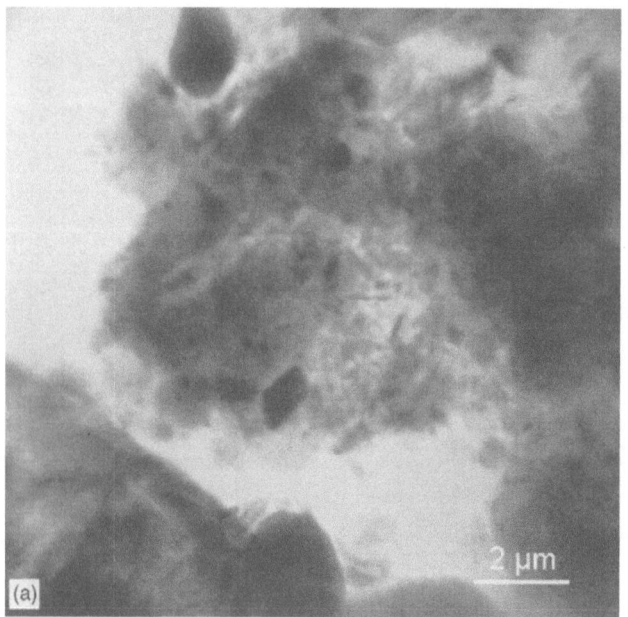

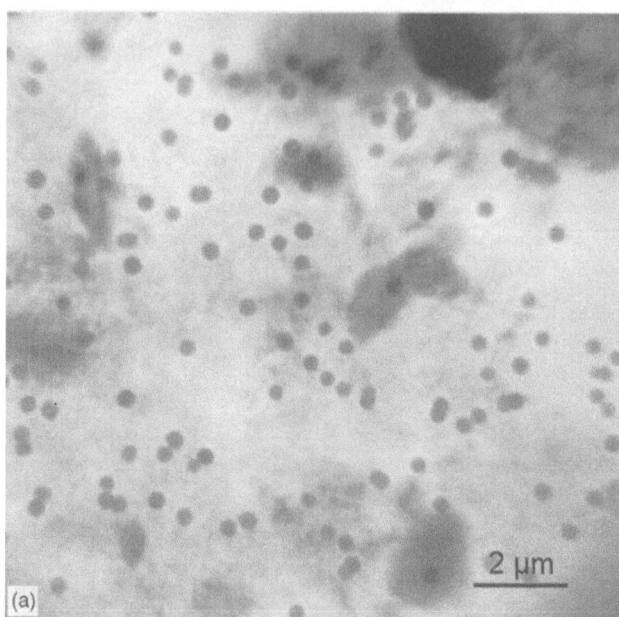

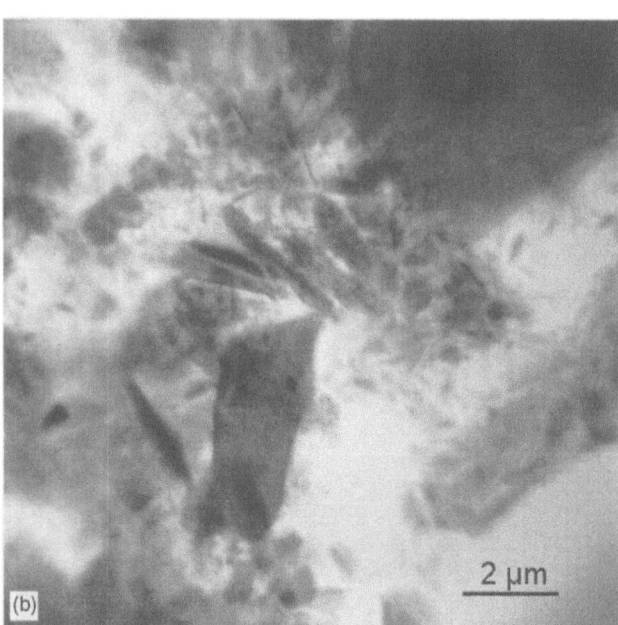

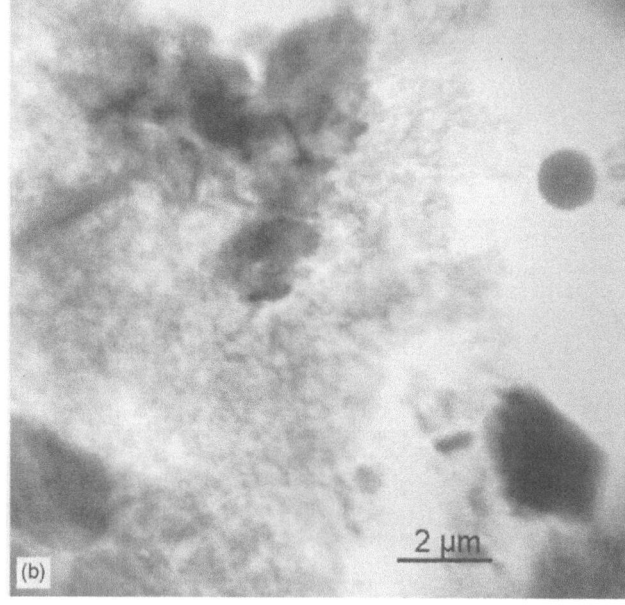

Fig. 8 Aggregates of colloidal particles from the A_H-horizon of a CAV in Rosdorf near Göttingen after the addition of the cationic detergent CTB, imaged with an X-ray microscope

Fig. 9 Structures within a colloidal dispersion from a CAV after the addition of the cationic detergent and humic substances, imaged with an X-ray microscope, revealing sphere-shaped particles with small (DTB-addition) and large (CTB-addition) diameters

all the detergents and none of this amount of DTB could interact with the CAV particles. This result can be seen in Fig. 6a and b.

Therefore, the amount of DTB has been increased in the next step. After the addition of $50\,\mu l$ of the 1% dispersion of humic substances $50\,\mu l$ of the 1% solution of DTB was added. Now the network structure of humic

substances seemed not to be able to adsorb all the detergents. Compactation of the soil particles occurred similar to that shown in Fig. 3. The coagulation force of the cationic detergent now was large enough to influence also the shape of the network. It appeared much denser than seen in Figs. 5 or 6. Both, the compactation and the increased density of the network is shown in Fig. 7a and b.

Progr Colloid Polym Sci (1998) 111: 193–201
© Steinkopff Verlag 1998

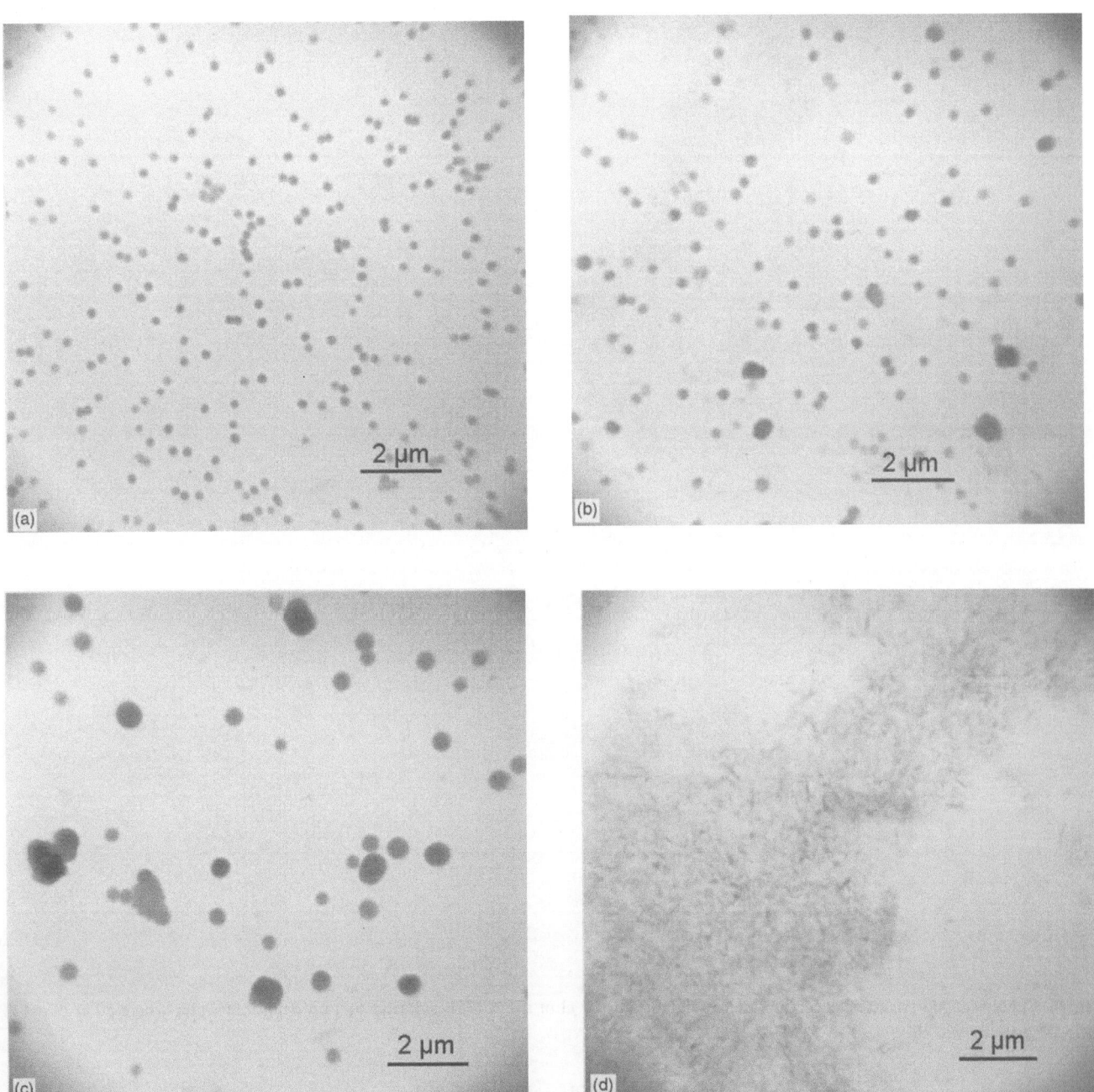

Fig. 10 Sphere-shaped particles, occurring after the addition of 1 μl (a) 5 μl (b) and 7 μl (c) of a 1% DTB solution to 1 ml deionized water where 50 μl of a 1% dispersion of humic substances was added, and a network-like structure appearing after the addition of 50 μl DTB

These experiments have been repeated with another cationic detergent, the CTB. During the preparation it was already clear, that CTB showed a much stronger influence as DTB on the structure of the soil colloids. Therefore, to a 1 ml aliquot of the CAV dispersion 1 μl of a 1% solution of CTB was added. This very small amount compared to DTB showed already the same influence on the visualized

structures. As can be seen in Fig. 8a and b, compactation of the aggregates took place, the particles appear much denser.

Again, humic substances and cationic detergent were added to the CAV sample. To the 1 ml aliquot of the CAV dispersion 50 μl of the humic substance dispersion was added, after which 1 μl of the 1% CTB solution was added. Network-like structures between the colloidal particles

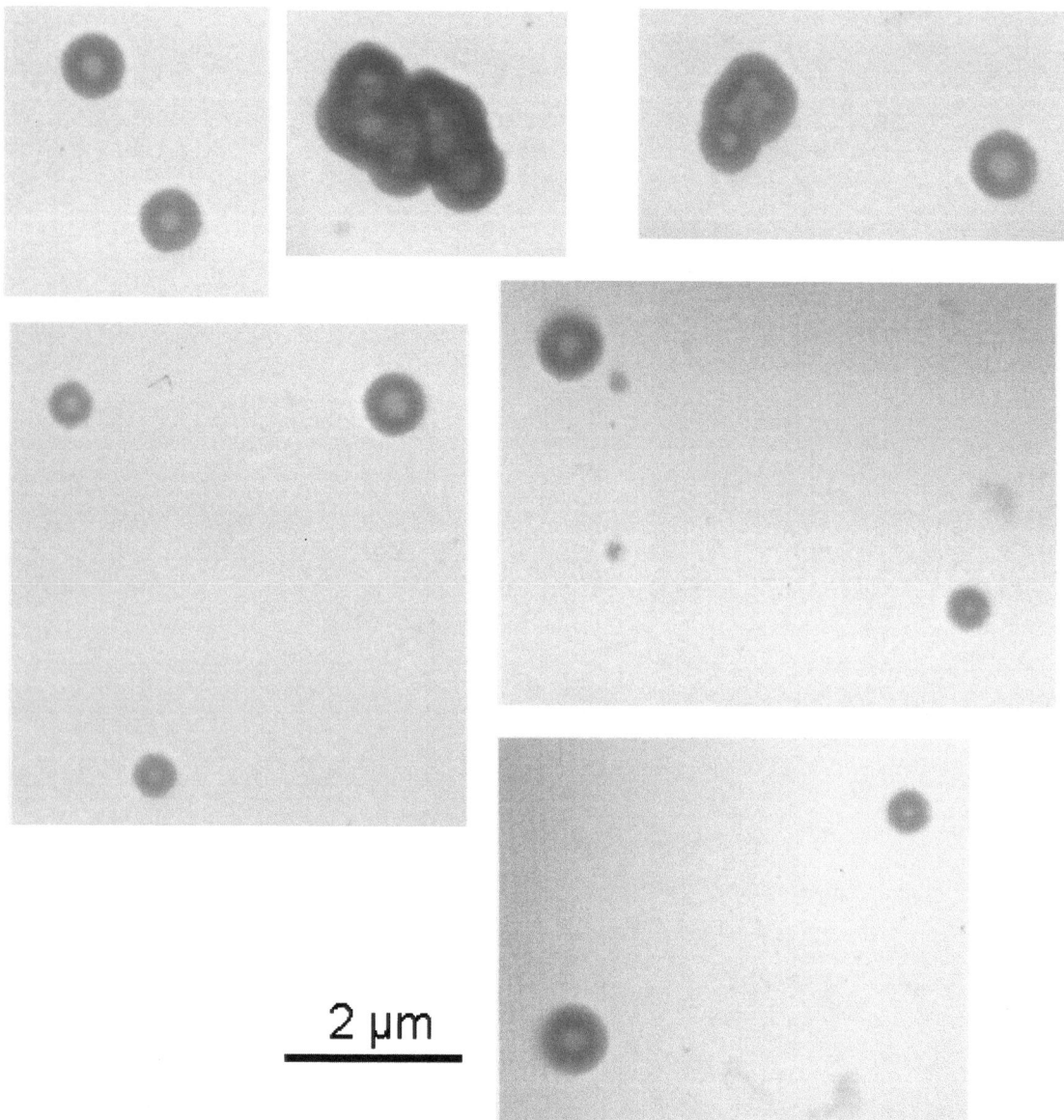

Fig. 11 Sphere-shaped particles, occurring after the addition of 1 μl of a 1% CTB solution to 1 ml deionized water where 50 μl of a 1% dispersion of humic substances was added

appeared, which were very similar to the structures imaged in Figs. 5 and 6. The aggregates were very large. However, the internal structure was comparable to that, when no detergent or small amounts of CTB were added.

As it seemed to be important for the appearance of the colloidal structure visualized by X-ray microscopy whether the cationic detergent reaches the colloids of the CAV or not, another experiment has been performed. To a 1 ml aliquot of the CAV dispersion 1 μl of the 1% solution of CTB has been added first before adding 50 μl of the 1% humic substance dispersion. Again, the network appeared in the images and the soil colloids appeared to be more compact. In addition to that small spheres appeared,

all nearly of the same size, see Fig. 9a. This experiment has been repeated several times to ensure that artifacts are avoided. The diameter of the spheres have been measured to $d = 360 \pm 22$ nm. A similar experiment was performed using DTB. To 1 ml of the CAV dispersion 50 μl of the 1% DTB solution was added before the addition of 50 μl of the 1% humic dispersion. A compactation of the soil colloids is visible and a dense network, comparable to Fig. 7. In addition, single large spheres appeared on the X-ray images. These spheres seem to be the result of the interaction of an excess of the cationic detergent and the humic substances. The spheres can be seen in Fig. 9a (CTB-addition) and b (DTB-addition).

Progr Colloid Polym Sci (1998) 111:193–201
© Steinkopff Verlag 1998

In order to study the interaction of humic substances and the cationic detergents in more detail, the next experiments have been performed without the CAV. To a 1 ml aliquot of deionised water 50 μl of the humic dispersion has been added. First, 1 μl of the 1% DTB solution has been added to that system. Small spheres appeared, showing a very small distribution of diameters, $d = 224 \pm 33$ nm as shown in Fig. 10a. An addition of 5 μl DTB to a 1 ml aliquot of deionized water, where 50 μl of humic dispersion has been added, has lead to an increase in the diameters of the spheres, to an increase in the diameter distribution and to the appearance of some aggregates, as seen in Fig. 10b. The particles showed diameters of 280–440 nm. A further increase, i.e. an addition of 7 μl, yielded larger particles with diameters from 370 to 590 nm and more aggregates, compare Fig. 10c. The sphere-shaped particles are not seen anymore when more than the amount of 7 μl of the 1% DTB solution was added to that system. The concentration of cationic detergent then seems to be too high for the creation of these particles. Instead, the network-like structure appears again, increasing in size with increasing amounts of DTB. Figure 10d shows a large network after the addition of 50 μl of the 1% DTB solution.

As already seen in the interaction with soil colloids, CTB showed again a much stronger effect as DTB. Here, the addition of 1 μl CTB to the system of 1 ml deionized water and 50 μl of the 1% humic dispersion showed much larger particles than 1 μl DTB could create. Furthermore, the spheres appeared to be much denser. The inside of the spheres showed a lesser absorption. This is an indication of a hollow sphere, where water or solution with a much lesser content of humic substances and detergents is inside. Very few particles, but large and dense, were found. Nearly all of them were aggregated in clusters. Figure 11 shows several of these particles, where the hollow centres can be seen.

Summary

The interaction of soil colloids with cationic detergents leads to an compactation of the structures. An indication for that behavior has already been found in the interaction of clay particles – which are very abundant soil colloids – with these deteregents [7]. However, in the colloid fraction from the A_H-horizon of a CAV soil, much more colloidal particles can be found. Nevertheless, did these colloids show a compactation due to the coagulating force of the cationic detergents.

The presence of humic substances in the colloidal fraction of the soil sample has lead to an tremendous increase in the inner surface available for reactions. This was demonstrated by the increase in the amount of cationic detergents that was necessary to introduce the already mentioned coagulation of the soil colloids.

Humic substances as anionic polyelectrolytes and cationic detergents coagulate forming sphere-shaped structures, where size, density and state of aggregation is a function of the concentration of the detergents. Even hollow spheres could be produced this way.

Outlook

X-ray microscopy experiments with other soil samples have to be performed in the next future. In addition, the interaction of humic substances with detergents has to be studied with other types of cationic detergents, and with anionic and nonionic detergents.

Acknowledgements This work has been supported by the Federal Ministry of Education, Science, and Technology, BMBF, under contract number 05 644 MAG, and by the Deutsche Bundesstiftung Umwelt under contract number 03149. We would like to thank the staff of BESSY for providing excellent working conditions.

References

1. Thieme J, Schmahl G, Rudolph D, Umbach E (eds) (1998) X-Ray Microscopy and Spectromicroscopy. Springer, Heidelberg
2. Thieme J, Guttmann P, Niemeyer J, Schneider G, David C, Niemann B, Rudolph D, Schmahl G (1992) Nachr Chem Techn Lab 40:562–563
3. Thieme J, Niemeyer J (1996) Geol Rundschau 85:852–856
4. Niemeyer J, Thieme J (1997) Z Pflanzenernähr Bodenk 160:93–95
5. Schmahl G, Rudolph D, Niemann B, Guttmann P, Thieme J, Schneider G (1996) Naturwissenschaften 83:61–70
6. Schmahl G, Rudolph D, Guttmann P, Schneider G, Thieme J, Niemann B, Wilhein T (1994) Synchrotron Rad News 7(4):19–22
7. Thieme J, Niemeyer J, Guttmann P, Wilhein T, Rudolph D, Schmahl G (1994) Progr Colloid Polym Sci 95:135–138
8. Mitteilungen der Bodenkundlichen Gesellschaft (1985) 42
9. Stevenson F (1994) Humus chemistry, 2 edn. Wiley, New York

Progr Colloid Polym Sci (1998) 111:202
© Steinkopff Verlag 1998

AUTHOR INDEX

Progr Colloid Polym Sci (1998) 111:203
© Steinkopff Verlag 1998

SUBJECT INDEX